The Middle Paraná River

M. H. Iriondo J. C. Paggi
M. J. Parma (Eds.)

The Middle Paraná River
Limnology of a Subtropical Wetland

With 128 Figures and 46 Tables

Springer

Prof. Dr. Martín H. Iriondo
CONICET-UNL
C.C 487
3100 Paraná, EntreRíos
Argentina
e-mail: rniriond@ceride.gov.ar

MSc. María Julieta Parma
Instituto Nacional de Limnología
(CONICET-UNL)
José Maciá 1933
3016 Santo Tomé, Santa Fe
Argentina
and
Facultad de Humanidades y Ciencias
(UNL)
3000 Santa Fe
Argentina
e-mail: julietaparma@datamarkets.com.ar

MSc. Juan César Paggi
Instituto Nacional de Limnología
(CONICET-UNL)
José Maciá 1933
3016 Santo Tomé, Santa Fe
Argentina
e-mail: juanpaggi@gmail.com

Cover illustration: Male of the drangonfly *Diastatops intensa* Montgomery (Odonata; Libellulidae) on a leaf of *Eichhornia crassipes* (Mart.) Solms (Pontederiacea), typical inhabitants of the Paraná River ecosystems. Photo by C.J. Debonis and P.A. Scarabotti, INALI (CONICET-UNL). Identification of the dragonfly by Dr.J. Muzón, ILPLA (CONICET)

Library of Congress Control Number: 2007925871

ISBN: 978-3-540-70623-6 Springer Berlin Heidelberg New York

Springer is a part of Springer Science+Business Media

springer.com

Editor: Dr. Dieter Czeschlik, Heidelberg, Germany
Desk editor: Dr. Andrea Schlitzberger, Heidelberg, Germany
Cover design: WMXDesign GmbH, Heidelberg, Germany
Production and typesetting: SPi

Printed on acid-free paper SPIN 11597070 31/3100 5 4 3 2 1 0

Contents

List of Contributors

Amsler, Mario L.
Facultad de Ingeniería y Ciencias Hídricas (UNL), CC 217, 3000 Santa Fe, Argentina, e-mail: mamsler@fich1.unl.edu.ar

Arzamendia, Vanesa
Instituto Nacional de Limnología (CONICET-UNL), José Maciá 1933 (3016) Santo Tomé, Santa Fe, Argentina, and Facultad de Humanidades y Ciencias, Universidad Nacional del Litoral (UNL). Maestría en Ecología, Facultad de Ciencia y Tecnología, UADER, Entre Ríos, Argentina

Beltzer, Adolfo H.
Instituto Nacional de Limnología (CONICET-UNL), José Maciá 1933, 3016 Santo Tomé, Argentina, e-mail: adolfohec2001@yahoo.com.ar

Collins, Pablo
Instituto Nacional de Limnología (CONICET-UNL), José Maciá 1933, 3016 Santo Tomé, Argentina, and FByCB-FHUC, Universidad Nacional del Litoral, Argentina, and FCyT, Universidad Autónoma Entre Ríos, Argentina, e-mail: pcollins@arnet.com.ar

Cordiviola, Elly
Instituto Nacional de Limnología (INALI). Consejo Nacional de Investigaciones Científicas y Técnicas de la República Argentina (CONICET) y Universidad Nacional del Litoral (UNL). Santo Tomé, Argentina

Depetris, Pedro J.
Centro de Investigaciones Geoquímicas y de Procesos de la Superficie (CIGeS), Facultad de Ciencias Exactas, Físicas y Naturales, Universidad Nacional de Córdoba, Avenida Vélez Sarsfield 1611, X5016DGA Cordoba, Argentina, e-mail: pdepetris@com.uncor.edu

Devercelli M.
Instituto Nacional de Limnología (CONICET-UNL) J. Maciá 1933, 3016 Santo Tomé, Argentina

de Domitrovic, Yolanda Zalocar
Centro de Ecología Aplicada del Litoral (CONICET), Corrientes, Argentina, e-mail: yzalocar@arnet.com.ar

Drago, Edmundo C.
Instituto Nacional de Limnología (CONICET-UNL). José Maciá 1933, 3016 Santo Tomé, Santa Fe, Argentina, e-mail: edmundodrago@arnet.com.ar

de Drago, Inés Ezcurra
Instituto Nacional de Limnología (CONICET-UNL). José Maciá 1933, 3016 Santo Tomé, Argentina, e-mail: inesezcurra@arnet.com.ar

de Emiliani, M.O. García
Instituto Nacional de Limnología (CONICET-UNL) J. Maciá 1933, 3016 Santo Tomé, Argentina

Giraudo, Alejandro R.
Instituto Nacional de Limnología (CONICET-UNL), José Maciá 1933 (3016) Santo Tomé, Santa Fe, Argentina, and Facultad de Humanidades y Ciencias, Universidad Nacional del Litoral (UNL). Maestría en Ecología, Facultad de Ciencia y Tecnología, UADER, Entre Ríos, Argentina, e-mail: alejandrogiraudo@hotmail.com

Giri, Federico
Instituto Nacional de Limnología (CONICET-UNL), José Maciá 1933, 3016 Santo Tomé, Argentina, and FCyT, Universidad Autónoma Entre Ríos, Argentina

Iriondo, Martín H.
CONICET/UNL, C.C. 487, 3100 Paraná, Entre Ríos, Argentina, e-mail: rniriond@ceride.gov.ar

Lajmanovich, Rafael C.
Instituto Nacional de Limnología (CONICET-UNL), José Maciás 1933, 3016 Santo Tomé, Santa Fe, Argentina, and Facultad de Bioquímica y Ciencias Biológicas (ESS-FBCB-UNL), Ciudad Universitaria, Pje. El Pozo s/n, 3000, Santa Fe, Argentina

Lallana, Victor H.
Facultad de Ciencias Agropecuarias, UNER, C.C. N° 24, E3100WAA Paraná, Entre Ríos, Argentina

López, Soledad M.
Instituto Nacional de Limnología (CONICET-UNL), José Maciá 1933 (3016) Santo Tomé, Santa Fe, Argentina

Marchese, Mercedes
Instituto Nacional de Limnología (CONICET-UNL), José Maciá 1933, 3016 Santo Tomé, Santa Fe, Argentina and Facultad de Humanidades y Ciencias (FHUC-UNL), Ciudad Universitaria 3000, Santa Fe, Argentina

Montalto, Luciana
Instituto Nacional de Limnología (CONICET-UNL), José Maciá 1933, 3016 Santo Tomé, Santa Fe, Argentina and Facultad de Humanidades y Ciencias (FHUC-UNL), Ciudad Universitaria 3000, Santa Fe, Argentina

Paira, Aldo R.
Instituto Nacional de Limnología (CONICET-UNL), José Maciá 1933, 3016 Santo Tomé, Santa Fe, Argentina, e-mail: alpaira@ceride.gov.ar

Paggi, Juan César
Instituto Nacional de Limnología (CONICET-UNL), José Maciá 1933, 3016 Santo Tomé, Santa Fe, Argentina, e-mail: juanpaggi@gmail.com

de Paggi, Susana José
Instituto Nacional de Limnología (CONICET-UNL), José Maciá 1933, 3016 Santo Tomé, Santa Fe, Argentina, e-mail: sjdepaggi@gmail.com

Parma, María Julieta
Instituto Nacional de Limnología (CONICET-UNL), José Maciá 1933, 3016 Santo Tomé, Santa Fe, Argentina and Facultad de Humanidades y Ciencias (UNL), Ciudad Universitaria 3000, Santa Fe, Argentina,
e-mail: julietaparma@datamarkets.com.ar

Pasquini, Andrea I.
Centro de Investigaciones Geoquímicas y de Procesos de la Superficie (CIGeS), Facultad de Ciencias Exactas, Físicas y Naturales, Universidad Nacional de Córdoba, Avenida Vélez Sarsfield 1611, X5016DGA Cordoba, Argentina

Peltzer, Paola M.
Instituto Nacional de Limnología (CONICET-UNL), José Maciás 1933, 3016 Santo Tomé, Santa Fe, Argentina, and Facultad de Bioquímica y Ciencias Biológicas (ESS-FBCB-UNL), Ciudad Universitaria, Pje. El Pozo s/n, 3000, Santa Fe, Argentina, e-mail: paolapeltzer@hotmail.com

Quiroga, Martin A.
Instituto Nacional de Limnología (CONICET-UNL), José Maciá 1933, 3016 Santo Tomé, Argentina

Rossi, Liliana
Instituto Nacional de Limnología (CONICET-UNL), José Maciá 1933, 3016 Santo Tomé, Santa Fe, Argentina and Facultad de Humanidades y Ciencias (UNL), Ciudad Universitaria, 3000 Santa Fe, Argentina,
e-mail: loyrossi@arnet.com.ar

Sabattini, Rafael A.
Facultad de Ciencias Agropecuarias, UNER, C.C. N° 24, E3100WAA Paraná, Entre Ríos, Argentina, e-mail: rsabatti@fca.uner.edu.ar

Williner, Veronica
Instituto Nacional de Limnología (CONICET-UNL), José Maciá 1933, 3016 Santo Tomé, Argentina, and FByCB-FHUC, Universidad Nacional del Litoral, Argentina

Introduction

Martín H. Iriondo[1], Juan César Paggi[2], and María Julieta Parma[2,3]

The great Paraná basin (2,600,000 km^2) covers a large portion of the Neotropical Realm, extending from the Andes to the Atlantic coast and connecting the central regions of South America along 2,200 km in a north–south direction. Most of the basin is under subtropical climates, which range from desertic in the west to humid in the east.

The Middle Paraná is a broad, complex floodplain extending 600 km in the heart of the interior lowlands of the continent. This system of river channels, shallow lakes, islands, and wetlands remains unoccupied by man and basically unaltered. From a systemic point of view, this area can be considered to be the most important link and collector of the fluvial chain that conveys huge volumes of nutrients, sediments, and salts from the continent to the South Atlantic Ocean. The influence of the Paraná upon the Atlantic is very important. It has been calculated that a sediment deposit 400 m thick and 1 million km^2 in area accumulated in the platform and nearby oceanic bottom over the last 2 million years, with a similar amount of dissolved salts contributed by the river to the ocean.

This book details the recent advances in the knowledge of the most important segment of the Parana River, organized according to the scientific structure of the National Institute of Limnology of Argentina (INALI-CONICET-UNL). It integrates contributions from specialists in physical geography, geomorphology, fluvial hydraulics, geochemistry and several branches of ecology: phytoplankton, macrophytes, zooplankton, benthos, littoral communities, fishes, amphibians, reptiles and birds. Such a diverse variety of approaches can hardly be synthesized by a unique word, although we authors believe that the classical term "limnology" still works.

The volume is organized into four sections. Part I contains the physical and chemical characteristics of the system and related regions. The first two chapters describe the physical geography of the river basin and the geomorphology of the floodplain. Chapter 1 describes the physical geography of the whole Paraná basin, with its geological and climatic components and location in

[1]CONICET/UNL, C.C. 487, 3100 Paraná, Entre Ríos, Argentina, e-mail: rniriond@ceride.gov.ar
[2]Instituto Nacional de Limnología (CONICET-UNL), José Maciá 1933, 3016 Santo Tomé, Santa Fe, Argentina, e-mail: juanpaggi@gmail.com
[3]Facultad de Humanidades y Ciencias (UNL), 3000 Santa Fe, Argentina, e-mail: julietaparma@datamarkets.com.ar

M.H. Iriondo, J.C. Paggi, and M.J. Parma (Eds.)
The Middle Paraná River: Limnology of a Subtropical Wetland

South America; limnological connections with the Amazon basin are developed. Chapter 2 synthesizes the geomorphology of the floodplain called the Middle Paraná, which is characterized by several well-defined geomorphological units, each of them with a different ecological meaning. Chapter 3 develops a characterization of shape, size, and depth for more than 1,500 lakes inside the plain by using techniques of quantitative geomorphology. It is really important to know the main parameters of a "mean lake" in this region with thousands of water bodies. Chapter 4 describes the complex dynamics of water in selected lakes of the plain during floods and low waters, a mechanism that reverses in some epochs and varies among years. Chapter 5 details the sediment budget, which discriminates Andean sources from other geological terrains and the main channel–floodplain interrelationships over decades. A comprehensive study of the geochemistry of the Paraná River and related systems is presented in Chap. 6, considering the influence of El Niño years as well as other factors.

Parts II and III address aquatic communities of vegetables and invertebrates related to the lentic and lotic environments of the floodplain and the main channel of the studied complex of ecosystems. Each chapter includes a synthetic description of the composition and abundance of these communities as well as an analysis of their functional and distributional variations in relation to the dominant environmental factors, meaning those directly or indirectly associated with the flood pulse. All these chapters deal with traditionally important conceptual areas of limnological knowledge but stress the particularities of these communities in a subtropical floodplain.

Chapter 7 is devoted to phytoplankton of the main channel and the alluvial plain and the spatiotemporal changes of this community associated to the hydrosedimentological regime. The essential role of the floodplain in maintaining algal populations and incorporating them towards or from the lotic environments is also analyzed.

The complex combination of lenitic and lotic environments with different conditions of flux and water quality located between the floodplain and the main channel were found to be closely associated with the composition, number of species, diversity, density, and biovolume of the assemblages inhabiting these environments. The more clear differences are shown during isolation periods.

In Chapter 8, the authors stress the important role of macrophytes, mainly *Eichhornia crassipes* and *Paspalum repens* due to their contribution to the productivity and to the physical structure of the system as habitat and refuge for a number of organisms. The importance of drifting vegetation as a dispersal mean and the distribution of nutrients is also discussed.

Chapter 9 details the realm of zooplankton and concludes that the assemblages of lotic environments of the Paraná River show characteristics similar to most other rivers of the world such as the dominance of small organisms and the low density, which is comparable to other South American rivers. The influence of the flood pulse on zooplankton of floodplain lakes is related to their degree of connection to a river. Between lotic and lentic environments there are at least two gradients in regard to diversity and abundance of zooplankton, one of them associated to the relative influence of the physical and biological control and the other related to the level of environmental heterogeneity. In the main channel, biological mechanisms to control zooplankton populations would be less important than physical mechanisms of control.

The composition and abundance of benthic invertebrates of the Middle Paraná River floodplain complex, which is a mosaic of habitats undergoing succession, is described and discussed in Chap. 10. Environmental factors, which mainly control benthic structure and composition, are flood pulses, groundwater seepage, hydraulic parameters, hydrological connectivity degree, macrophyte cover, and water quality. In spite the great amount of organic matter accumulated on the bottom of floodplain environments, the "gathering-collectors" functional group is dominant, being the "shredders" present only in the floodplain. In the central strip of the higher channels, the assemblages are characterized by the dominance of species endemic to the Neotropical region, mainly *Narapa bonettoi*.

The littoral communities of floodplain environments of the Middle Paraná River are then treated in Chap. 11, which include a group of complexes of high biological diversity and abundant populations. Macrocrustaceans constitute an important group and have active participation in the community structure and biomass. Dominant species of this group are prawns of the genera *Macrobrachium*, *Palaemonetes*, and crabs belonging to the family Trichodactylidae. Biological and ecological aspects are then discussed, particularly growth, reproduction, spatial distribution, and trophic ecology. Co-adaptation of the assemblages of macrocrustaceans are also analyzed under the functional context imposed in the Paraná River from its formation, habitat characteristics, and stability degree, these factors being the product of a joint organic evolution.

Part IV addresses vertebrate animal life. Chapter 12 shows the high fish diversity and productivity in this section of the Paraná River and summarizes information about migratory displacements, reproductive strategies and life histories, participation of the fish species in the complex trophic webs, and data of metabolism and responses to hypoxia.

The objective of Chapter 13 is to present information about the diversity and natural history (tadpole and adult characteristics, habitat use, and reproductive traits) of anurans commonly found in riparian areas of the Middle Paraná River. The factors that determine the presence of anuran species in ponds are part of a complex network of relationships that act synergically.

Chapter 14 then analyzes the reptile community living in the floodplain, including aspects of species richness, ecological traits, and environmental variables. A total of 71 reptile species were recorded in the Middle Paraná River with snakes being the richest reptile group with 49 taxa, and they were found to occupy all type of habitats and substrates, including various species adapted to live in the water.

In conclusion, Chap. 15 presents data on the ecology of birds in the middle section of the river in relation to trophic and functional groups, seasonal dynamics, and characteristics of reproductive biology.

Acknowledgements. The editors are very grateful to the National Council of Scientific and Technical Research (CONICET) of the Argentine Republic for the financial assistance from the Program PIP N° 02195 under the direction of MSc. María Julieta Parma.

Part I
The Physical and Chemical Environment

1 Physical Geography of the Basin

Martín H. Iriondo[1] and Aldo R. Paira[2]

1.1 Introduction

The Paraná River drains a continental-sized basin formed by several different, even contrasting, regions. Such regions cover areas with hundreds of thousands of square kilometers in extension. The Middle Paraná is a type of fluvial synthesis of a huge territory covering an area of 2,600,000 to 2,800,000 km^2 (according to different criteria of definition), which includes a large diversity of rainforests, mountain deserts, and savannas. Basically, however, the fluvial net of the Paraná basin can be considered as a great machine of continental size that collects, modifies, and transports water, nutrients and other inorganic and organic products from tropical environments to the middle latitudes of the South Atlantic Ocean.

The Paraná basin is about 230 times larger than the lower reach of the fluvial system analyzed in this book and is known as the "Middle Paraná". Hence, the influence of the geology and climate of some regions of the basin on this (comparatively) small area is undoubtedly decisive, although such influence mostly remains not yet understood.

1.2 Geology

The Paraná basin is composed of several geologic regions that influence to different degrees the geochemistry and sedimentology of the collector. The geological map of the area reveals that both the Brazilian shield and the Andes cordillera have only modest areas in the basin (Fig. 1.1). This fact suggests that the direct influence of both geotectonic systems is probably not as important as traditionally thought. The geological composition of the Paraná basin is as follows:

Brazilian shield: This formation shapes an irregular belt at the northeastern corner of the basin, comprising 7.4% of the total area. Gneisses and other

[1]CONICET/UNL, C.C. 487, 3100 Paraná, Entre Ríos, Argentina, e-mail: rniriond@ceride.gov.ar
[2]Instituto Nacional de Limnología (CONICET-UNL), José Maciá 1933, 3016 Santo Tomé, Santa Fe, Argentina

M.H. Iriondo, J.C. Paggi, and M.J. Parma (Eds.)
The Middle Paraná River: Limnology of a Subtropical Wetland

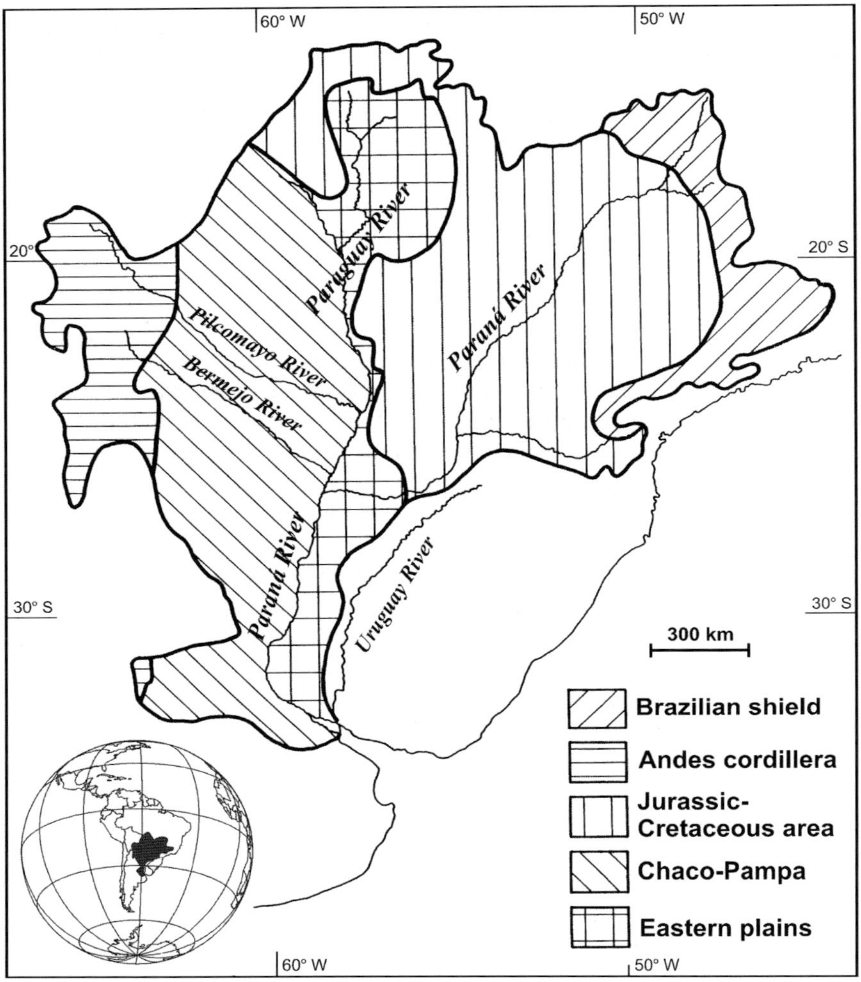

Fig. 1.1 Geology of the basin

Proterozoic metamorphic rocks predominate; red soils cover the region. The shield is a rich source of quartz and kaolinite.

Andes cordillera: The Andes sector of the Paraná basin is formed by the Sierras Subandinas and the Atacama Highplain. This area describes a rectangle of about 170,000 km^2 surface (approximately 7.4% of the total area) at the west of the basin. It is composed of a sequence of N–S folded chains of Tertiary age. Numerous rock types can be found in the region, among them predominate lutites, phyllites and other fine-grained sedimentaries. The region provides abundant fine sediments to the hydrographic net, mainly silt and illite.

Jurassic–Cretaceous region of the Upper Paraná: This region comprises almost 29% of the total surface of the basin. The rocks are tholeiitic basalts and siliceous sandstones originated in desert and fluviatile environments. Basalts and sandstones cover similar extensions in the region. The surface is formed of red soils and the area provides quartz and kaolinite to the rivers.

Carboniferous area of the Upper Paraná: Glacial and periglacial rocks of the Upper Carboniferous, mainly tillites, sandstones and siltstones are characteristic. They form two belts bordering the Jurassic–Cretaceous area, comprising 5.6% of the total area of the basin.

Chaco–Pampa plain: This is the largest geologic region of the Paraná basin (29.8% of the total area) and is composed of Quaternary sediments, mainly fine solid, silt, and associated illite. The sediments originated to the west and south of the basin. Brown soils are typical. Most of the sediments contain soluble salts. The main contribution of the Chaco–Pampa plain to the Paraná is dissolved salts, through small rivers and groundwater seepage.

Eastern plains: The eastern plains are composed of two areas located at the left side of the Paraguay-Paraná line. They cover 10.9% of the total basin. The northern area is the Mato Grosso Pantanal, which is composed of large alluvial cones; its collector is the Paraguay River. The southern area, known as "Mesopotamia", is located in northeastern Argentina. Montmorillonite clays and quartz sand compose the upper levels of the geologic column. The sediments covering both areas are Quaternary in age.

Other geological regions: 10% of the basin is formed by miscellaneous rocks and sediments, predominantly Paleozoic sedimentaries.

1.3 Climate

The climatic pattern of South America is controlled by the general circulation of the atmosphere superimposed on an isolation of the continent inside large oceanic extensions. An excess of solar radiation occurs in the zone extending from the Equator south to the latitude of 40/42°, hence, including the whole Paraná basin. From this line to the South Pole, the surface of the Earth radiates more heat that it receives, which results in large transfer processes of energy and matter (winds, water vapor, etc.) to the south, tending to maintain the thermodynamic balance of the globe. A major system in the atmosphere is the Intertropical Convergence Zone (ITCZ), a low-pressure belt characterized by abundant precipitation and high turbulence. Its mean position is located at the Equator; the belt migrates to the north in July and the south in the austral summer, reaching the upper basin in January. South America is flanked by anticyclones in both oceans (Atlantic and Pacific) at tropical latitudes. The South Atlantic Anticyclone, which circulates in a counter-clockwise direction, directly influences the climate of the Paraná basin by sending warm and humid air masses, mainly in the summer. Dry and cold southerly winds, ultimately originated in the South Pacific Anticyclone, mark the winter weather, together with rare polar spells. The result in the Paraná basin is the following: Tropical-subtropical climate with a north–south gradient in temperatures and east–west gradient in precipitation. The occurrence of a seasonal anticyclone in the Chaco region somehow complicates this simple pattern.

The Paraná basin extends from 15° S to 35° S, covering a long latitudinal range. The distribution of altitudes above sea level is also important in an east–west direction. The climate develops on this topographically complex scenario by following a few main structures. In first place, the General Circulation of the atmosphere forces the Intertropical Convergence Zone (ITCZ) to the south in the austral summer (December to February), producing rains in most of the basin (Fig. 1.2). The ITCZ itself reaches only the northernmost region of the basin; other areas are covered by spells of humid air masses migrating from the ITCZ to the south. Winters are drier.

A second significant structure is the occurrence of seasonal blocking anticyclones above the Bolivian Altiplano in summertime (Necco 1989; Iriondo 1999), which hinders the general circulation of air masses in summer. Some years the anticyclone extends to the east and south, covering the whole Paraná basin and provoking general dries in the region. It is noteworthy that a climatic alteration occurred in the decade of 1970, which produced an increase of about 200 mm yr^{-1} in precipitation and 1°C in temperature in central and southern sectors of the basin; the opposite tendency occurred in the northwest corner.

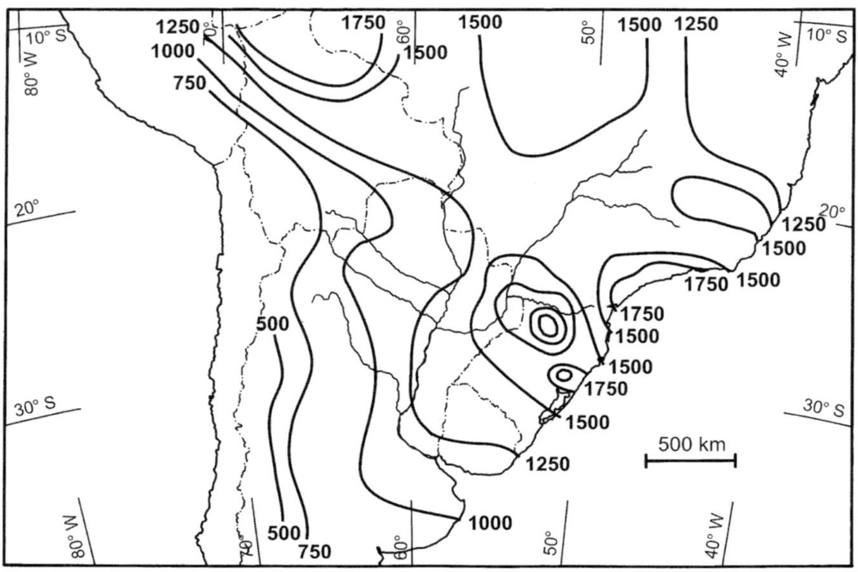

Fig. 1.2 Mean annual precipitation

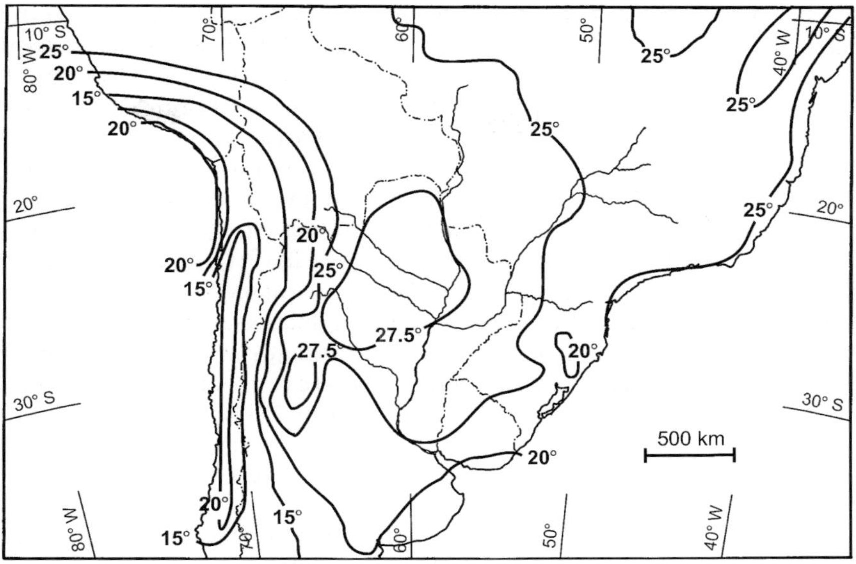

Fig. 1.3 Middle fields of temperature in January

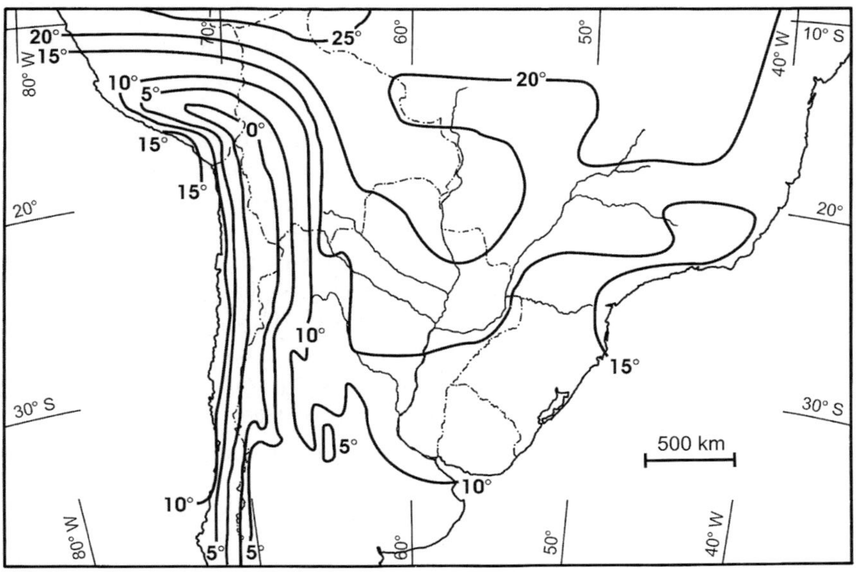

Fig. 1.4 Middle fields of temperature in July

The general climatic characterization of the basin is as follows: humid tropical in the northeast (15°S) shifting to humid subtropical/temperate in the south (35°S), and desertic subtropical "mountain climates" in the west. Frost and snow are nonexistent or irrelevant. Mean temperature fields in January indicate that the Chaco region is the warmest area in South America, above 27.5°C (Paoli et al. 2000, Fig. 1.3). Mean temperatures in July are between 20 and 25°C in the north and from 10 to 15°C in the south (Fig. 1.4). Maximal annual precipitation occurs in the northeast of the basin (more than 2,250 mm) owing to a combination of factors: the location of the ITCZ in summer and the influence of the South Pacific Anticyclone in winter. A second maximum (above 1,750 mm) occurs at the flanks of the Altiplano, produced by the ITCZ plus mountain rains. Minimal registers, below 250 mm yr^{-1}, occur in the southern Altiplano (SW Bolivia/NW Argentina) (Fig. 1.2).

1.4 Hydrography

The hydrographic net of the river upstream the middle reach of the collector is composed by several sub-systems: The Upper Paraná, the Paraguay River, the Chaco tributaries, the Carcaraña River and its minor tributaries in the southeast (Fig. 1.5).

1.4.1 The Upper Paraná River

The Upper Paraná River basin is a large region characterized by old landscapes developed in rocks of Proterozoic to Cretaceous ages. Such landscapes were basically erosive and were generated during three phases in Cretaceous and Tertiary times: large planation surfaces formed by pediments and peneplains with meandering channels incised in the rocky substratum. The Paraná basin did not exist during that period millions of years long, basins tend to derive to the Atlantic Ocean. In present times, some channels are partially preserved and occupied by tributaries of the river, in a different topographic and hydrological pattern. The river flows more than 2,500 km in that geomorphological unit, until entering in the megafan near Posadas (27° S, 57° W) (Fig. 1.6).

The northernmost first-order channels of the Paraná basin are located in the Serra dos Preneos (Brazil) at 15°30'S, originating the Sao Bartolomeu River, which joins downstream other channels, derived from the Piloes and Río Claro ranges, forming the Paranaíba River that already is the Paraná. Another important hydrographic sub-system (the Río Grande) originates near the Atlantic Ocean and joins the Paranaíba after a length of 1,306 km at the latitude of 20°S. At that point the river is named "Paraná".

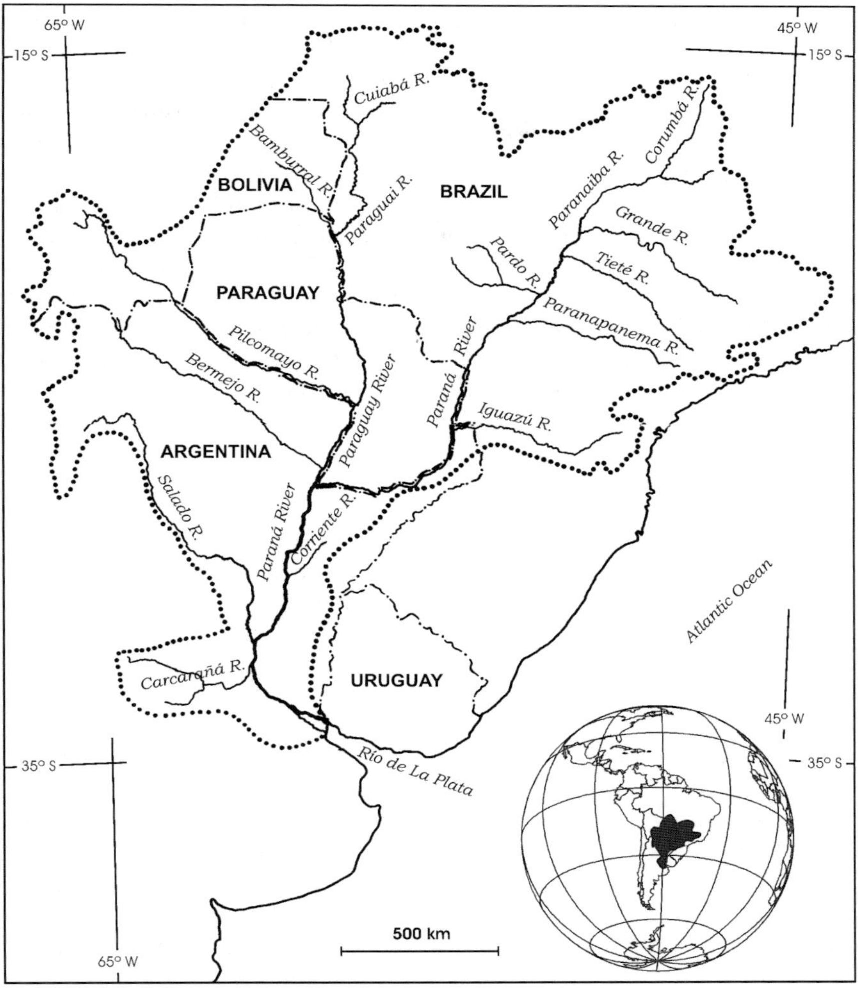

Fig. 1.5 Hydrographic net of the basin

1.4.2 The Megafan

Leaving the basaltic meset, the Paraná River enters the Argentine plains. In the first segment, it generates a large alluvial fan (or megafan), which covers the NW half of the Corrientes Province in Argentina and the southern part of eastern Paraguay. The fan is 260 km long and 500 km wide in the distal line, and is located at the Paraguay-Paraná left bank. The mega-fan has been stable for most of the Upper Quaternary. Inside it, the river channel

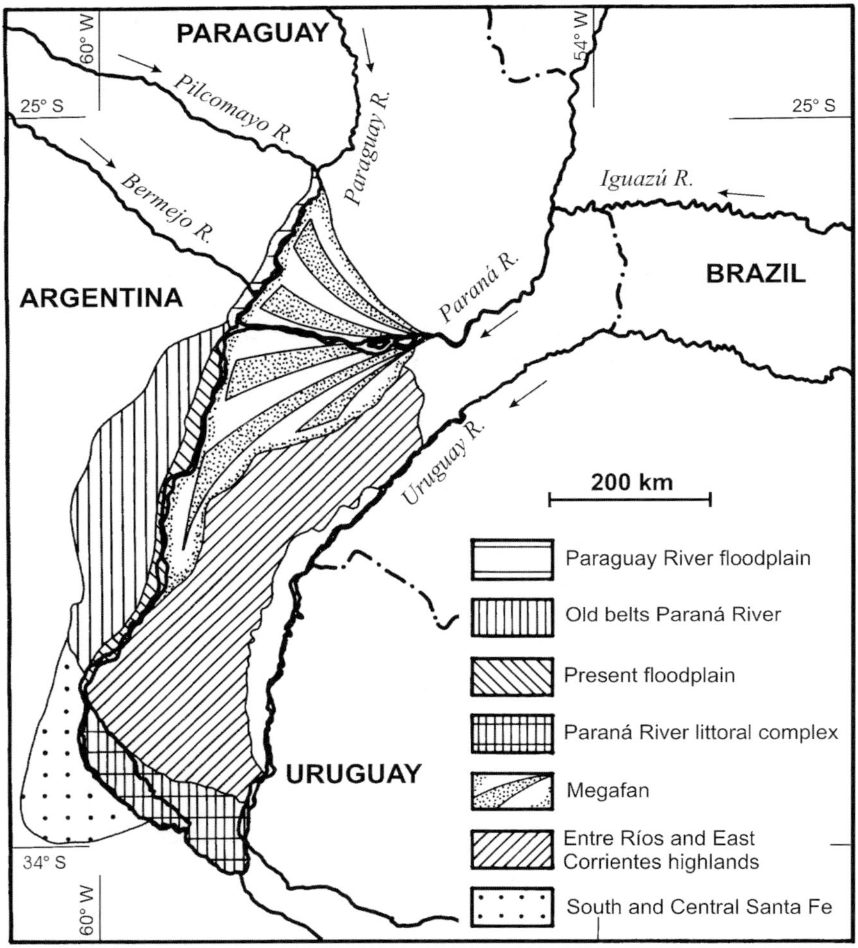

Fig. 1.6 Regions limiting the Middle Paraná

built relatively stable belts, occupied during some hundreds or thousands of years, eventually abandoned by the main stream and replaced by large swamps. A sequence of such intervals formed the present megafan. A few extensive deposits produced by spill outs during dry climatic phases in the Upper Quaternary are intercalated among the abandoned belts. Also during dry phases, important deflation of sand occurred in the abandoned belts, generating up to 80-km-long and 5-km-wide dune fields.

The present river belt crosses the fan in an east–west direction; the channel there is braided, with frequent elliptical islands and sand banks. The youngest abandoned belt in the megafan is Holocene in age and runs along the eastern limit of the system and includes the Iberá and Luna lakes. The ecological system of the megafan is similar to that of the Middle Paraná.

1.4.3 The Chaco

The Chaco is a large tropical plain located in the interior of South America, which consists of sectors of Argentina, Paraguay, and Bolivia, and covers an area of 840,000 km^2 and is characterized by forests, savannas, and extensive swamps, which give it a marked climatic and biogeographic identity. The plain encompasses five huge alluvial fans built by the major rivers that cross the region: the Salado, Bermejo, Pilcomayo, Parapetí and Grande Rivers. The fans are composed of several sedimentary units deposited during different times of the late Quaternary under diverse climates. Two fluvial terraces appear at the apex of each fan; the older one is probably late Pleistocene in age and the second was formed in postglacial times. Humid climates, such as the present one, favored the generation of soils and stable fluvial belts; drier climates led to widespread sedimentation along small ephemeral channels and large spill-outs. During two intervals, one in the Late Glacial Maximum and one in the Late Holocene, dry climates occurred in the region, leading to the formation of dune fields and loess mantles (Fig. 1.7).

1.4.3.1 Western Chaco

The Western Chaco is dominated by the fluvial dynamics of allochthonous rivers flowing from the humid Sierras Subandinas. The rivers cross the mountains through deep transverse canyons and carry well-sorted, fine quartz sands. In spite of the high transport capacity of the rivers, coarse sediments are absent in the fluvial deposits. The only exception to this sedimentological pattern is the Bermejo River, which conveys large pebbles and boulders in the upper reach. The fluvial systems have developed two terraces on the eastern slope of the Sierras.

Large rivers cross the region to the east along complex alluvial belts composed of long old channels, oxbow lakes, swamps, and other minor landforms built by the channel and related geomorphic agents. Occasionally, a major avulsion occurs. About 100 years ago, the Bermejo River abandoned its former channel and shifted to a different location along a stretch of 200 km (Iriondo 1974). The Pilcomayo River has been undergoing a similar process since 1980. Such types of avulsion take several years until completed.

During past dry climates, the Chaco rivers were smaller and less stable than they are at present. Numerous small paleochannels can be observed in the plain, forming large-scale distributary patterns, which is especially evident in the Pilcomayo and Salado systems. Two arid episodes occurred in the

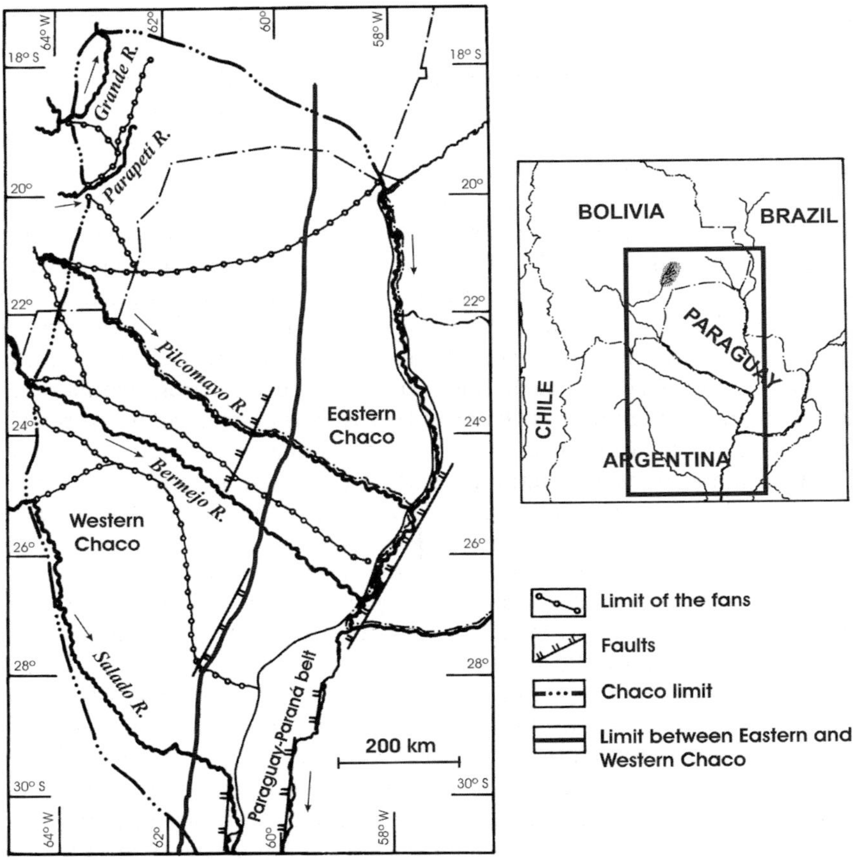

Fig. 1.7 Main units of the Chaco

late Quaternary, during which strong, dry northern winds formed dune fields and loess mantles in western Chaco.

1.4.3.2 *Eastern Chaco*

The eastern Chaco is formed by the distal areas of the alluvial fans of the major rivers: they are basically swamp environments, crossed by old fluvial belts of the rivers. The climate is subhumid to humid, and the floods of the allochthonous rivers discharge large volumes of water across the landscape. Infiltration is practically nil because the terrain is composed of 12 to 20-m-thick impervious silty clays. The extremely low regional slope does not favor runoff of the excess water and, as a consequence, permanent and ephemeral swamps, densely covered by floating and paludal vegetation, cover broad areas in the region.

Evapotranspiration in the swamps is several times higher than the potential evaporation, and organic matter accumulates in the anaerobic environment at the base of the swamps. Neiff (1986) described two types of swamp:

(a) Temporary swamps ("bañados") are formed by rainwater and do not accumulate sediments, water remains periodically for several months, forming a 30 to 40-cm-deep layer on the surface of the plain. A short period of oxygen deficiency appears at the beginning of the flooding, originated by the decomposition of vegetation. (b) Permanent swamps ("esteros") have well-defined borders and are larger and deeper. The roots of the vegetation are adapted to survive anaerobic conditions for long periods of time. A natural mechanism in the swamp is the periodic occurrence of fires. Production of organic matter is high in these tropical wet environments, and is estimated at 20 t ha^{-1} year (dry weight) according to Neiff (1986). Approximately 70% of the organic matter accumulates at the bottom, forming a peat horizon that degrades very slowly due to the oxygen deficiency and to a low C/N ratio.

The major swamps, 100–200 km long and 3 to 10 km wide, are typically less than 1 m deep. The central area of the swamp is often free of vegetation and forms an irregular belt of slowly flowing water (Iriondo 1989). The Quaternary paludal deposits and present swamps cover an area of 125,000 km^2 in the Bermejo and Pilcomayo fans.

Large swamps are the headwaters of local river networks, which have developed on the large fans during the present humid climate. The collectors of such systems are small tortuous channels, 2–5 m wide (Orfeo 1986). They flow along abandoned channels of the major allochthonous rivers, which are normally bordered by levees. They are a special type of underfit rivers (Cucchi 1973; Iriondo 1974), with the former channel belonging to a river basin several times larger than the present one. When major floods cover the whole region, only the old levees remain above the water level.

A significant percentage of the sedimentary loads of local rivers are colloids, which form as much as 75% of the total weight of some samples. Pedrozo and Orfeo (1986), in their study of several river systems, found between 40 and 52% suspended solids composed of particles with diameters ranging from 1.2 to 0.45 μm. Colloids are composed of clay minerals and organic matter in different stages of evolution, probably resulting in the formation of organic-clay complexes.

1.4.3.3 The Paraguay Belt

The Paraguay is one of the long rivers of the world. Its large basin (1,095,000 km^2) begins at the Preneos Ranges 3,000 m a.s.l. and in the upper section it drains the Mato Grosso meset. The most interesting sector of the river is the Pantanal do Mato Grosso, a 60,000-km^2 wetland that controls the runoff and delays the floods for 1–2 months. The lower reach (220 km long) crosses a sector of the Paraná floodplain. The Paraguay channel there is very active geomorphologically, resulting in a deep modification of the geoforms into a young meander plain.

The primary component of the sediments is clean, fine-grained quartz sand, originated in the erosion of Cretaceous sandstones in southern Brazil and Paraguay. Silt and illitic and montmorillonitic clays are also found in minor proportions. Sedimentary deposits are formed of coarse strata of channel sand extending to several tens of meters in depth. Local relief is moderate to low.

1.4.3.4 The Chaco Tributaries

The large tributaries crossing the Chaco have formed alluvial fans, which are complex systems composed of numerous sedimentary and geomorphological units that formed at different times and under different climatic conditions. Although their general characteristics are similar, each shows unique features. From south to north, the fluvial systems are the following: the Salado, Bermejo, Pilcomayo, Parapetí and Grande.

The Río Grande River system. The Río Grande alluvial fan system enters the plain at Puerto Camacho (Fig. 1.8). It forms a large alluvial fan, 65,000 km^2 in area, 30% of which forms the northwestern corner of the Chaco, and the rest belongs to the Amazonic Llanos. The Chaco sector of the system is described as follows:

Two river terraces are conspicuous on the flanks of the Sierras. The oldest one is 50 m above the present river level, its surface is modified by a tropical half-orange morphology (characterized by rounded hills) and covered by a yellowish brown loess. The terrace is equivalent to similar geoforms in the Pilcomayo and Parapetí systems.

The lower terrace is 10 m high at the fan apex, near the mountain front, diminishing to 3.30 m 700 m downstream. The stratigraphic section is composed of channel sand at the base, and a sequence of a silty abandoned channel infilling, with a brown soil at the top. Both beds correlate with the lower terrace of the Pilcomayo. That sequence is buried by massive soft silt, clearly younger, corresponding to overflow sediments deposited simultaneously with a Holocene sand mobilization occurred in the region.

The major bed of the river, which is active during floods, is 300 m wide at the foot of the western mountains and 5 km wide 100 km downstream. It is covered by ill-defined bars of very fine sand, which is composed of more than 95% quartz. In some sectors, small accumulations of pebble-sized quartzites are frequent. Minor overflow channels, partially filled with silt and [page no 6] clay, are common. During the dry season, large desiccation polygons up to 1 m in diameter are formed in the silt. Transverse and diagonal bars of quartzite pebbles and blocks are visible every 100–300 m in the dry channels. Except for the

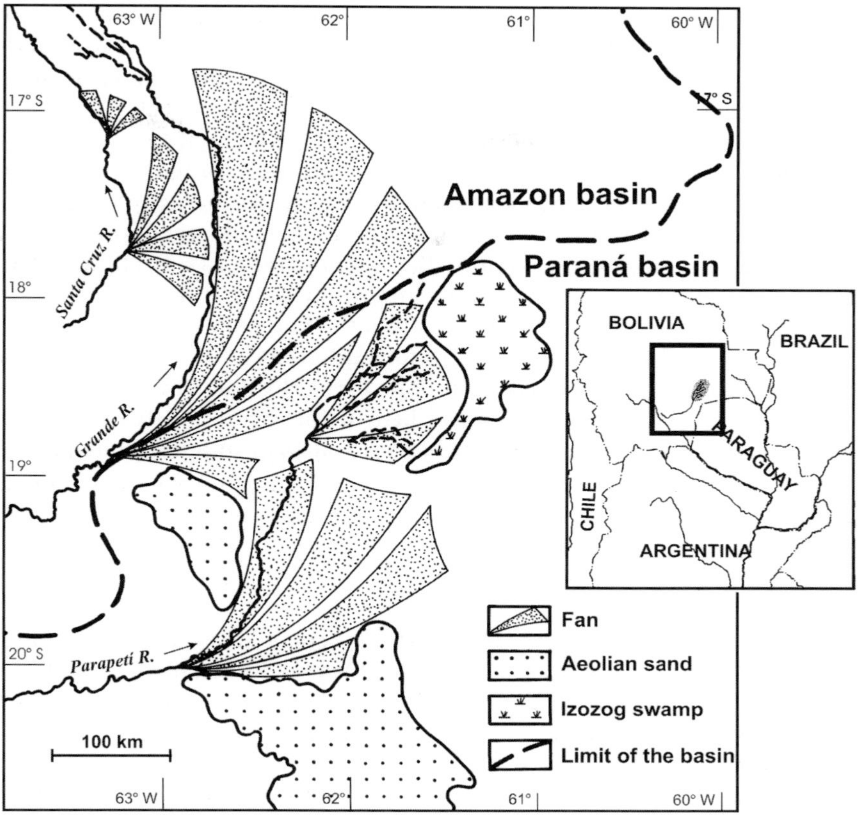

Fig. 1.8 Río Grande fan and divisory with the Amazon basin

pebble bars, the sedimentary characteristics of the major channel are similar to those of the basal sand in the lower terrace. The minor channel is 80 m wide and has a meandering pattern.

The river at Pailón, some 200 km downstream, is similar to the upstream section. The major bed is composed of very fine sand and large planar bars. A large amount of tree trunks, branches, and other vegetative debris is buried in the sand. Overflows and paleochannels of the Río Grande cover the northwestern corner of the Chaco. Overflow deposits are composed of brown quartz silt with subordinate illite; they extend to the very foothill of the Precambrian Shield, at the northern boundary of the Chaco. Even as far as San José de Chiquitos, Río Grande silts fill the valley, without any sedimentary contribution from the surrounding low mountains. The unit is equivalent to the Parapetí loess and the western Formosa general surface. Several paleochannels of the Río Grande incise the surface and convey clean water during the rainy season. They are 40–80 m wide, 2–3 m deep, and join the

San Pablo River to the north. Such channels are locally named "cañadas", the most important of them are Los Aceites, Los Papagallos, and Kolla Muerto; they have the same age as the paleochannels in the Bermejo and Pilcomayo alluvial fan systems.

The Parapetí River system. The Parapetí River has formed an alluvial fan in Bolivia and Paraguay, which has a surface area of several tens of thousands square kilometers. Part of it is located in the Paraná basin, the remainder lies within the Amazon basin. At present, the permanent channel flows into the Izozog swamp, and eventually into the Mamoré River, a tributary of the Amazon. During the rainy season an important water transfluence occurs, forming the Timané River, which flows into the Paraguay River.

According to Huamán et al. (1975), four different units compose the surface of the Parapetí alluvial system in Bolivia: (a) old alluvial deposits; (b) the aeolian plain; (c) the present alluvial belt of the river; and (d) the Izozog swamp (Fig. 1.9). Two terraces were formed at the fan apex, equivalent to

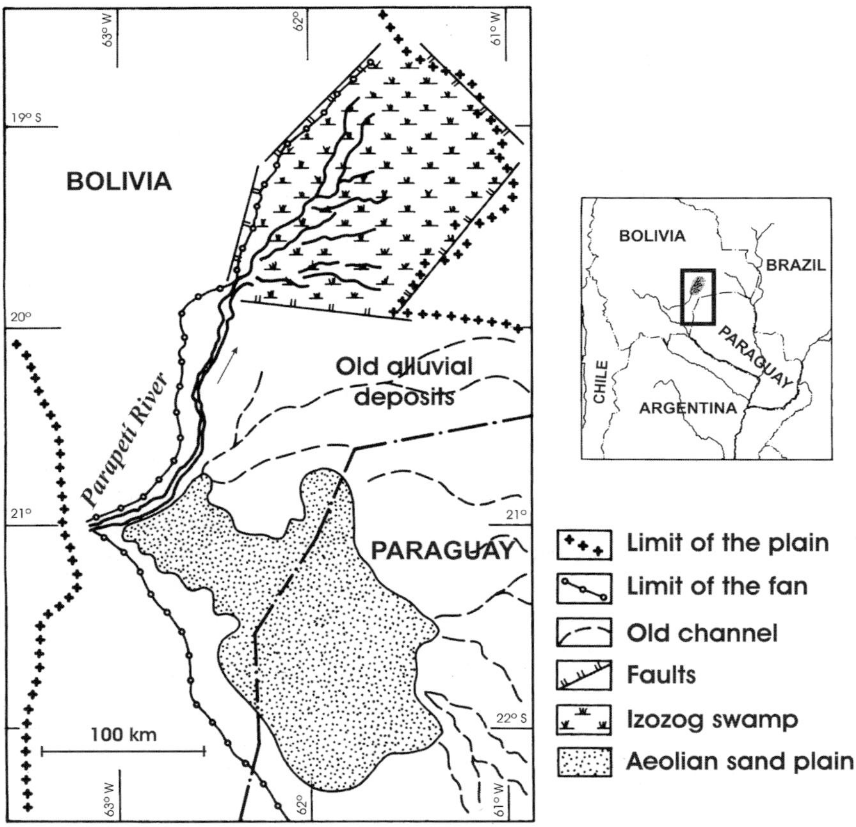

Fig. 1.9 Internal units of the Parapetí fan

those of the Pilcomayo system in Villa Montes. The old alluvial deposits form a large plain with slopes below 1%, located between 62°15′ W and 62°30′ W longitudes. Water drains very slowly and the surface is crossed by numerous paleochannels. The sediment is clayey to sandy loam and is composed of quartz and illite. In general, soils on this unit are deep and well developed, with a Bt horizon; in the subsoil $CaCO_3$ precipitates form pseudomycelia and small patches. The sediment has been partially affected by local deflation. The old alluvial deposits are equivalent to the western Formosa surface of the Pilcomayo alluvial fan system.

The aeolian sand plain covers a surface area of 25,000 km^2 in eastern Bolivia and northwestern Paraguay and is formed by large parabolic dunes oriented in a north–south direction. The individual dunes range up to 10 km in length and are composed of fine sand (98% quartz, 2% felspars). The unit was formed during a general aeolian mobilization in the late Holocene, between 3400 B.P. and 1400 B.P. (Servant et al. 1981). The present alluvial belt of the Parapetí River is formed by the river channel and a series of old channels abandoned by avulsion. It is 10 km wide at 63° W longitude, where it turns to the north. When it reaches the Izozog swamp, its width has increased up to 20 km. The wind forms sand shadows and small mounds at the southern bank of the channels during the dry season. The unit is discordant with the aeolian plain, hence, it is younger than 1,400 years.

The Izozog swamp has an area of 6,800 km^2 and is located on a tectonic depression. As the rivers enter the swamp, its channel divides into numerous distributaries flanked by levees. The sedimentary load of the Parapetí River is deposited in the swamp, where it is locally redistributed by the wind during the dry winter. Most of the water infiltrates and is lost by evapotranspiration, the rest slowly migrates to the Amazon River. In local depressions, the sedimentary fill is composed of fine-grained stratified mud. The levees are composed of coarser particles.

The Pilcomayo River system. The upper basin of this 1,070-km-long river is developed in the Cordillera de Los Frailes, where the channels receive most of their water from the melting of winter snow. Downstream, the valley crosses the Subandean ranges to the east forming large incised meanders, in a clear antecedent relationship. Hence, a Tertiary age, probably early Pliocene, can be proposed for the Pilcomayo River. The Subandean tributaries, on the contrary, are subsequent, young, and scarcely developed.

Entering the plain, the Pilcomayo River constructed the most important alluvial fan in the Chaco. It is the almost unique case of an active alluvial fan of a major South American river (Fig. 1.7). The total area is 210,000 km^2. The river reaches the plain at Villa Montes, forming the fan apex. The fluvial deposits of the area are preserved in two terraces. The older one has a variable altitude above the present channel (40 m in the upstream valley and 20 m in the piedmont); it is covered by 3-m-thick reddish aeolian silt with a well-developed soil on the top. The younger terrace covers a large surface in the region. In the upstream valley, near the road bridge (5 km from Villa

Montes), it is 6 m thick and composed of two units. The lower one is composed of pebbles and blocks, gray in color. The upper unit of the section is composed of smaller pebbles and boulders in a red sandy matrix. The Pilcomayo River presently transports very fine gray quartz sand at Villa Montes. The sand accumulates in large bars, which also include small deposits of isolated rounded pebbles. The maximum discharge of the channel is 800 $m^3\ sec^{-1}$, 45 times higher than the minimum; a really extreme condition for a large river. Downstream, in the piedmont area, the lower terrace is 2 m high and is formed by coarse and very coarse planar strata with sharp contacts. The continuous surface of the terrace extends farther to the east, forming a wide area in the Western Chaco. In west Formosa, 300 to 400 km to the southeast, an older unit is present; it contains quartz silt and very fine quartz sand. The grains are covered by hematite and large plates of illite. The sediment is yellowish brown in color, moderately plastic and compact. It is geomorphologically characterized by numerous ephemeral channels, 5–15 km in visible length and with an irregular pattern. This unit extends to 60° W longitude, forming a regional fan pattern.

An older sedimentary unit, probably equivalent to the upper terrace in Villa Montes, appears on the surface to the east. It is composed of 10 to 20-m-thick paludal silty clay, covered at the present by temporary and permanent swamps. Fluvial belts, formed by paleochannels and levees, cross the area in west-to-east and northwest-to-southeast directions; they transport waters of local fluvial basins (Monte Lindo, Pilagá, Melo, and others) (Cucchi 1973; Iriondo 1974), whose main features are similar to those described for the Los Amores basin. Such a unit composes the whole distal region of the Pilcomayo fan, limited by the Paraguay-Paraná belt along 650 km. In western Chaco, the Pilcomayo River changes its channel several times during the present humid climate. Abandoned channels are filled with fine silty sand, without clay or coarse clastics. Its internal relief is about 1 m. In some cases, the channels convey considerable quantities of water during rainy seasons (Rabicaluc 1986).

Over the last four or five centuries, the Upper Pilcomayo River discharged its water into a 15,000-km^2 tectonic depression, the Estero Patiño, located at the Argentina/Paraguay border, 250 km upstream from the confluence with the Paraguay River (Cordini 1947). The main channel deposited huge volumes of silt and very fine sand in the area. Cordini (1947) measured very high concentrations of suspended solids during floods (up to 40,000 ppm). Dikes of vegetation debris, which accumulate during periods of high water, acted as sediment traps, building up plugs of silt and sand, which frequently caused channels to shift. Eventually, the wood and leaves disappear through oxidation, resulting in a collapse of the overlying sand. Such a process produces a landscape of "hoyales" (holes), which is characterized by numerous pits up to 5 m in diameter and 1.5 m in depth. In 1980, the river completed a century-long sedimentation process by entirely filling the depression. Consequently, the Pilcomayo began an avulsion process, depositing fine sand and silt in the lower reach of its channel, and with water overflowing into the surrounding plain. In successive years, it occurs a channel filling migrating upstream, leaving behind a rapidly enlarging sector abandoned by the main stream.

The velocity of retreat is 10 to 35 km yr^{-1}, with a total of 160 km in 7 years. The overflowing water in Argentina formed a swamp 250 km long and 7–12 km wide, with paludal dynamics and geomorphology. Approximately the same volume of water spilled to the north, into the Republic of Paraguay, forming a similar swamp.

A few well-preserved paleochannels, cutting the general surface of the region, were mapped in western Formosa. A well drilled in one of them showed that the channel fill is composed of well-sorted very fine quartz sand, reddish in color. It is most probable that the same sedimentary process active at present times in the Pilcomayo filled the former channels during the Holocene.

As a first approximation, the Pilcomayo alluvial fan system was produced by two contrasting processes: (a) development of stable fluvial belts during humid periods, such as the present one, and (b) generalized sedimentation by spillouts and ephemeral channels during periods of dry climate. In both cases, swamps formed in tectonically sunken blocks.

The Bermejo River system. This river system is 1,450 km long, draining a basin of 135,000 km^2. The sources are located in the Eastern Cordillera in Bolivia and Argentina and in the Argentine Puna; one of the tributaries, the Iruya River, contributes 70% of the suspended load of the collector. The apex of the Bermejo fan is located near Embarcación, in Bolivia. The system extends 650 km to the east and southeast, as far as the Paraguay-Paraná belt. In the western and central Chaco, a large number of Holocene paleochannels can be observed in the plain. Their general orientation is northwest to southeast. In the south, a large spillout lobe reaches the province of Santa Fe. Farther east, in the Charadai area, swamps cover the region, probably since Pleistocene times. The sediments are composed of 6–12 m of thick silty clays, greenish-gray and reddish-brown in color, deposited in a depression formed by a 5,000-km^2 tectonically sunken block. The Bermejo River presently flows in a several-kilometer-wide belt, which at Las Lomitas is discontinuous and 10 km in width. It is characterized by numerous minor abandoned channels in different degrees of preservation. The oldest ones are hardly visible, obliterated with overlying sediments. Several deflation depressions are located at the central line in such paleochannels. The youngest channels are well preserved, 3–4 m deep, and transport water during floods. A hard reddish clayey silt composed of quartz and illite forms the substratum of the belt. The young deposits are formed of loose brown clayey silt with the same mineralogical composition as the substratum. The paleochannels described here are formed in 1 to 2-km-wide fringes composed of young deposits and separated from each other by areas of reddish substratum.

In this area, the river is actively eroding a 4-m-high bank where the local sedimentary column is well exposed. The upper section of the bank is composed of sand bars deposited during floods and paleochannels filled with dark clayey silt containing 15–25% of very fine laminated sand. The silt is rich in decomposed organic matter and leaf molds. Strata are plane, concordant, 0.40 to 1 m thick and 50–300 m long. The lower section outcrops less than 1 m above the mean water level. It is composed of reddish silty sand, more compact than the upper section, with gray mottles and black patches of manganese minerals; it probably represents the substratum of the present fluvial belt.

The river flows through a very tortuous 1-km-wide channel with a high load of suspended sediments and very fine sand transported as bedload. The sand accumulates in planar bars several hundred meters long; numerous trunks and branches of "palo bobo" (*Tessaria integrifolia*) are included in the sand mass. Large desiccation polygons to 1 m in diameter and similar depth appear in the sand when dry. Such contraction structures, common in all Chaco rivers, are probably caused by the high percentages of colloids in the sediments. Such colloids are probably composed of clay-organic complexes.

Several local fluvial networks were developed in the eastern region of the Bermejo alluvial fan system: Tapenagá, El Rey, Los Amores, Negro, all having similar characteristics. The Los Amores River will be described as typical: The upper basin, approximately 600 km^2 in area, belongs to the western Chaco. The area is underlain by pervious brown loess. Because it is dominated by infiltration, with little runoff, there is virtually no contribution of sedimentation to the channels. The middle basin is located in the eastern Chaco. It covers an area of 4,000 km^2, consisting of hard, impervious, grayish-green and reddish-brown clays, paludal in origin. It is a slightly sunken block, covered by large permanent swamps, and dense vegetation. Runoff is very slow, resulting in generalized accumulation of organic matter and development of tropical peat in some places. The region does not contribute clastic sediments channel, but the volume of colloids entering the system is large.

The lower basin of Los Amores, 3,500 km^2 in area, drains a sector of old Paraná sediments. The relief is moderate to low and the fluvial system is composed of a comparatively well-developed net of tributaries and an active collector channel. The runoff is high, with bank erosion and sediment transport (Iriondo 1987a).

The Salado River system. The southernmost tributary in the Chaco is the Salado River, with an important basin of 247,000 km^2 and 1,500 km longitude. The upper basin is formed by numerous mountain creeks draining the southern Puna (a 4,000 m high meset, characterized by salt lakes and playas). The river crosses downstream the folded Subandean Ranges and forms a complex alluvial fan in the plain. The fan system is 650 km long and 150 km wide in the distal region and composed of several subsystems: minor fans, fluvial belts, lacustrine, and swamp deposits. Silt, illite clay, and fine sand are the more important clastic sediments; chlorides dominate among the soluble salts. The minor fans were formed during dry periods on tectonically sunken blocks. Such blocks are several thousand square kilometers in area (Iriondo 1987b).

The uppermost fan of the Salado system is located in the provinces of Salta and Chaco (Argentina). It is formed by a series of small, well-preserved paleochannels cutting across late Pleistocene loess deposits in western Chaco without accumulation of significant volumes of sediments. The most probable age for the fan development is late Holocene, because channels erode the late Pleistocene loess and the early Holocene was characterized by channel stability in the region. At present, the segment of the Salado crossing the region is a permanent, well-defined channel until it reaches the latitude of 25°40′ S, where it enters into a large swamp (Bañado de Copo, 300 km^2 in area).

The swamp affects the general characteristics of the Salado system. Evapotranspiration reduces the water discharge by 50%, causing the diminution of the grain size of the sediments from coarse silt upstream to fine silt at the edge of the swamp where the river lives the depression. Downstream, the river is reconstituted by successive contributions of small channels originated inside the swamp.

The southeastern corner of the distal region of the Salado system, located in the province of Santa Fe, is composed of a series of large, stable paleochannels, formed during a late Pleistocene humid period. The channels are buried by a 6 to 8-m-thick loess carpet; however, they are still visible in the field and in aerial photographs. Their primary features are a well-sorted sand content, marked tortuosity, and a width of up to 200 m. Paleochannels of different ages and directions can be seen partially superimposed and crossed in an irregular pattern. The sand in the paleochannels constitutes reservoirs of low salinity groundwater, a fact of high significance in a region with serious water shortage.

The rest of the distal region is formed by clayey silt of swampy origin containing high percentages of colloids. The area is flooded during several months in humid years. The only features emerging in a very flat, tall grass landscape are small surfaces covered by dense forest, named "isletas" (little islands). Such mounds are built by the biological activity of ants (gen. *Atta*). A colony of *Atta* ants can build a mound several meters wide, with a system of chambers and galleries reaching as much as 3 m under the surface (Bonetto 1959). The volume of sediment removed and altered in each ant hill can total several cubic meters, resulting in a soft and damp material composed of rounded clods smaller than 2 mm in diameter and having high porosity. The accretion of successive colonies can result in 40 to 80-m-long landforms providing a very good substratum for trees and shrubs.

1.4.4 The Pampa

The Pampa is a large plain of eolian origin. It is formed by a sand sea (outside of the Paraná basin) surrounded by a wide loess belt (Fig. 1.10). The loess,

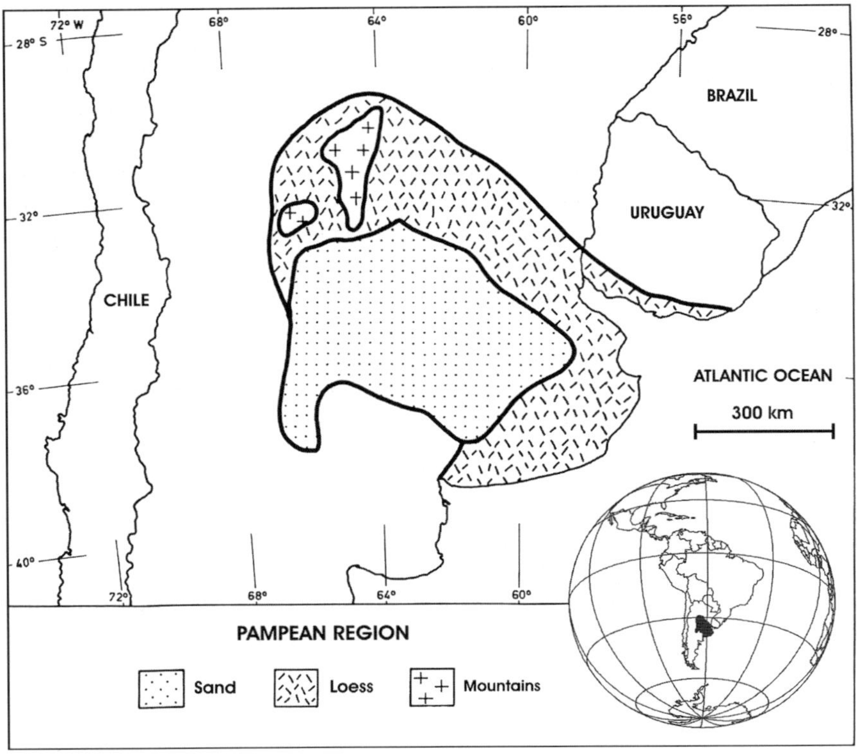

Fig. 1.10 Pampa region

partially located in the basin, is rich in quartz, volcanic glass, and illitic clay; it contains frequent carbonate concretions. It is poorly drained by river channels, the only important system is the allochthonous Carcarañá River. On the other hand, this region is dominated by very interesting cases of "non-classical" fluvial nets, characterized by diffuse divides, swampy low-order waterways and other anomalous elements. The scenario was described as follows (Iriondo and Drago 2004).

1.4.4.1 First Element - Water Divides

The nature of divides is probably the most complex and confusing problem linked to water bodies in plains. The small slopes and poor divide definition of the catchments sometimes allow that water masses flow towards one or another drainage basin. In some cases, it is possible to define the water divides according to the classical methods of geomorphology and hydrology. In other cases, water divides are not well-defined borders on the terrain because they are located in large and flat surfaces. Therefore, water flows in one or other direction according to wind, vegetation type, or hydraulic gradient due to the differences in the rain distribution. This is a classical case of a "fuzzy class", i.e., the transition between a class and other (neighbor) class is not neatly in one line but gradual in one long series, which ranks from zero to one. From a point of view of surface hydrology, such areas have a probabilistic structure, meaning the surficial waters at different points flow with a higher frequency to one or another basin according to the distance to the definite basin; the relative frequencies change from a border of the divisory area to the opposite border.

Another type of undefined water divide is characterized by slight slopes that are independent of a particular catchment, such areas are inclined planes (shared by two adjacent catchments) that dip parallel to the collectors. An example of this type is the divide of the Cululú River, a tributary of the Salado River located in the central area of the Santa Fe Province.

Within this basic hydrological pattern, a typical case in sandy areas (such as the Pampean Sand Sea) is the rapid infiltration of rainwater in the relatively elevated sectors, followed by a sub-superficial flow to neighbor depressed areas, which result afterwards very slowly flooded by seepage.

This process is a very important phenomenon in the hydrological behavior of the Pampean Plain. In 1983, owing to a large El Niño, several thousand square kilometers of the northern part of Santa Fe northwest of Buenos Aires and south of Córdoba were flooded by this process. Studies made in the province of Santa Fe (Iriondo 1983), allowed to group the water divides into three types: (a) Clearly defined, linear segments; (b) High, plane and homogeneous undefined areas, without dells or any kind of irregularities, mainly composed of loessic sediments, that is, highly permeable materials. Due to this fact, there is no water storage on the surface, but a rapid infiltration; (c) Small closed areas of recent tectonic origin.

1.4.4.2 Second Element - Transient Marshes (Bañados)

Drainage networks in the Pampa begin in a kind of non-permanent wetlands, locally named "bañados". These elements have a marked identity in the plain and are well known in the region for centuries. They are relatively low and flat surfaces periodically covered by water during larger or shorter time spells. Such water bodies occupy large, gentle depressions of different origins, most of them originated in the past eolian landscape. The flooding of large areas (e.g., 20,000 km^2) during large periods (weeks or months) is one of the most important characteristics of the Pampean Plain.

The bañado is the typical case of unchanneled flow on the plain. It is a temporary aquatic environment, mainly fed by local rains with an undefined and variable perimeter and without its own sediments and a maximum depth less than 1 m. Such a phenomenon occurs at large intervals;

recurrences have a minimum of several years; it is linked to extreme climatic events such as the as El Niño phenomenon. A striking characteristic of these inundations is that they leave no indications in the landscape once they disappear. A typical bañado moves very slowly downwards along the gentle relief, in terms of days or weeks, undergoing changes in surface area. It can be described as a *"transient marsh"*.

An unchanneled flow has the following hydrological phases:

a) **A water storage phase**, with a very slow flow velocity (usually centripetal). Here, the water stagnation predominate over the drainage.
b) **A water drainage (first phase)** arises after storage, lowering the residence time in the storage areas.
c) **A flood phase**, results in flooding of the areas located downslope of the storage areas.
d) **A water drainage (second phase)**, which occurs when temporary inundated areas are drained.
e) **A channelization phase** means that the residual waters are conveyed along dells and fluvial channels.

Frequently, these phases show a partial simultaneity in this flat landscape.

1.4.4.3 Third Element – Dells

The water masses flowing slowly on the plain eventually occupy geomorphological elements named dells (or "cañadas" in Spanish). Dells are shallow linear depressions, most having straight and well-defined limits containing permanent or temporary waters. Some dells include a small channel at the center, but such a channel always is a minor feature, strongly subordinated to the rest of the depression in the general hydrodynamics.

In many cases, dells are tectonic lineations located at parallel and regular intervals. Most of them are asymmetrical in cross section; such a characteristic is imperceptible to direct observation in the field, but it can be detected by observing the different widths of the vegetation belts at both sides of the central depression. The deeper dells can contain permanent water and paludal vegetation. At both sides are located semi-permanent wetlands with tall grasses ("pajonal"). Frequently, the wetland on one side of the depression belt is two or three times wider than that at the opposite side, composing an asymmetrical cross section. Such a pattern is attributed to a tilting of small tectonic blocks.

The size of dells is variable, ranging between 200 and 400 m in width and reaching usually more than 5 km in length, with maximum lengths up to 35 km. In some areas in north Santa Fe appear parallel dells with 1.5–2.5 km equidistance. Dells are scarce or absent on other areas. In Santa Fe Province (Argentina), dells underwent the influence of hydric and eolian processes of variable importance. Three different types can be recognized:

a) **Simple dells.** Generated by neotectonic processes, without influence of exogenic agents.
b) **Dells with deflation hollows.** Dells were favorable environments to the formation of deflation hollows during arid climates. At present, the deflation hollows (seasonally transformed in ponds) form long rows aligned along the central line of the dells.
c) **Dells with fluvial channels.** This class has undergone a more advanced hydric modelation. The channel is normally considerably narrower than the dell width; it has little influence in the local drainage dynamics.

Dells collect the local surficial and hypodermic flow. The majority have temporary or intermittent regime, conveying water only during the rain season or after important storms. From a hydrological point of view, dells are characterized by a high roughness, originated by dense mats of tall grasses as *Panicum prionitis* ("pajonal") or *Spartina densiflora* ("espartillar") among others. The high roughness and the low longitudinal slope strongly hinder the movement of water. Second-rank differences among dells are visible. Each type of dells show a different hydrological behavior; type "c" allows a comparatively rapid drainage whereas type "b" (with deflation hollows) shows the lowest drainage capacity. The later have larger retention times, originating larger infiltration, evaporation, and evapotranspiration rates.

During the present humid climate, dells tend to evolve into a fluvial landscape, integrating in some degree the drainage network. Such a process is considerably evolutioned in the Cululú River basin (located in the center of Santa Fe Province, Argentina), where dells developed to true fluvial channels completely integrated in the Carcarañá fluvial net. On the other side, in the north of the province, the process is less significant or absent; the basin of the San Antonio River includes several dells directly connected with the collector and other completely isolated. The Ceres-Tostado area (NW of the Santa Fe Province) is characterized by a large number of dells without any hydric evolution.

Frequently, dells function as headwaters of autochthonous rivers, which means that they are the first-order channels (sensu Horton 1945; Strahler 1952) in such basins. The drainage pattern is parallel, without a dendritic development, because the tectonic origin of the morphological elements. In general, the total length of the small dells is commonly far larger than the total length of the major ones.

1.4.4.4 Fourth Element – Fluvial Channels

A fluvial channel is easily distinguished from the above described elements; it is a trench incised by the downslope flow of water. In an active channel, the water velocity varies by some dm s^{-1} to several m s^{-1}. Fluvial channels in the aggradation plain are shaped by the water which conveys. The hydraulic geometry of the fluvial channel (cross section, longitudinal profile, depth, banks, etc.) as well as the evolutionary mechanisms is very different to those in dells. The nature of a channel is represented by the volume of water and sediment conveyed by processes of erosion and sedimentation in low gradients.

1.4.5 Connections with The Amazon Basin

The contact (sporadic or seasonal) of the Paraná and Amazon fluvial nets is real. The divisory between both basins is around 2,000 km long, crossing several geomorphological units of different characteristics. Some of these landscapes favor transfluences, captures, floods and spillouts to one or other side. Two examples can illustrate this condition. The most important one is located in eastern Bolivia, where the 65,000-km^2 alluvial fan of the Río Grande is crossed by the present divisory, 70% of them (including the main stream) belongs to the Amazon, and the rest to the Paraná basin (Fig. 1.8). Overflows and paleochannels of the Río Grande form the Paraná portion of the alluvial fan; they extend to the foothill of the Bolivian Precambrian Shield, at the northern border of the Chaco. Paleochanels are 40–80 m wide and 2–3 m deep, dimensions suggesting considerable discharges.

The second example is provided by the sources of the Paraguay River. Small first-order channels drain a meset of Cretacic/Tertiary age, where an old landscape is preserved in hard rocks. Paleochannels of this epoch are partially occupied by tributaries of both sides of the divisory. The rivers Aguapehy (Paraná basin) and Alegre (Amazon basin) flow in a parallel direction along 40 km, separated by a distance of only 1,000 m (Soldano 1947, Fig. 1.11).

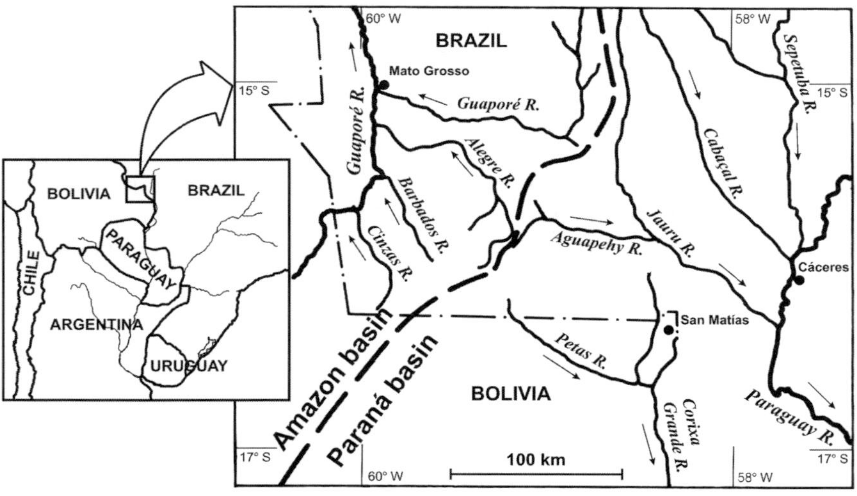

Fig. 1.11 Upper Paraguay-Amazon divisory

References

Bonetto A (1959) Las hormigas “cortadoras” de la Provincia de Santa Fe. Dirección General de Recursos Naturales, Ministerio de Agricultura de Santa Fe. Rosario, 83 pp

Cordini R (1947) Los ríos Pilcomayo en la región del Patiño. Anales I, Dirección de Minas y Geología. Buenos Aires, 82 pp

Cucchi R (1973) Los ríos fuera de proporción de Formosa y su significado paleoclimático. Revista de la Asociación Geológica Argentina 28:588–590

Horton RE (1945) Erosional development of streams and their drainage basins: hydrophysical approach to quantitative morphology. Bull Geol Soc Am 56:275–350

Huamán A, Ballivián G, Méndez O, Algañaraz J (1975) Levantamiento integrado de los recursos naturales del sector occidental de la provincia de Cordillera, parte llana. Corporación de Desarrollo Regional de Santa Cruz. Santa Cruz, 30 pp

Iriondo M (1974) Los ríos desajustados de Formosa. Una hipótesis alternativa. Revista de la Asociación Geológica Argentina 29:136–137

Iriondo M (1983) Geomorfología de la cuenca inferior del Río Salado como base para la investigación hidrológica. En: Coloquio Internacional de Grandes Llanuras CONAPHI Olavarría

Iriondo M (1987a) Geomorfología y Cuaternario de la provincia de Santa Fe. D’Orbignyana, Corrientes 4:1–54

Iriondo M (1987b) Mapa de las llanuras de agradación argentinas. XII International Congress, INQUA Ottawa

Iriondo M (1989) The Quaternary lakes of Argentina. Palaeogeogr Palaeoclimatol Palaeoecol 70:81–88

Iriondo M (1999) Climatic changes in the South American plains: records of a continent-scale oscillation. Quaternary International Pergamon Press, Oxford, pp 57–58, 93–112

Iriondo M, Drago E (2004) The headwater hydrographic characteristic of large plains: the Pampa case. Ecohydrol Hydrobiol 4(1):7–16

Necco G (1989) Extratropical weather systems in South America. In: Annals 3rd Meteorological and Oceanographic Conference of the Southern Hemisphere. American Meteorological Society, C.A.M. Buenos Aires, pp 149–155

Neiff J (1986) Sinopsis ecológica y estado actual del Chaco Oriental. Ambiente Subtropical, Corrientes 1:5–36

Orfeo O (1986) Evaluación de sólidos suspendidos en algunos ríos del noreste argentino. Revista de la Asociación de Mineralogía, Petrología y Sedimentología 17:61–66

Pedrozo F, Orfeo O (1986) Estudio sedimentológico de ambientes fluviales del Chaco Oriental. Ambiente Subtropical, Corrientes 1:60–72

Rabicaluc H (1986) Situación del río Pilcomayo. Informe 32H86. Dirección de Recursos Hídricos de Formosa, 8 pp

Paoli C, Iriondo M, García M (2000) Características de las cuencas de aporte. In: Paoli C, Schreider M (eds) El Río Paraná en su Tramo Medio, Universidad Nacional del Litoral, Santa Fe, pp 27–68

Servant M, Fontes J, Rieu M, Saliege J (1981) Phases climatiques arides holocenes dans le sud-ouest de l’Amazonie. Comptes Rendues Academie des Sciences, París 292(II):1295–1297

Soldano F (1947) Régimen y Aprovechamiento de la Red Fluvial Argentina. Editorial Cimera, Buenos Aires, pp 214

Strahler AN (1952) Hypsometric (area-altitude) análisis of erosional topography. Bull Geolog Soc Am 63:1117–1142

2 Geomorphology

Martín H. Iriondo

2.1 The Surrounding Regions

The Middle Paraná floodplain (Fig. 2.1) is surrounded by several regions that influence the environment of the system in various ways. All of them, except for one (the delta at the Río de la Plata) are located upstream, a condition that reinforces the character of contributors that deliver water, sediments, salts and heat into the plain (Fig. 2.2).

2.1.1 The Lower Paraguay River

The lower Paraná River flows inside an abandoned segment of the Paraná floodplain which is 220 km long and 5 to 15 km wide (Drago 1975; Iriondo 1993). The intense morphogenetic activity of the Paraguay re-shaped the old internal units of the plain; Drago (op. cit.) described three geomorphological units: present meander plain, old meander plain, and hindered dranage plain, all of them Holocene in age. The hindered dranage plain is the oldest, and is characterized by numerous permanent and non-permanent swamps. It has been partially eroded, remaining only a few relicts of the original surface. The old-meander plain is located preferently on the right side of the valley, which suggests a secular migration of the channel to the east. This river has a particular hydrological behavior owing to important natural perturbations occurring in the basin. The annual flood has a long and shallow wave because the attenuating and delaying effect in the runoff provoked by the interference of the Pantanal. The Pantanal is a 60,000 km^2 wetland formed by drowned eolian dunes, paleochannels of alluvial fans, and other obstacles. A general delay of 1–2 months is registered in the peak.

The lower Paraguay receives two important tributaries that can duplicate the discharge of water: the Bermejo River at the right bank, and the Tebicuary River at the left bank; the Bermejo, besides, contributes with most of the discharge of suspended sediment in the whole Middle Paraná system. The backwater effect of the Paraná River at the confluence modifies the hydraulic slope at the Paraguay

CONICET/UNL, C.C. 487, 3100 Paraná, Entre Ríos, Argentina, e-mail: rniriond@ceride.gov.ar

M.H. Iriondo, J.C. Paggi, and M.J. Parma (Eds.)
The Middle Paraná River: Limnology of a Subtropical Wetland

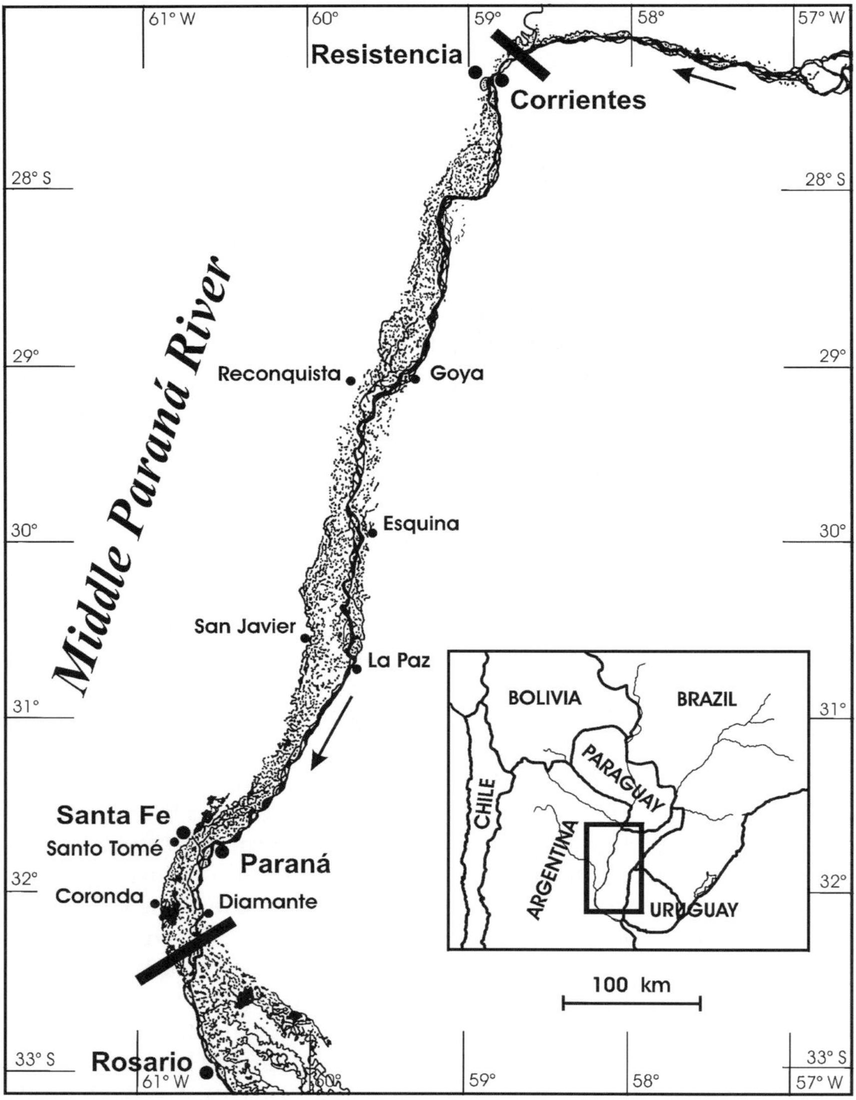

Fig. 2.1 Locations of the Middle Paraná floodplain

mouth with variable intensity, affecting all the segment in extreme cases (Paraná flood with low waters in the Paraguay). The bed-load of the lower Paraguay is formed by fine and very fine sand composed of more than 99% quartz. A terrace 4 m high above high waters appears at the right bank of the floodplain that is composed of plastic grayish green clay covered by 30 cm of gray eolian silt.

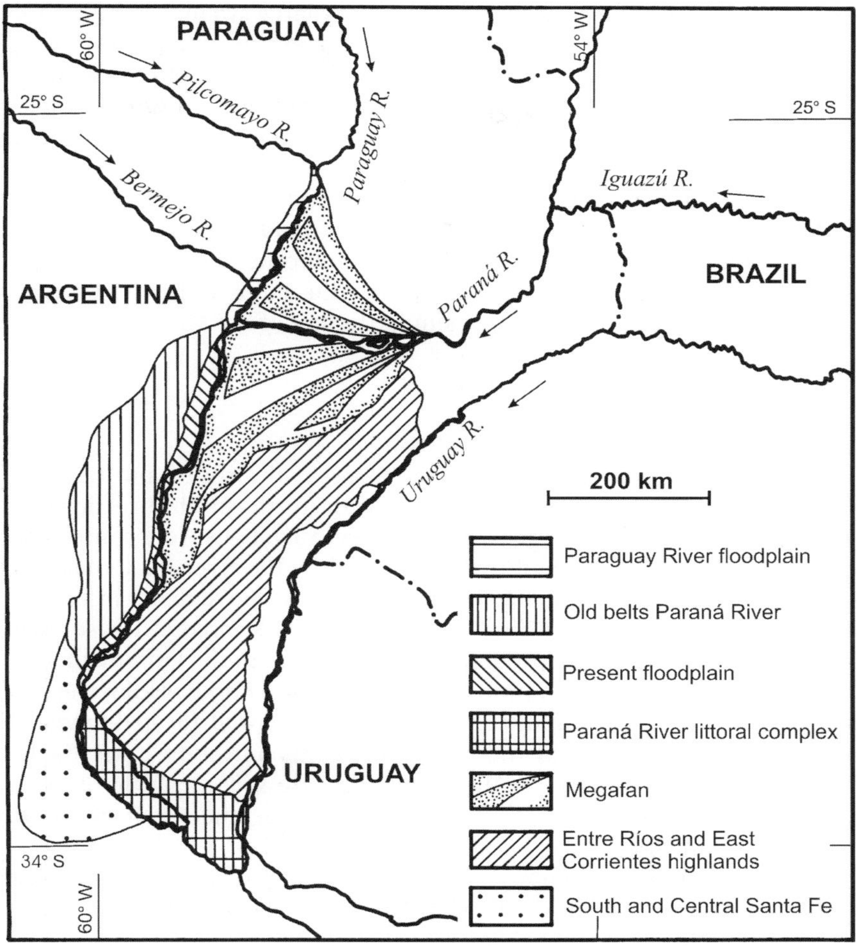

Fig. 2.2 Neighboring regions

2.1.2 The Megafan

Leaving the basaltic meset, the Paraná River enters the Argentine plains. In the first segment, it generated a large alluvial fan (or megafan), which covers the NW half of the Corrientes province in Argentina and the southern part of eastern Paraguay. The fan is 260 km long and 500 km wide in the distal line and is located at the Paraguay-Paraná left bank. The megafan has been stable for most of the Upper Quaternary. Inside it, the river channel built relatively stable belts, occupied during some hundreds or thousands of years, eventually abandoned by the main stream and replaced by large swamps. A sequence of such intervals formed the present megafan. A few extensive deposits produced by spillouts during dry climatic phases in the Upper Quaternary are intercalated among the abandoned belts. Also during dry phases, important deflation of sand occurred in the abandoned belts, generating up to 80-km-long and 5-km-wide dune fields.

The present river belt crosses the fan in an east–west direction; the channel there is braided, with frequent elliptical islands and sand banks. The youngest abandoned belt in the megafan is Holocene in age and runs along the eastern limit of the system and includes the Iberá and Luna lakes. The ecological system of the megafan is similar to that of the Middle Paraná.

2.1.3 The Chaco Old Belts

Downstream of the large fan, the river turns to the south and begins the Middle Paraná reach. Several floodplains with north–south directions were built by the stream along its geological history. Such belts were controlled by tectonic fractures. Each position was occupied by the river for thousands of years and eventually abandoned owing to tectonic movements in relatively short-time shifts called "avulsion". Old belts are preserved at the surface in the SE of the Chaco region (province of Santa Fe) up to the latitude of 31°40'. The region was subject to a humid climate during the middle Holocene (8,500 to 3,500 years B.P.). Local alluvial systems developed well-formed fluvial nets on the inherited landscape abandoned by the Paraná. In some cases, the local rivers flow in paleochannels, others have structural control; the most frequent case is a sequence of both types of control. The major valleys are from 2,000 to 2,300 m wide.

The Chaco old belts compose a broad stripe along the right bank of the Middle Paraná with a maximal width of 90 to 100 km; the downstream extreme is located at the city of Santa Fe. This geomorphological unit is characterized by paleochannels of different sizes formed by the main channel and minor branches. Surficial paleochannels are Upper Pleistocene in age; most of them have a meandering pattern (meaning different to the present main channel) and have suffered variable degrees of destruction and infilling with local sediments. The growing of tropical peat, although in lesser scale than in the fan, is also common in swamps and paleochannels of this unit.

A few dune fields were accumulated by the wind at the fringe of the floodplain during a dry climatic phase that occurred in the Late Holocene (3,500 to 1,000/1,400 years B.P.); the northernmost (near the town of Alejandra), having a 17-km N–S longitude and 5 km width, is the largest one. The eolian dynamics has also produced deflation hollows several hundreds of meters in diameter, in some cases flanked by lunettes or "clay dunes". Frequently, such depressions are transformed in shallow lakes.

2.1.4 Eastern Highlands

This area is formed by old Quaternary sediment of the Uruguay River and Last Glacial loess generated by SW winds. The Quaternary sediment is represented by the Hernandarias Formation, playa deposits characterized by montmorillonite clays and very fine sands composed of mature quartz. Abundant carbonate and gypsum precipitates appear in the sediments, particularly in the lower section of the formation and the thickness is between 20 and 40 m. The dominant colors are gray, green, and reddish brown. The landscape is flat, covered by vertic soils and is prone to saturation by rains. This system covers about 30,000 km^2 in the Paraná basin and extends to the east into the Uruguay River basin.

The Last Glacial loess is formally named the Tezanos Pinto Formation. It covers a landscape of hills and valleys downstream from the former area. The loess is composed of a loam formed by illitic clays mixed with felspars and quartz. The color is pale brown and deep brunizem soils were formed at the surface of the terrain. The original wood vegetation of the hills has been replaced by agriculture in most of the area. This loess covers a surface of 20,000 km^2 and is a sector of the Pampean sediments described in Chap. 1 of this book.

2.1.5 The Pampa in South Santa Fe

This sector of the Pampa region limits the southwest half the Middle Paraná floodplain at the right bank. An extended description of the Pampean geology and hydrology can be found in Chap. 1 of this book. The area immediate to the Middle Paraná is the northeastern corner of the system and is characterized by an Upper Pleistocene sequence of loess and swamp sediments with a thickness of 15 to 40 m. The relief is scarce, and is dominated by tectonic tilting of small blocks. Few and small rivers reach the food plain in this area, the largest among them is the Carcarañá. The contribution of chlorides and others salts via fluvial and seepage waters into the Paraná system is probably important.

2.1.6 The Littoral Complex at the Paraná Mouth

The Paraná River reaches the Atlantic Ocean in a 300-km-long and 80-km-wide littoral complex. Most of its evolution occurred during the Holocene. The development of this complex comprises four phases: (1) A fluvial period represented by river flood deposits; (2) a marine ingression with development of a sand barrier, a lagoon, minor tributary deltas and estuaries and well-developed regression deposits; (3) an estuarine phase characterized by extensive tidal deposits in the central area; and (4) the present fluvial period, with channel deposits and deltaic deposits advancing into the Río de la Plata. The present dynamics of the complex is dominated by the Paraná floods, the floods of the Uruguay and Gualeguay Rivers, tides, and inundations produced by a southeast wind.

2.2 Neotectonics

Plains are generally influenced by tectonic movements. Not only the regional features but also minor forms (direction of channels, asymmetry of small basins, etc.) frequently have structural control. The vertical component of the movements is very small (no more than 6 m in the Paraná floodplain, for example); however, it is significant because of the near horizontality of the plains of South America.

Large rivers with high morphogenetic energy tend to make uniform the inherited geological features. Therefore, most of the visible structural elements in the large floodplains are the consequence of present-day or recent neotectonic processes. On raised blocks, the floodplain is narrow, the channel straight, and the older sedimentary deposits appear as terraces. On sunken blocks, the longitudinal slope is very small, the floodplain is wide, the channel shifts, and old sediments are permanently submerged. Evidence of this tilting can be observed as in the case of the channel eroding only one of the valley cliffs, and terraces on one side of the river, etc. (Iriondo and Suguio 1981). Both the Amazon and Paraná Rivers are clearly influenced by neotectonics.

The Paraná River flows across a region composed of several structural blocks elongated in a N–S direction, and most of these are tilted to the east. The blocks are superimposed on an old structural system characterized by NW–SE and NE–SW alignments. In fact, the pattern is similar to the Amazon region, but with the N–S components being more developed. In the Pampa plain, west of the Paraná valley, small grabens with square or rectangular shape are frequent. Typical dimensions for these structures are 50–200 km^2 in area and 2–5 m topographic difference in elevation.

The Paraná River flows 600 km from north to south along a tectonic lineation. The lineation, which is probably Pleistocene in age, is a complex element that crosses through minor blocks suffering independent vertical movements. Most of the blocks are tilted to the east, and, consequently, the main channel tends to flow along the eastern cliff of the floodplain. In contrast to the Amazon, in the Paraná we could not find any differences between the present and other Holocene movements. In general, the upper reach of the floodplain (from Corrientes to Bella Vista) is in process of subsidence. The lower reach, between Santa Fe and Rosario, tends to rise. The intermediate sector is composed of several minor blocks with a non-uniform behavior.

The hydraulic gradient of the river is composed of a sequence of segments of different slope.

2.3 Climate

The important point to understand in this issue is that the Paraná floodplain shapes its own climate, owing to the dominant influence of the huge volume of water flowing in a north–south direction in channels, and in lakes and swamps. Particularly, the large river discharge amounting to about 20,000 m^3/s with water flowing near 1 m/s produces a significant transport of heat from the tropics to the temperate region. In synthesis, the climate of the Middle Paraná is humid subtropical with hydric excess during the entire year, without forests, 1 to 5 storms per month, and moderate cloudiness (100–120 days/year).

2.4 Climatic Changes in the Quaternary

Important climatic changes that occurred over the last few thousand years provoked modifications in the region and influenced the present ecology of the Middle Paraná (in both physical characteristics and biological pattern). The Quaternary was a period that produced strong climatic changes over the entire earth. In the great South American plains, located at a low latitude, wet and dry climates alternated. During the Pleistocene, the Argentine plains were subject to two long humid periods followed by dry intervals (Iriondo 1984). Data are lacking about the temperatures of each climate, but very probably humid climates were warm and the dry intervals cooler than the present.

Large variations in precipitation also occurred in the Holocene. During the humid periods, the fluvial nets developed and wide floodplains appeared in the major tributaries and during the semiarid intervals, large alluvial fans and swamp deposits developed. Extremely dry conditions also produced aeolian erosion and deposition (Iriondo 1982).

During the late Pleistocene, the Argentine section of the Paraná River was under an arid climate, with generalized loess deposition and disintegration of the fluvial nets in the surrounding plains. In the northeastern provinces, an area not reached by the loess sedimentation, the depressions were filled by swamp environments (Iriondo 1984).

The lower and middle Holocene, including the Hypsithermal Episode, was characterized by a humid climate that reactivated the hydrographic nets. Deep soils developed on the interfluves, with well-defined B-horizons, indicating good vegetal cover and landscape stability (Iriondo 1987).

Between 3500 B.P. and 1400 B.P. an arid episode occurred that provoked wind erosion of the A-horizon of the soils and the sedimentation of a loess mantle a few decimeters thick (San Guillermo Fm.). The aeolian sedimentation advanced 300 km more to the northeast than the Pleistocene loess. The dune fields located in the Paraná floodplain near Reconquista and northern of Esquina also developed during that period.

The present subtropical humid climate stabilized around 1000 B.P. The isohyets have a general NW–SE direction; the available information (Iriondo 1984, 1987) suggests a similar disposition during the upper Pleistocene.

2.5 The Sediments

In the Middle Paraná (Fig. 2.3), the river flows on a broad and thick carpet formed by its own sediments in the typical pattern of floodplains. The bulk of sediments is composed of well-sorted, fine quartz sand with grain sizes between 0.2 and 0.3 mm. The sand was originated in the destruction of Mesozoic eolian sandstones in the upper basin of the Paraná, so the high degree of sorting of the sand is the result of the fluvial dynamics, but it is inherited from the former sedimentary cycle. The thickness of the sand in the floodplain is about 40 m, carved in erosive discordance on older floodplains.

Fine sediments have contrasting sources: most of them arrive from the Cordilleran/Puna realm along the Bermejo River and other tributaries and are formed by illitic clays. Another volume composed of kaolinite is conveyed by the upper Paraná, and a small proportion dominated by montmorillonite is contributed by small basins in the SE, developed in terrains derived from the Uruguay River system.

Surficial sediments which compose the different geomorphological units of the Middle Paraná are dominated by four sedimentary facies: (A) Channel facies, which are composed of fine and very fine sand generally in 10 to 40-cm-thick horizontal strata with diagonal structures; (B) levee facies, which are characterized by silty-clayey very fine sand with intermediate plasticity; ochre to greenish gray in color, mottled, with numerous pores, tubes and root molds. Gley processes are frequent; (C) lake and swamp facies, composed of dark gray silt with abundant organic matter in different degrees of decomposition. It is compact, with a low porosity; sand and clay are scarce; and (D) facies of aeolian dune, very fine sand, without visible internal structures, brown in color; it appears in isolated spots in the northern half of the floodplain.

2.6 Geomorphology

The morphology and sedimentary deposits of floodplains are determined by the flow regime, which in turn depends on the hydrological pattern of the basin. Two groups of mechanisms operate in floodplains: channel processes and flood processes. Channel processes deposit sand and pebbles in elliptic of arcuate

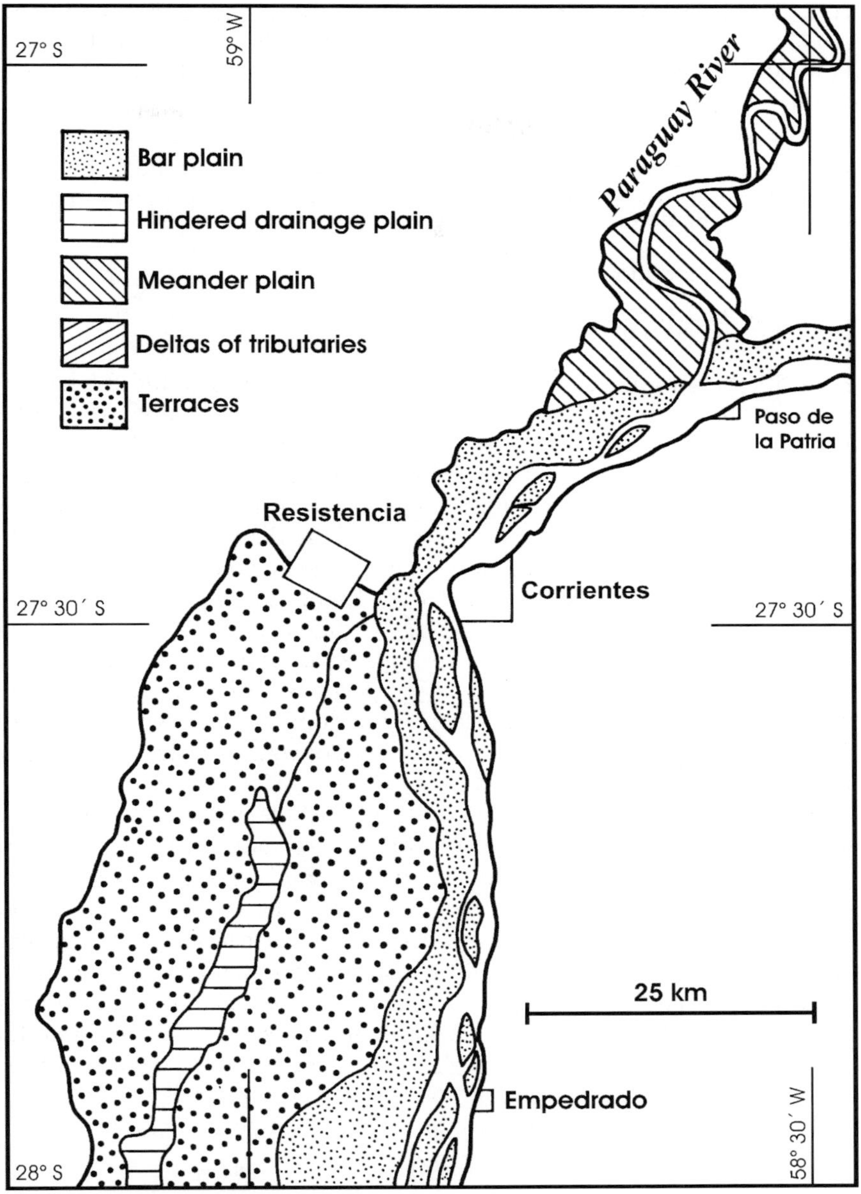

Fig. 2.3a–d Geomorphological map of the floodplain

(Continued)

banks; they develop at maximum in humid climates. Typical flood deposits, on the other hand, derive from the deposition of fine sediments transported as wash load and form a landscape with swamps and lakes. Flood deposits develop better in semi-arid climates characterized by uneven distribution of rains.

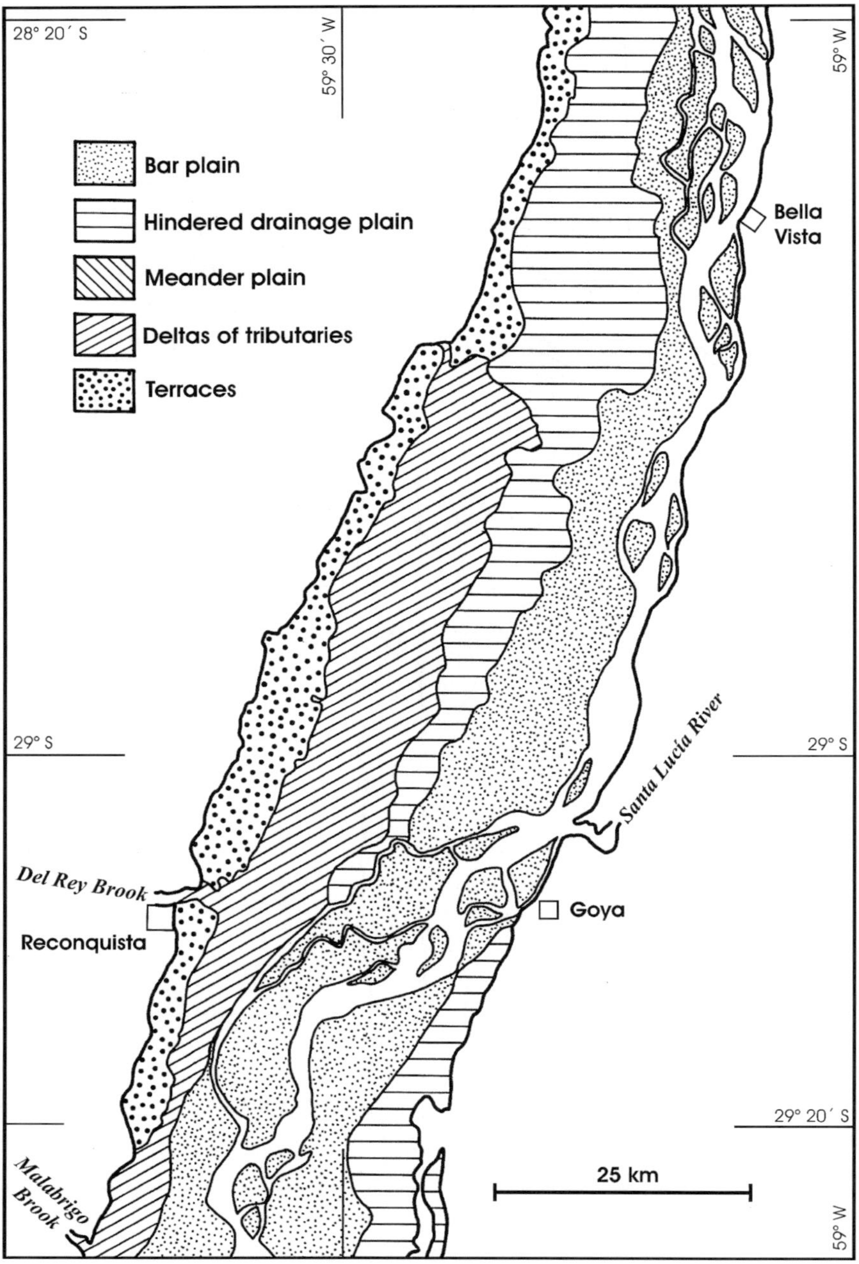

Fig. 2.3a–d (cont'd)

(*Continued*)

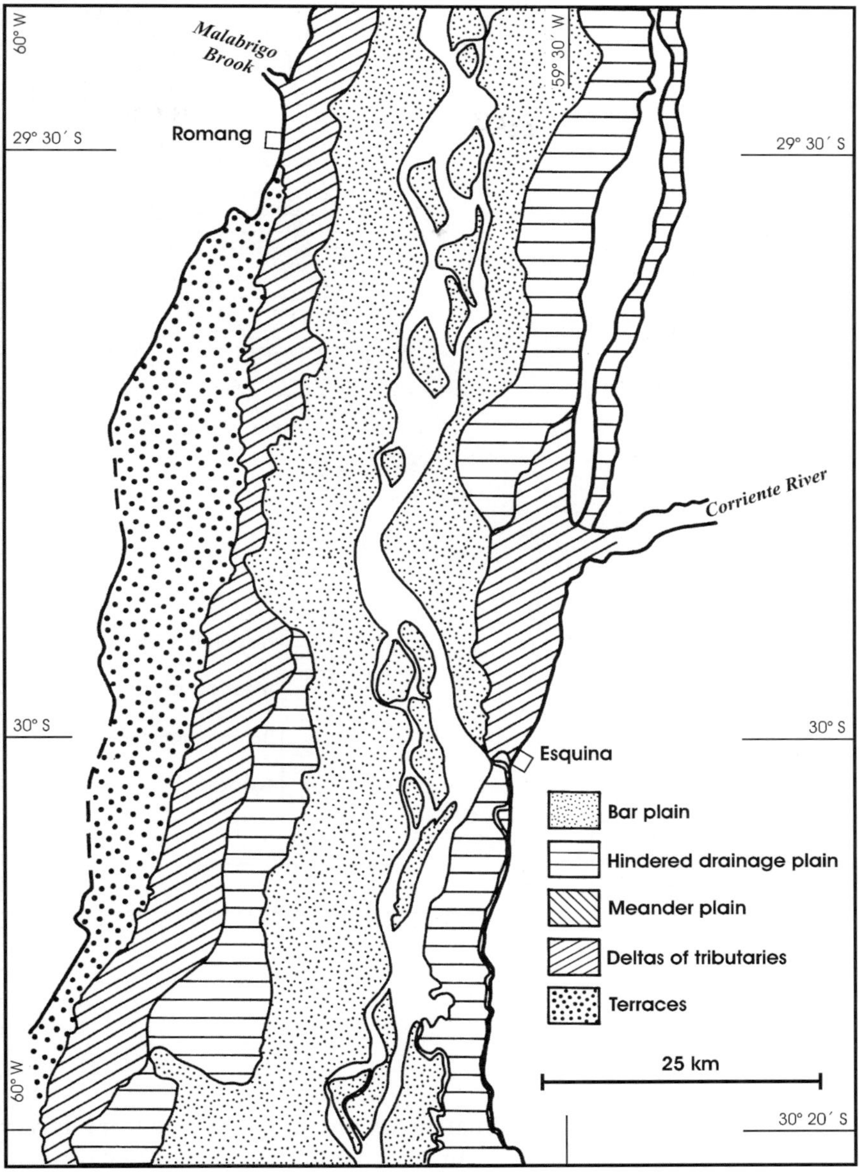

Fig. 2.3a–d (cont'd)

(Continued)

The most striking point in the comparison between the Amazon and Paraná Rivers is the similarity of both floodplains. Their general evolution during the Holocene and the local characteristics of the geomorphological units are very similar, with only second-order differences.

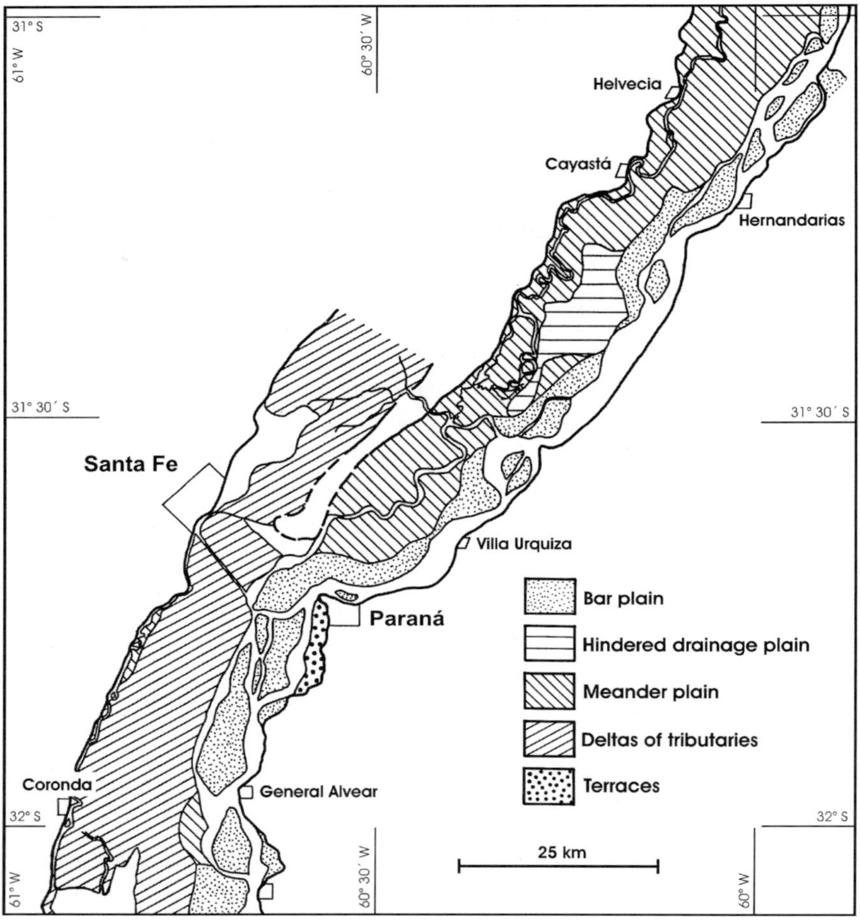

Fig. 2.3a–d (cont'd)

2.6.1 Geometry

The Paraná floodplain is approximately 600 km long, reaching from the confluence Paraná-Paraguay, 27°25′ S to 33°S, south of Rosario. Downstream it passes transitionally to a well-developed Holocene delta. The general direction of the plain is north–south, crossing perpendicularly or diagonally several tectonic blocks. This relationship has generated local segment within the floodplain, with different directions and slopes.

From the confluence of Paraná-Paraguay until 28°10′ S, the Paraná flows 75 km on a sunken block, widening the floodplain from 13 to 45 km; the hydraulic slope of the channel is 4.45×10-5. The segment that follows is 150 km long and steeper, with a hydraulic slope of 5.80×10-5; the width of the plain varies between 25 and 35 km. From Goya to Esquina (30°S), the plain passes through an elevated block with a $6.37 \times 10\text{-}5$ slope. Downstream from here, the floodplain passes across several minar blocks and narrows from 40 to 8 km at Santa Fe, a section which owes its origin to a regionally important fracture.

In most of the reach, Confluencia-Santa Fe the main channel flows along the left cliff of the plain, indicating a west–east tilting of the tectonic blocks. From Santa Fe to the delta apex, on the contrary, the channel crosses the floodplain and flows along the right margin. The plain widens to 40 km at the delta apex and the total area of the floodplain is 19,290 km^2.

There are two major groups of geomorphological units inside the floodplain; one of them directly linked to the main channel and the other formed or influenced by the tributaries. Those directly depending on the main channel are the channel itself, the bar-plain, the hindered-drainage plain and meander-plain of the Colastiné branch. Units influenced by the more important tributaries form delta-like areas at the confluences of the rivers Salado, San Javier, Corrientes, El Rey, and others.

2.6.2 The Eastern Cliff

The left cliff of the floodplain is an imposing geographic element 1,200 km in length and 30–50 m in height. It begins in the apex of the alluvial fan (27°30'S and 56°W), crosses the entire Middle Paraná, and ends at the littoral complex of the Paraná mouth at 34°S. The cliff is formed by a sequence of geological formations. The long-term evolution of this feature is the retreat by means of mass movements, triggered during sporadic long rainy seasons. Each geological formation has a particular "equilibrium profile", which is the maximal angle that can resist the weight of the own formation plus the above column of sediments. When the slope is higher than this, a mass movement occurs until the equilibrium profile is reached.

The active factor in the Middle Paraná is the main channel of the river, which flows along the left margin of the floodplain; the stream slowly erodes the base of such equilibrium profile, provoking a rejuvenation of the retreating process. Sectors of high instability are intercalated with other more stable ones, depending on the different clay minerals included in the sediments.

A second mechanism of cliff retreat with important ecological and biogeographic influence is the development of gullies. Those grow by rill erosion caused by the rain in the neighboring high terrains. A typical gully is 500–2,000 m long, 100–300 m wide, and 10–30 m deep, with a plane bottom and vertical margins. Such gullies serve as warm ecological niches where the tropical vegetation advances to the southern temperate latitudes.

2.6.3 The Main Channel

The main channel of the Paraná is typically braided and composed of a sequence of wide segments that are characterized by two or more branches with lateral erosion and sedimentation activity forming an unstable island. Such segments are limited by short and well-defined contractions, in which the river flows in a single channel. The contractions, determined by transverse fractures, are planimetrically stable sections where the channel has not shifted over the last 80 years.

The wide segments are 20–30 km long and 4–8 km wide including islands and sand bars. The summed width of the channel branches is 2–5 km and the typical depth oscillates between 5 and 10 m. In the contractions, the width of the single channel is smaller, measuring between 0.5 and 1.5 km; the depth oscillates between 15 and 30 m. The total surface of the main channel is 1,803 km.

2.6.4 The Bar Plain

A belt of elliptic bars composed of fine sand encloses the main channel along the entire floodplain. This belt is permanently modified by erosion and sedimentation and has a width of 2–7 km and a total area of approximately 4,150 km. The bar-plain is formed by islands inside the major channel and sand bars developed at the channel margins. Most islands began to form and grow as a juxtaposition of individual bars, which are 500–1,000 m long and 200–500 m wide. The islands are up to 5–6 km long and 2–3 km wide in the area of Bella Vista, and larger in the south (up to 35 km long and 7 km wide). This difference in size suggests a tendency towards a neotectonic sinking in the south. The islands are crossed by minor active channels, most of them with intense lateral migration, resulting in irregular belts of small meanders.

At both sides of the channel, this geomorphological unit has the same pattern as the islands, with irregularly juxtaposed bars and active minor channels originated in the collector and that join it downstream.

2.6.5 Large Branches

The main channel originates a few large branches that flow on the right side of the floodplain, such as the San Javier and Colastiné “rivers”. These convey large discharges on the order of 1,000 m^2/s and produce an intense morphogenesis.

The San Javier branch flows near the west cliff of the floodplain from Reconquista (28°20’S) to the south, along more than 200 km. It crosses low-lying, almost permanent drowned terrains, forming an entanglement of small anastomosing channels linked to 200 to 300-m-long shallow lakes. Near Romang, such lakes are well defined, and in other places they pass in a transition to swamps. The channels are fixed, without lateral migration, and are 100 m wide. This unit is characterized by organic sediments and clays, forming a 4 to 5-km-wide belt.

The Colastiné branch is 39 km long and appears at the same latitude as Santa Fe city. The discharge of the channel is 1,400 m^3/s. It is a typical meandering channel with intense morphogenic activity; some meanders migrate laterally several meters per year. The channel constructed a 60-km-long and 6 to 8-km-wide meander plain, characterized by 0.8–1.5 m deep meander scrolls, transformed in shallow lakes and seasonal swamps. Meanders cut and form ox-bow lakes in processes that take several years of progressive capture of the water discharge and closing of the old segment (Iriondo 1975). The sediment is fine, well-sorted quartz sand and 0.2–0.3 mm in diameter. The channel is formed by a regular sequence of riffles and pools with a depth ranging from 7.3–11 m. The grain-size of the sediment in the ox-bow lakes is clearly finer, with a mean between 10 and 20 μm (silt) in some points.

2.6.6 The Meander Plain

Along the right margin of the floodplain, on the opposite side of the main channel, flow-active secondary channels of the meander type (Fig. 2.4) are develop significant meander plains in several areas. The largest meander plain is located north of Santa Fe city, where an important branch of the river (the Colastiné "river") flows for 60 km. The plain is 60 km long and 6–8 km wide. This geomorphologic unit is composed of fine silty sand at the surface. The total area of the meander plains is 2,917 km^2 (Iriondo 1972).

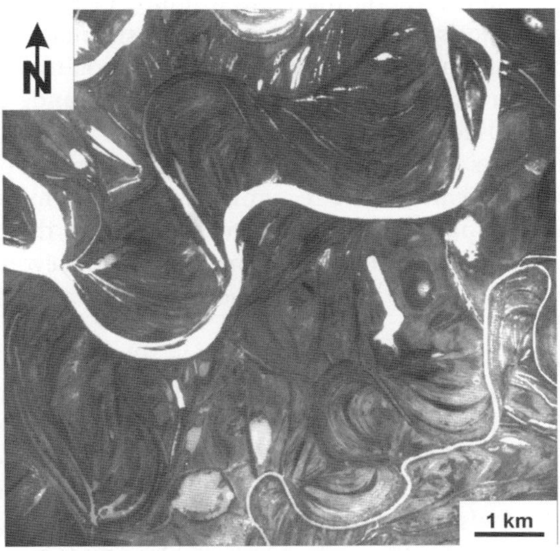

Fig. 2.4 Meander plain

2.6.7 Hindered Drainage Plain

This unit is characterized by a morphology of flat areas with numerous shallow ponds, swamps, and small adventitious channels. Such channels are active only at the beginning and end of the floods. Flood deposits are located at the right bank of the main channel along the entire floodplain. The intricate topography hinders the downstream flow. Water moves very slowly, practically in laminar flow, and follows minor tracts, sometimes in different directions depending on the state of the channel. Some areas of this type were slightly sunken by tectonics. The numerous shallow lakes are irregular in shape and 400 to 1,000 m long; adventitious channels are tortuous and about 100 m wide. They are also numerous, and were generated in the past under a hydric regimen different than the present one. The channels are interconnected in a chaotic way: some of them have well-formed levees at the banks, a feature that impedes the flow even more. This geomorphological unit forms wide belts at both sides of the bar plain. The sediments are mainly composed of two sedimentary facies: (a) levee facies, characterized by very fine sand with silt and clay admixtures, an ochre and greenish gray color, mottled, with abundant pores, tubes and root moulds. It shows an intermediate plasticity and gleying processes are evident; (b) pond facies, composed of dark gray silt with a high percentage of organic matter at different degrees of decomposition. It contains small percentages of clay and very fine sand. The sediment is compact, with low porosity, except in very recent deposits.

An important area is located south of Goya city, with a longitude of 50 km and 6–8 km in width. At the right side of the system, a similar surface appears at the latitude of Alejandra with 8–10 km width and extends for dozens of kilometers to the south; there, the lakes are larger (1–3 km long) and less numerous. Adventitious channels are numerous, small, and frequently interconnected; some of them flow in an east–west direction transversal to the valley. Most of the surface is covered by swamps and ponds characterized by vegetation decomposed in various degrees.

The flood deposits cover an area of 6,048 km. The mapping at 1:100,000 and 1:20,000 scales permitted the recognition of several minor units, each of them representing a particular state of evolution from the original channel morphology towards a definite flood morphology. The evolution shows two diverging lines, one of them in areas subject to tectonic rising and the other in subsident blocks.

The first state of the flood morphology is the same in both lines. The minor channels crossing the bar plain are isolated from the general water circulation and begin to widen in some places, appearing as small ponds. In the raised blocks the channels gradually disappear, and the ponds change according to lenitic dynamics, resulting in the following sequence: areas with widened channels/areas with ponds and lakes/elevated areas with lakes.

In the subsident blocks the surface covered by ponds and swamps gradually increases. The morphological sequence is as follows: areas with widened channels/areas with lakes and associated channels/permanently submerged areas. One important case is the area crossed by the "Río San Javier", a branch of the main channel.

2.6.8 The Ancient Deposits

Ancient deposits of the Paraná form terraces that flank both sides of the floodplain in the upper half of the Middle Paraná. These terraces are characterized principally by a flat surface with extremely low relief mostly not visible in the field. Only scarce swamps 100–300 m in diameter and irregular in shape are detectable in such areas. The ancient deposits forming the eastern terrace are composed of medium-gray green sand with calcareous concretions; small pores and tubes are common. Gray sandy silt intercalations with salt efflorescences can be observed in outcrops.

The ancient deposits are present from Confluencia (where the Middle Paraná begins) reaching a latitude of 30°40′S and forming two discontinuous belts on both sides of the floodplain. In the northern sector (up to a latitude of 28°30′3S) they are sunken, with the ancient surface at lower levels than the modern geomorphological units. In the southern sector, the ancient deposits have been raised up, so that they presently form a terrace. The difference in level between both sectors is more than 4 m.

The ancient sediments were deposited during a hydrological phase drier than the present one in which the Paraná had a reduced discharge and a water salinity higher than today. The age of the ancient deposits can be estimated as middle to upper Holocene, probably contemporaneous to the amazonic ancient geomorphological units, since the aeolian sediments covering the fluvial surface are dated from 3000 to 1000 B.P. (Iriondo 1981).

The Reconquista-Villa Ocampo terrace extends 60 km along the west cliff of the floodplain. It is 2–5 km wide with a flat, generally swampy surface and scarce morphological features. A gentle tectonic tilting originates the existence of small shallow lakes on the west side. The Goya terrace is located at the left bank, a sector of an elevated old floodplain, partially eroded. At present, it is not reached by the river dynamics, but undergoes floods and sedimentation originated in small local basins developed in the eastern areas. This terrace is different and looks older than the Reconquista counterpart. The Goya terrace extends for about 100 km, is 5–11 km wide, is crossed by meandering paleochannels of 2–4 km wavelength, which are quite large features. Such paleochannels belong to the Santa Lucía river system, an important tributary that drains an area of several thousand square kilometers in the Corrientes province. In the southern half of the terrace, several minor local basins contribute with modest volumes of water and sediments during sporadic floods. Permanent water bodies are absent, probably owing to the sandy subsoil. Such subsoil sediments were originated in the large Iberá Wetlands located at the center of the Corrientes province.

The Goya terrace is crossed by a younger branch of the Paraná main channel. This branch is now abandoned and forms a 45-km-long, 1.5 to 2-km-wide swampy depression that ends at the mouth of the Corrientes River. The existence of this feature suggests a tectonic elevation of the terrace (which is several meters high and also tilted) synchronic with the fluvial activity of the river branch.

2.6.9 Deltas of Tributaries

These are areas of sedimentation formed inside the floodplain during a dry climate that has recently occurred (probably from 3500 to 1000/1400 years B.P.). Water discharges were lower than today in all the system and tributaries did not reach the collector, deposing the sediment loads in lateral positions inside the floodplain. The Paraná was also smaller then than it is today. That resulted in relatively short and wide delta-like sedimentary bodies. The basins of western tributaries belong to the Chaco region, and convey a different type of sediments than the main channel; hence, the deltas have a definite signal in the Middle Paraná. Reworking and modifications of the original geoforms were important in some of them. Larger discharges presently allow a permanent connection with the main channel. In consequence, the deltas are now inactive as sedimentation systems, but undergo several fluvial processes. Each of them has particular characteristics.

The Arroyo del Rey River (right bank) generated a 45-km-wide and 8-km-long delta. It is formed by low-lying, swampy terrains with scarce well-defined shallow lakes. Small channels cross the area; the channel of Arroyo del Rey (50 m wide) is characterized by contorted curves about 100 m long. An abandoned channel of similar characteristics formed a narrow belt. Both channels finish into an area of lakes located along the Correntoso branch of the main channel, which is separated from the lakes by a wide levee.

The Los Amores River enters the Middle Paraná floodplain south of Villa Ocampo and flows near the right bank of the plain for 50 km inside a narrow (500–1,000 m wide) belt. The channel forms closed curves and is flanked by numerous oxbow lakes. The sediment in the belt is predominantly sand; it ends in a small delta 20 km wide and 5 km long at the latitude of the Reconquista. The delta is occupied by several large, well-defined lakes 600 to 1,500 m long, which are connected by numerous well-formed adventitious channels. It is also markedly sandy.

The Corrientes River drains a large area of the Paraná alluvial fan along 280 km and joins the floodplain at the left bank. It formed a fan-shaped deposit 46 km wide in north–south direction, and 6 km wide in east–west direction. The area is a meander plain crossed by channels originated in the Corrientes River and also in the Paraná main channel. The sediment is dominated by well-sorted fine quartz sand, kaolinite, montmorillonite, and organic debris. A few eolian dune fields were formed in this area in recent times. The largest one is 12 km long and 2–3 km wide; it extends in a north–south direction. Dunes are several meters high and are formed by loose, well-sorted fine sand originated in the Corrientes River.

References

Chiozza E, Figueira R (1990) Atlas Físico de la República Argentina. Volumen 2. Centro Editor de América Latina, Buenos Aires, 511 pp

Iriondo MH (1972) Mapa Geomorfológico de la Llanura Aluvial del Río Paraná entre Helvecia y San Nicolás (República Argentina). Revista de la Asociación Geológica Argentina 27/2:155–160

Iriondo MH, Drago E (1972) Descripción cuantitativa de dos unidades geomorfológicas de la llanura aluvial del Paraná medio. Revista de la Asociación Geológica Argentina, Bs. As 27(2):37–42

Iriondo MH (1973) Mineralogía de las arenas de la Formación Yupoí. Revista de la Asociación de Ciencias Naturales del Litoral, Santa Fe 4:87–96

Iriondo MH (1975) Geomorfología y sedimentología del río Colastiné. Revista de la Asociación Geológica Argentina, Bs. As 30(4):349–359

Iriondo MH (1979) Origen y evolución del río Paraná. In: Trabajos presentados a las 2das. Jornadas del Paraná Medio, Santa Fe, Argentina

Iriondo MH (1980) El Cuaternario de Entre Ríos. Revista de la Asociación de Ciencias Naturales del Litoral, Santa Fe 11:125–144

Iriondo MH, Musetti M y Real G (1983) Geomorfología del área de Coronda. Folleto Univ. Nacional del Litoral–Ministerio de Agricultura y Ganadería, Santa Fe, Argentina

Iriondo MH (1982) Estudio geomorfológico de la cuenca del río Salado en Santa Fe como base para la investigación hidrológica. – Convenio UNL–INCYTH, 4 tomos, Santa Fe, Argentina

Iriondo MH (1983–1985) Mapa Geomorfológico de la Llanura Aluvial del río Paraná. – Geren cia de Estudios y Proyectos Paraná Medio, Agua y Energía Eléctrica, Santa Fe, Argentina

Iriondo MH (1984) The Quaternary of northeastern Argentina. In: Rabassa J (ed) Quaternary of South America and Antarctic Peninsula, vol 2, pp 51–78

Iriondo MH (1987) Geomorfología y Cuaternario de la Provincia de Santa Fe. D'Orbignyana, Corrientes 4:54

Iriondo MH (1988a) A comparison between the Amazon and the Paraná fluvial systems. Mitteilungen des Geologisch–Paläontologisches Instituts der Universität Hamburg, SCOPE/UNEP Sonderband, Hamburg, H. 66, S 77–92

Iriondo MH (1988b) Guía de Campo Nro. 4: Llanura aluvial del Paraná , Simposio Internacional sobre el Holoceno en América del Sur, INQUA, 9 pp

Iriondo MH (1987) Los últimos 20.000 años en el departamento General López. – Informe inédito Subsecretaría de Cultura, Municipalidad de Venado Tuerto, 12 pp

Iriondo MH (1989) Major fractures of the Chaco–Pampa plain. Bulletin of the INQUA Neotectonics Commission, Estocolmo, Suecia 12:42

Iriondo MH (1993) El Litoral. En: El Holoceno en la Argentina, M. Iriondo (Ed.) Cadinqua. Paraná, 2:1–22

Iriondo MH (1999) Climatic changes in the South American plains: records of a continent-scale oscillation. Quat Int 57–58:93–112

Iriondo MH (1999) The Neogene of the Llanos-Chaco-Pampa depression. Episodes (IUGS) Special. International Union of Geological Sciences. Pekín 22(3):226–231

Iriondo MH (2004) Large wetlands of South America: a model for Quaternary humid environments. Quat Int 114:3–10

Iriondo MH (2004) The littoral complex at the Paraná mouth. Quaternary Intenational, Pergamon P, 114:143–154

Iriondo MH, García N (1993) Climatic variations in the Argentine plains during the last 18,000 years. Palaeogeogr Paleoclimatol Palaeoecol 101:209–220

Paoli C, Iriondo MH, García M (2000) Características de las cuencas de aporte. In: Paoli C, Schreider M (eds) El Río Paraná en su Tramo Medio. Universidad Nacional del Litoral, Santa Fe, pp 27–68

Soldano F (1947) Régimen y Aprovechamiento de la Red Fluvial Argentina. Editorial Cimera. Buenos Aires, 214 pp

3 Origin, Evolution, and Types of Floodplain Water Bodies

ALDO R. PAIRA[1] AND EDMUNDO C. DRAGO

3.1 Introduction

The interest from geomorphological and limnological standpoints is the analysis of the origin, evolution, morphometry, and abundance of floodplains lakes, which provides relevant information about the management of these wetland ecosystems. The Middle Paraná River contains a heterogeneous mosaic of lotic and lentic water bodies with changing shapes and areas according to the variable dynamics of the fluvial processes and the short- and long-term evolution of the different portions of the river floodplain. Thus, floodplain lotic and lentic environments can be eroded, filled, or enlarged at each flood event. The field data obtained from more than 30 years of the last century to the beginning ones, and the interpretation and analysis of the morphological structure of the floodplain lakes were accomplished using aerial photographs scaled from 1:20,000 to 1:50,000, Landsat satellite images and topographic charts (Paira 2003; Paira and Drago 2006). The aerial photographs and satellite images used were taken during mean and low water stages of the Paraná River. The aim of this chapter is to summarize a basin-wide characterization of the origin, shape, and morphological evolution of the floodplain lakes and its relationships with the aquatic and terrestrial vegetation and river dynamics.

3.2 The Middle Paraná Floodplain

3.2.1 Floodplain Characteristics

The areas geomorphologically more active of the composite fringing floodplain of the Middle Paraná River are those located near the main channel or along the larger secondary channels. Due to the braided pattern of the main channel, the adjacent floodplain areas usually depict a morphology classified by Iriondo (1988) as the *modern sand bar-meander plain* (Fig. 3.1A). The evolution of this morphology is the *evolved bar-meander plain unit* (Fig. 3.1B), which is very

Instituto Nacional de Limnología (CONICET-UNL). José Maciá 1933, 3016 Santo Tomé, Santa Fe,
[1]e-mail: alpaira@ceride.gov.ar

M.H. Iriondo, J.C. Paggi, and M.J. Parma (Eds.)
The Middle Paraná River: Limnology of a Subtropical Wetland

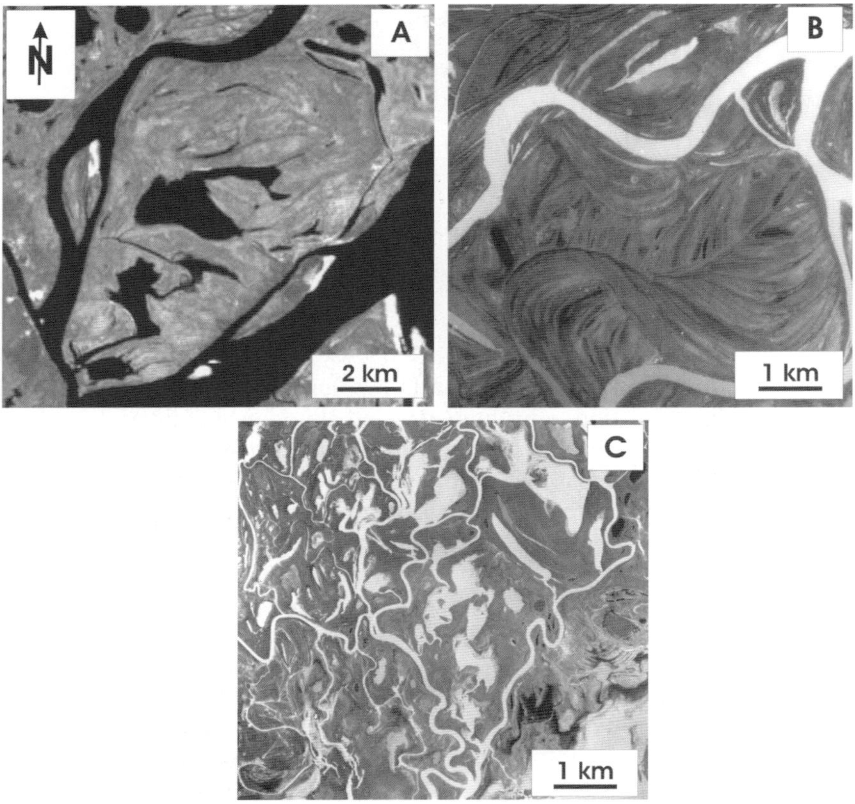

Fig. 3.1 Examples of the three main geomorphological units. **A** Modern sand bar-meander plain; **B** Evolved bar-meander plain; **C** Anastomosed plain. Note the differences in shape types and densities of the floodplain lakes for each unit

recent and was also probably formed during the present water regime (Iriondo 1988). The old portion of the floodplain is composed by the *anastomosed unit* (Fig. 3.1C), which has been built by flood deposits (Drago 1976; Iriondo 1972, 1988) located on the right margin of the alluvial valley, far away of the main channel. In short, the modern and evolved bar-meander plains have been built by channel deposits and are representative of the younger stages of the floodplain, while the anastomosed unit represents the late phase of the alluvial terrain. Moreover, the anastomosed unit is characterized by areas affected by neotectonic deformation (Iriondo 1988; Ramonell et al. 2000). Thus, there is a sequence of raised and subsided blocks, each one displaying different hydrographic patterns. The uplifted blocks show a major development of large and small meanders together with a great number of scroll lakes and irregular lakes (Fig. 3.2A), whereas the sunken blocks display a complex anastomosing drainage pattern with a high density of large and irregular interconnected lakes (Fig. 3.2B).

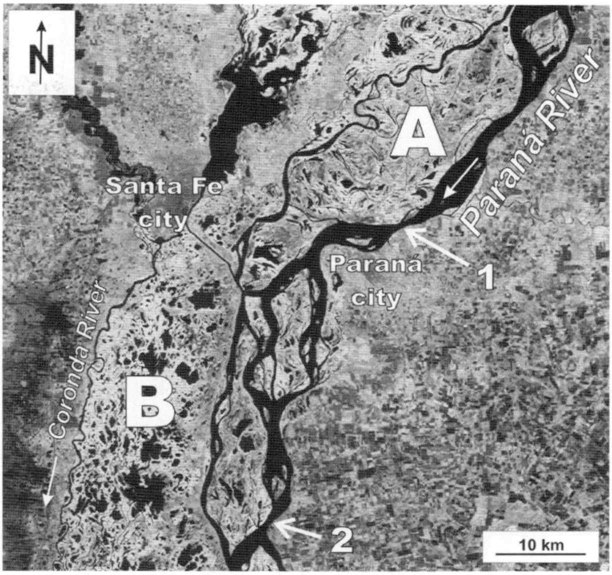

Fig. 3.2 Satellite image of different floodplain water networks in uplifted block (**A**) and sunken block (**B**) sectors and nodal points (*1* and *2*)

3.2.2 Main Channel

Channel shape is a contributing factor in the distribution of depth and velocity of the water and is a function of the overall basin hydrology and geology. The main channel of the Middle Paraná River has a braided pattern showing a sequence of wide segments characterized by two or more anabranches and narrow and short reaches in which the river flows in a single channel. The channels divided by islands and bars within the main channel are termed anabranches (Drago et al. 2003), whereas secondary channels form the floodplain drainage system. Thus, the floodplain channels branch and rejoin, but each is a distinct channel bounded by the surface of the alluvial plain ("anastomosing", sensu Schumm 1971) showing sometimes different plan form patterns (Table 3.1, Fig. 3.2).

The braided reaches are the most unstable and wide cross sections of the main channel (Drago 1977a, 1990; Figs. 3.3 and 3.4). Channels islands and sand bars are unstable and sometimes change their location quickly (Drago 1977a, 1990). The channel division is associated with increased width of water surface, increased slope, and decreased depth. The ratio of depths in the divided reaches to depth in the undivided reaches ranges from 0.3 to 0.9 (mean 0.6), similar to the data given by Leopold and Wolman (1957) for natural rivers (0.6–0.9). The main channel and its anabranches carry the 80–85% of the total discharge flowing throughout the valley. Measurements of the

Table 3.1 Different floodplain drainage structure according to their location in old floodplain area (*OFA*) and in a new floodplain area (*NFA*). The data were obtained from an area of 4,000 km²

Parameter	OFA	NFA
Channel number per km²	2.5	1.0
Drainage density (km/km²)	2.3	1.0
Mean channel width (km)	0.03	0.06
Number of channel junctions per km²	30	12
Angles of channel junctions (°)	85	45
Radius of curvature (km)	0.32	2.5
Amplitude range of channel bearings (°)	160	77
Main vector bearing (°)	240	247

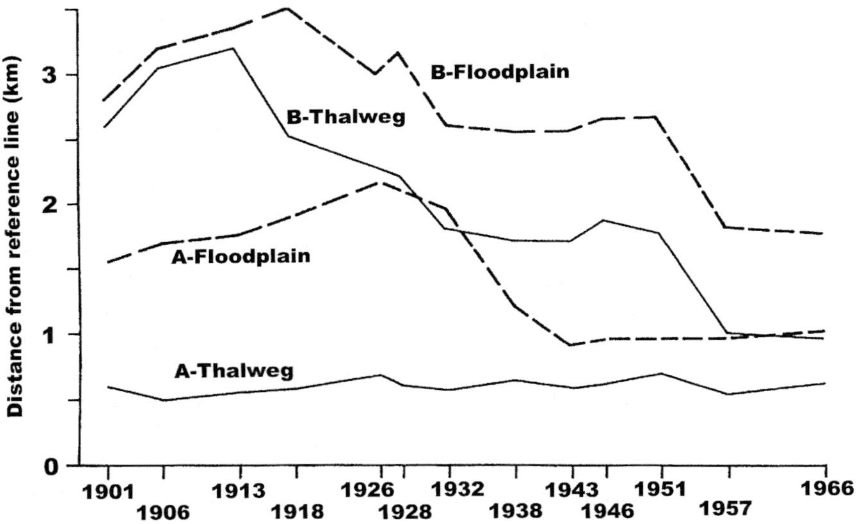

Fig. 3.3 Shiftings of the thalweg and floodplain bankline in two cross-sections. **A** narrow section or nodal point (Toma de Aguas Corrientes, upstream Paraná city, *1*); **B** wide section (see also Fig. 3.4)

number of anabranches in the main channel of the Middle Paraná River performed every 20 km along a reach of 800 km showed reaches with maximum division of six channels and a mean of two anabranches in the overall length of main channel. The channel narrowing is called nodal points or primary control points (Coleman 1969). At these points, the Paraná main channel is narrow (700–1,500 m in width) and quite deep, ranging the maximum depths between 30 and 45 m, in contrast to less than 30 m in the wide or non-restricted areas. At a nodal point, the bed water velocity measured over a 5-year period ranged from 0.44 to 1.11 m s^{-1}. Nodal points are the most stable

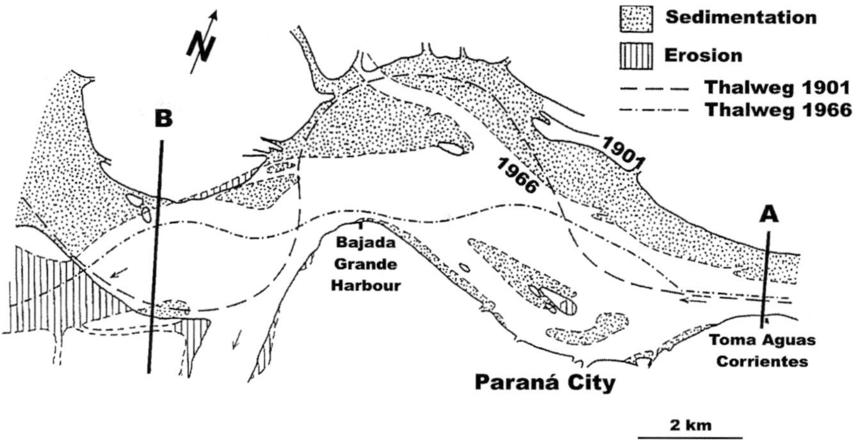

Fig. 3.4 Bankline migration, shifting of the thalweg and main sites of erosion and deposition during the 1901–1966 period. **A** narrow cross section or nodal point (Toma de Aguas Corrientes cross section); **B** wide cross section. The reach is just located downriver of the Aguas Corrientes nodal point (see Fig. 3.3)

cross sections and usually remain rather fixed through long periods of time. An example is the Toma de Aguas Corrientes cross section, located 3 km upriver from the Paraná city, which has showed only small variations since 1847 (Figs. 3.3 and 3.4; Drago 1977a, 1990). The Tertiary sandstone deposits on the left margin have not allowed the river to migrate as freely in this area as in others, where the main channel flows between alluvial banks. In most of the reach from the Paraguay River confluence to Paraná city, the main channel flows along the left cliff of the plain, indicating a west–east tilting of the tectonic blocks (Iriondo 1988, see Chap. 2; Fig. 3.1). Nodal points having some of the deeper points of the main channel could be also related to neotectonic processes, as fault systems and sunken blocks (Fig. 3.2: points 1, 2). Usually, the deep holes of tectonic origin are more than 30 m in depth in contrast with the shallow scour holes originated by fluvial processes, which show depths lower than 20 m (Drago unpubl.). Ephemeral sand bars exist in the nodal points and channel cross sections and are either U-shaped or asymmetrically V-shaped. The narrow and wide cross sections of the main channel show also two clearly different areas, the central strip and the bank strip along the banklines. The former occupied 90–95% of the wetted perimeter, being the more barren and homogeneous area of the river, with mobile sandy bed materials and where the river bottom is seldom flat. Physically, the central strip habitat changes little with the season or river stage. Current speeds are always high, ranging from 0.5–2.6 m s^{-1} under low to moderate discharges and often exceeding 3 m s^{-1} during high flows. The bank strips are the zones adjacent to the central strip and the boundary between these habitats is not always easily determined. Slopes of natural banks are usually >25–35° and are often nearly vertical in the upper portions of the banks. In the wide cross sections, the steep banks as well as low ones sometimes are associated with sand bars, being areas with different physical pattern in water depth, current speed and substrate. Furthermore, the bankline can be associated with high cliffs (>5 to 80 m) or with low cliffs (<5 m). Large segments of the main channel left bank are faulted-raised cliffs (Iriondo 1988), where sometimes very narrow beaches are developed but usually there are great depths immediately offshore of the bankline. Drago (1977a, 1990) measured maximum bankline extension and retreat averages of 139 m $year^{-1}$ and 78 m $year^{-1}$, respectively. Usually, the maximum bankline migrations are detected on the floodplain border and the minimum on the high fault-raised cliffs composed of Tertiary sandstone (Figs. 3.3, 3.4). As much as 20,000 m^3 of sediment may be involved in a single bank failure of 200 m length, changing drastically the onshore strip area (Drago unpubl.). The complexity of the bank structure is due to several factors, such as different high-low water level fluctuations, scalloping caused by block slumping, irregularities caused by the differential erodability of bank materials, fallen trees and brush, and mixing of allochthonous and autochthonous sediments (bad-sorting sediment). The accumulation of woody debris (mainly large snags) accelerated the sedimentation in the mouths of the anabranches, also fixing the bedforms and originating the joining of channel sand bars and islands on the banks. This process results in the loss of a portion of channel and the formation of a floodplain lake (Drago 1976, 1989, 1990).

The bedforms of the Middle Paraná River were classified by Drago (1977b) in four groups: ripples, with wave height <0.30 m; megaripples, have a wave height ranging from 0.3–1.5 m; dunes, the third category, range in wave height from 1.5–7.5 m; and sand waves, which have a wave height of over 7.5 m. At high water stages the dunes move downriver at a mean velocity of 12 m d^{-1} and the average rate of movement of the superimposed ripples and megaripples reaches 37 m d^{-1}. During low stages, the rate of movement decreases to 3.3 m d^{-1} for dunes, and 9.9 m d^{-1} for the superimposed ripples and megaripples (Lima et al. 1990; Fig. 3.5). Substrates at the channel center uniformly consist of sand (Table 3.2) and are constantly shifting. The bedload movement near Paraná city (Entre Ríos Province) ranged between 20 and 30 million metric tons per year (Amsler and Prendes 2000). Based on data covering 66 years of surveying in the same area, Drago (1977a, 1990) calculated a mean annual deposition of 5.5 million m^3 of sediment (Fig. 3.4). The wash load transported by the river averages 100 million metric tons per year, with most of the transport occurring in the main channel (Drago and Amsler 1988).

3.2.3 Floodplain Network: Secondary Channels

As in many alluvial large rivers, the Middle Paraná shows elements of both braiding and anastomosing at the same time and in the same reach. The floodplain channels or secondary channels conformed the drainage network of the floodplain complex and their hydromorphological dynamics originated a large heterogeneity at the riverscape scale (Drago et al. 2003; Fig. 3.2).

The drainage pattern of the Paraná floodplain depicts a predominant anastomosing pattern that includes a large number of sinuous, low-energy channels following flow-paths across the alluvial terrain that operate independently over considerable distances (Figs. 3.1C, 3.2B). These secondary channels usually meander, often with the highly sinuous, even tortuous plan form associated with low-energy streams flowing through predominantly cohesive sediments. Except in the major secondary channels where mid-channel and point bars are common and the small islands are uncommon, in the minor meandering courses the typical aggradational forms are the point bars (Fig. 3.1A, B). Dunes and ripples are mainly associated with the largest floodplain channels, i.e., in courses with mean discharges over 800–1,000 $m^3\ s^{-1}$ (Tables 3.2, 3.3; Drago et al. 2003). The mobile sand bed configuration is similar to the one detected in the main channel, but showing dunes with lower wave heights less than 3 m (Fig. 3.5). They also shows deep scour holes and meander scour pools, which ranges from 5 m to 17 m in depth at

Table 3.2 Mean (*x*) and standard deviation (*s*) of river bed sediments in the Middle Paraná River hydrosystem

	Paraná River main channel		Main secondary channels		Minor secondary channels	
	x	s	x	s	x	s
Sand (%)	98.97	1.671	97.50	2.646	17.25	14.033
Silt (%)	0.70	1.922	1.50	1.290	51.50	12.014
Clay (%)	0.19	0.431	1.00	1.414	31.25	9.179

Table 3.3 Hierarchical/functional classification of the floodplain channels according to their mean annual discharge (Q_{SC}) in relation to the mean annual discharge of the Paraná main channel (Q_{MC}=17,000 $m^3\ s^{-1}$) and its degree of intermittency

Channel classification	% of Q_{MC}	Example	Q_{SC}($m^3\ s^{-1}$)	Months without flow
Permanent channel	>5	Colastiné River	2,200	0
		Coronda River	1,125	0
Low-intermittency channel	>1–<5	Tiradero Viejo River	500	1–2
Middle-intermittency channel	<1	Correntoso River	94	2–4
High intermittency	<0.5	Yacaré River	35	>4

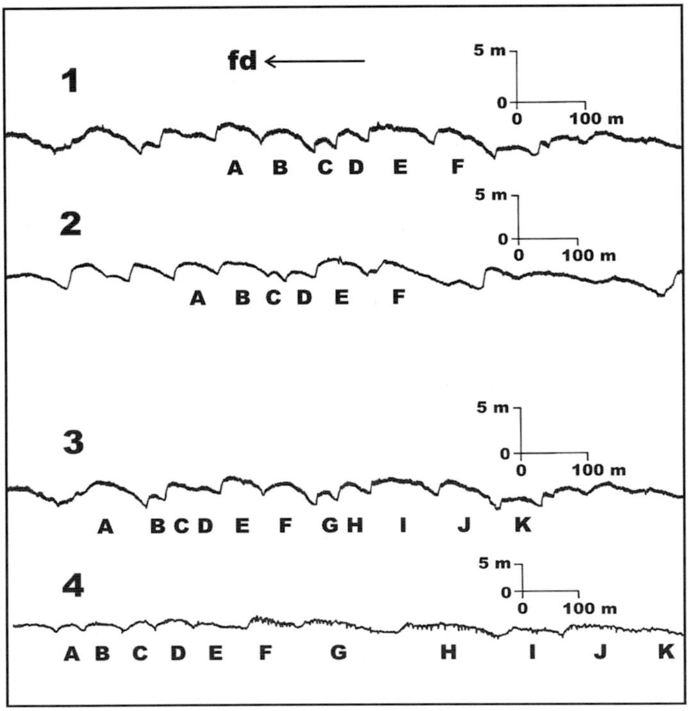

Fig. 3.5 Reproduction of hydrograph tracings from the center of the main channel showing well-developed dunes and their migration in a 30-day period (*1–2*), and during a rising river stage, showing the strong change in the bedforms in a 73-day period (*3–4*) (redrawn from Lima et al. 1990); *fd* flow direction

mid-water river stage. We found some of the deepest scour holes immediately located downstream of the entrance mouth of several floodplain channels when they flow to the floodplain from others majors secondary channels. We stated that their origin is due to the hydraulic vortices formed when the water flow across the scoured submerged levees during the floods.

The lower hierarchy channels with mean discharges less than 200 $m^3 s^{-1}$ (Table 3.3) sometimes present ephemeral small ripples (wave heights <0.30 m) along their straight reaches. Conversely, the minor hierarchy channels have a dominant silt-clay size composition in their beds (Table 3.2) with inserted large or small sandy reaches on which small ripples are common (Fig. 3.5). The characteristic feature along the longitudinal profiles of the minor channels is the alternation of shallow reaches and meander- and confluence-scour holes (Fig. 3.5).

Some areas of the alluvial terrain show a marked ridge and swale topography as a product of the meandering channels, whereas other areas display a very complex drainage pattern of lentic water bodies associated with the anastomosing-meandering network (Figs. 3.1, 3.2). The floodplain streams show a wide range in their mean discharges as well as in their degree of intermittence. Drago et al. (2003) have distinguished floodplain channels from the main channel and its anabranches by defining floodplain channels that have several order of magnitude smaller discharge than the main channel flow. Channels with higher discharges are annually active or perennials, whereas the minor hierarchies channels are not active during low-water periods. Thus, the floodplain channels of the Middle Paraná valley have been classified according to their mean annual discharges (Q_{SC}) in relation to the mean annual discharge of the main channel cross section (Q_{MC}) and their degree of intermittence (Drago et al. 2003; Table 3.3). According to our experience, the discharge is one of the main physical factors both in relation to the permanency-intermittency degree of the channel flow and also in the distribution, for example, of the benthic macroinvertebrates (Marchese and Ezcurra de Drago 1992; Marchese et al. 2002). Floodplain channels with mean annual discharges over 850 $m^3 s^{-1}$ have a permanent flow throughout the year, while the minor channels (mainly those with less of 100 $m^3 s^{-1}$ in annual average) show a strong reduction of discharges with temporal dry reaches and ponded water in the scour holes and meander pools during the lowest stages of severe droughts. Permanent secondary channels maintain the annual lotic connectivity between the parent river and its floodplain. On the contrary, during drought phases, the smaller intermittent channels ceased their flow, beginning the short-or long-term "lentification processes" according to the duration of the low-water phases (Drago et al. 2003).

Scour holes and meander pools are the lotic zones that display the maximum depths during the lower water level phases (about 5–8 m), showing physical behaviors similar to that of the adjacent floodplain lakes. The same effect is also detected in the shallow reaches of the minor channels where the flow ceased temporally. These areas show a temporary lentification process, which is maintained until the beginning of the flood phase (Drago et al. 2003). Furthermore, during the drought phases, we have found strong physical bars (temperature, suspended sediment, transparency) in the junctions between flowing and no-flowing streams (Table 3.4).

Table 3.4 Typical physical/chemical bar at the confluence of a low intermittent channel (Correntoso River) and an high intermittent channel (Yacaré River) during low water stage. *V* Water velocity, *SD* Secchi disk, *SSC* Suspended sediment concentration, T_W Water temperature, *K* Conductivity, (*1*): Water condition during high level (see Table 3.3)

River	V ($cm\ s^{-1}$)	SD (m)	SSC ($mg\ l^{-1}$)	T_W (°C)	K ($\mu S\ cm^{-1}$)
Correntoso	<2	0.22	81	29.0	487
Yacaré	0	0.90	17	23.0	344
Correntoso (1)	34	0.17	130	27.7	90
Yacaré (1)	20	0.15	100	27.6	90

In the old floodplain areas far away from the main channel, the number of channels per square kilometer is twofold greater than in areas near the mainstem channel, showing a greater degree of the fluvialscape dissection (Drago et al. 2003; Figs. 3.1C, 3.2B). This fact is due to the evolution of the floodplain drainage network, which shows from the main channel to the old areas, a progressive change in their hydrographic structure, from braided and meandering to anastomosed patterns (Fig. 3.1). The average channel width at the old floodplain is lower (27 m) than those closer to the Paraná mainstem (55 m). These narrower channels can be completely covered by aquatic macrophytes during the summer. The vegetation in turn changes the hydraulic conditions of these lotic habitats, driving thus the rate of the erosion-deposition processes of the channels, the vertical and transverse water velocities and the distribution of the channel bed sediments. These features explain the differences in the benthic communities (Marchese and Ezcurra de Drago 1992; Marchese et al. 2002; see Chap. 10). The increase of the drainage network complexity according to the increasing the distance between the main channel and the alluvial terrains is described in Table 3.1. Thus, in the large river-floodplain hydrosystems the spatio-temporal heterogeneity of the lotic-lentic connectivity increases with the increasing of the distance from the main course. We define as *hydromorphological complexity concept of the riverine floodplains* (HCC) to this relationship between the major and permanent running waters pathways and the minor floodplain lotic and lentic waters within a large alluvial river valley (Fig. 3.2). This concept is strongly linked to the flood pulse concept (FPC; Junk et al. 1989) and also with the expansion-contraction cycles or "flow pulses" described by Tockner et al. (2000).

The hydraulic conditions in the floodplain channels are reflected in the different erosion rates of their banks and in the diverse textural composition of their channel beds. The high-hierarchy channels present mobile sandy beds with a predominance of medium and coarse grain sizes. Sand-silty or silt-sandy patches are found in the slackwater areas, mainly in that located upstream or downstream of the point bars. Sand fractions always total more than 90% of the samples and silt and clay show similar percentages, although the silt contents are usually higher (Table 3.2). The grain size distribution of bed sediments along the low-hierarchy channels is quite variable, according to the different intermittence degree, position in the drainage network and aquatic vegetation cover of the streams. The intermittence degree is closely related with the position of the channel in the floodplain network. Thus, the directly connected with the larger courses or the main channel present higher percentages of sand or directly sandy reaches, while those connected with the minor floodplain streams, will have silt-clayed beds (Table 3.2). Therefore, the hydrological strength of the connection between the floodplain channels is a very important factor in the distribution of the bed sediments from the main channel to the floodplain channels. Despite that it is possible to remark the existence of a decreasing gradient in the bed grain-sizes from the main channel to the minor floodplain channels, such gradient it is not present as a continuum from the main stem to the more distant floodplain channel. In fact, the floodplain dynamics and evolution creates different network structures along the river where other factors, as tectonic influences or nodal points, determinate different floodplain channel densities and patterns and consequently, different hydraulic connections between lotic water bodies. These features exert a noticeable influence on the channel morphometry and bed sediment distribution. The typical sequence of shallow reaches and scour holes in the minor channels is depicted in Fig. 3.6, where it is possible to detect different bed grain-size composition according the channel morphology. Thus, in the first scour hole nearest to the connection with a major channel the amounts of sand are predominant as well as at the downstream confluence with a larger channel. In the last case, the large proportions of sands are related to the sand bar formed at the mouth of the confluence. On the contrary, in the scour holes and bendway pools as well as in the intervening straight shallow reaches, the bed texture changes to a marked silty-clayed

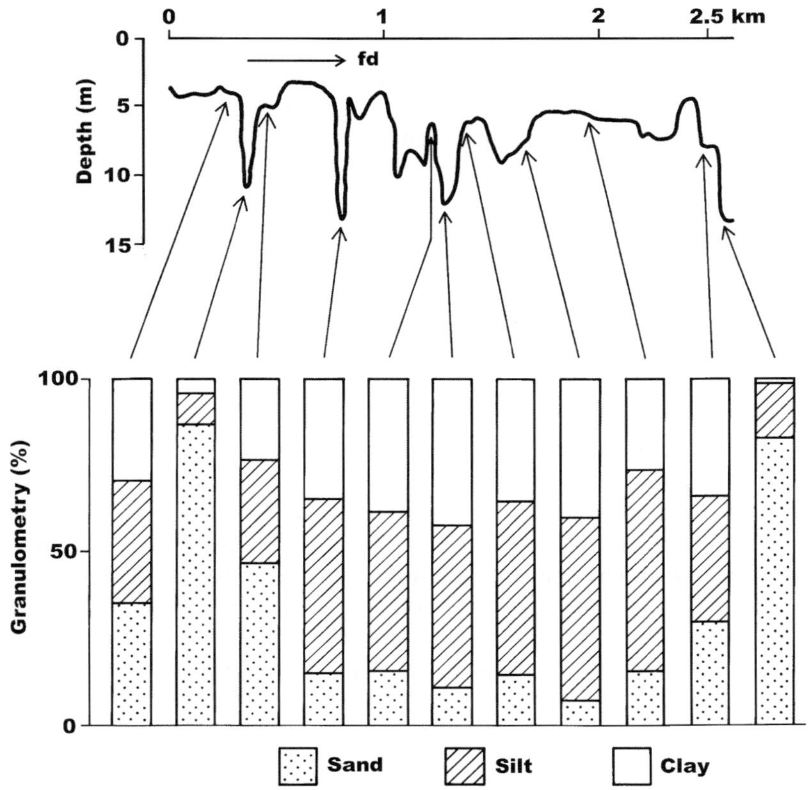

Fig. 3.6 Longitudinal echogram of the secondary channel Correntoso River showing the hole-step sequences, depth, and bed granulometry variation in a mid-water stage; *fd* flow direction

substrate. This change is generated by the current velocity diminution and the competence loss of the channel. Channels of minor hierarchies fed by streams similar to the example previously mentioned, display silt-clayed beds with small scattered sand patches (Table 3.2).

The particulate organic matter that is carried on the beds of the intermittent channels, is trapped in large quantities before it can conveyed into the permanent larger channels. The sand bars built at the outlets of the minor floodplain streams with larger channels are efficient traps for the coarse organic matter flowing from the floodplain. Nevertheless, during highest flows, a large portion of that organic matter accumulations is transported downriver. A part of this organic material forms the organic beds of the abandoned channels (e.g., oxbow lakes, meander scroll swale lakes, etc.).

3.3 Origin and Shape of Lentic Water Bodies

The information on the origin, shape, and evolution of the floodplain lakes has been gathered from several published and unpublished studies on floodplain lakes morphology (Drago 1973, 1976, 1981, 1989; Drago and Paira 1987; Drago et al. 2003; Paira 2003; Paira and Drago 2006). These works encompass morphometrical analysis of more than 2,000 lentic water bodies scattered on the alluvial areas from the main channel islands to the older zones of the Middle Paraná river floodplain. Nearly 50% of the floodplain surface was embraced in these studies and the selected alluvial sectors were located in the different geomorphological units classified by Iriondo (1972, 1988; Sect. 3.2.1). In this way, the 59.2% of the surveyed lakes were located in the modern sand bar-meander unit, the 28.7% in the evolved bar-meander unit and the 12.1% in the anastomosed unit. In these studies, special emphasis was given to the longitudinal and lateral distribution of the lakes (Paira 2003; Paira and Drago 2006; Fig. 3.1).

Floodplain lakes of the Middle Paraná River are defined here as permanent or temporary basins that may present their surfaces free or covered totally or partially by emergent vegetation (Drago et al. 2003; Paira 2003; Paira and Drago 2006). The assemblage of braided islands and the consequent isolation of the anabranches in the main channel originated the large and irregular lakes located in the modern sand bar-meander plain (Fig. 3.1A, B; Sect. 3.2.1). The three floodplain units developed numerous oxbows and scroll lakes due to the migration of the secondary channels. These elongated channel-shaped basins do not have the dimensions of the Paraná main channel because the major and minor floodplain channels created most of the meander scrolls and their associated lentic water bodies (Paira and Drago 2006).

The Middle Paraná, with its high sediments load has nearly filled up its valley, forming a very complex floodplain with island, bars, levees, secondary channels, shallow lakes and swamps (Drago 1989, 1990; Drago et al. 2003;

Iriondo 1972, 1988). About 40% of the alluvialscape is interspersed with different permanent and temporary lentic water bodies during mean river stage, being also composed by a complex lotic drainage pattern, dense fields of aquatic and palustrine plants and alluvial forests.

More than of 90% of the lakes in the Paraná floodplain have been active channels that developed into lakes by the abandonment of the channels (e.g., oxbow lakes) or by the isolation of a channel reach by avulsion and bars and islands assemblages. The latest process is common in the main channel and floodplain border, while the formers are typical in the floodplain streams. A very low number of temporary lakes had been formed by the accumulation of water in small and shallow depressions built on the floodplain by the uneven aggradation during floods.

Here we also differentiated between external and internal floodplain lakes. External lakes are those basins formed in the channel islands as well as along the floodplain-main channel border. On the other hand, internal lakes are water bodies originated within the floodplain (e.g., oxbows and scroll lakes). This division was based on the predominance of the evolution of the channels in the generation of floodplain lakes as well as in their hydrological connectivity (Paira 2003; Paira and Drago 2006; Sects. 3.2.2 and 3.2.3). Thus, lakes associated with the main channel support the strongest hydrosedimentological changes. Moreover, external lakes may change to internal lake conditions due to the floodplain evolution.

The origin of some kinds of floodplain lakes can be recognized through their basin shape. In the lake population analyzed in the three geomorphological units, we divided the lake basins in the following main shapes: *round-oval*, *elongated*, and *irregular* (Table 3.5). *Round-oval lakes* encompassing circular, elliptical and "drop" shape basins. We called "drop" shape-lake to the basins formed in the single young channel-islands, similar to the islands built at the front of the river deltas. A large number of round-oval lakes are located in the older areas of the floodplain (Figs. 3.2B, 3.7A1). However, many recent round-elliptical and drop lakes are formed in the main channel islands of the braided reaches, here called "island levee lakes" or "island lakes". Despite that these basins are generated by the same processes that form the deltaic levee lakes (Hutchinson 1957; Timms 1992), the location and hydrology of the basins are different. Old round lakes are usually associated with swamps, showing shorelines with very gentle slopes covered with aquatic and

Table 3.5 Classification of lake shapes in the Middle Paraná floodplain

1. Round-oval	
2. Elongated	2.1. Thin elongated
	2.2. Wide elongated
3. Irregular	

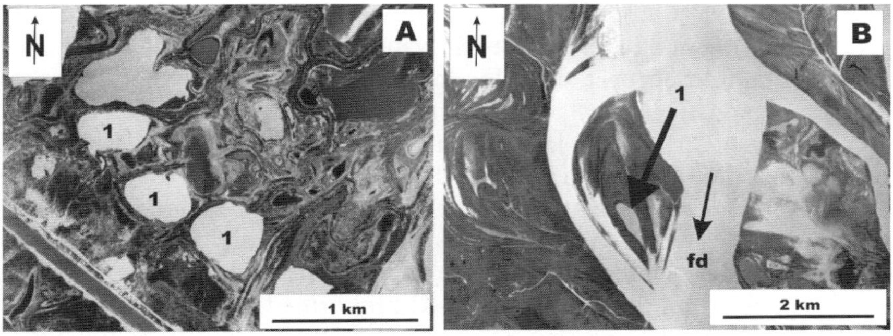

Fig. 3.7 Round lake shapes. **A1** Circular and elliptical lakes shapes in older areas of the floodplain; **B1** Drop lake shape in main channel island; *fd* flow direction

palustrine vegetation (Fig. 3.7A1). They are isolated from the running water for long periods of time. On the contrary, the island lakes are located in the channel islands and are entirely surrounded by levees, therefore their isolation phases are short (Fig. 3.7B1). *Elongated lakes* are those with channel-shaped basins that show "wide" and "thin" basins. They are well-defined channel-shaped basins formed by the abandonment of braided reaches of the main river or stretches of the floodplain channels (Fig. 3.1A). Meander loops (wide basins) and others abandoned straight and sinuous secondary channels, maintain their channel shape over long periods of their morphological evolution, mainly in that wide channels with higher depths. There are numerous lakes formed in abandoned river channels that did not originate from abandonment of a single meander (Fig. 3.8A1, 2). Such channels represent former sinuous or meander courses abandoned following siltation and the cutting of new channels and are common within the anastomosing complex. Some of these genetical shallower lakes are much subject to sedimentation as small oxbow lakes, but are for less sinuous and are often scoured during floods when they have their main axis paralleling the active secondary channels. On the contrary, very shallow scroll lakes lost their channel shape more rapidly due to their low depths and also by the fusion of several swale depressions, usually belonging to the same meander scroll set. Due to their plan form pattern, we call the longest and narrowest scroll lakes "fishbone lakes" (Fig. 3.1B). As noted by Mertes (1985) for the Amazon floodplain, the dimensions of the oxbow and scroll lakes in the Paraná floodplain are also a consequence of the floodplain channels shifting more than of the main channel dynamics. Meander scroll topography is very common in the three morphological units forming the alluvial terrain where small and medium elongated lakes frequently occur in the swales between concentric levees. A particular fact is that the "wide elongated" and "thin elongated" lake basins of the Paraná riverscape have been created by the abandonment of floodplain meandering channels in an average proportion of three scroll lakes by each

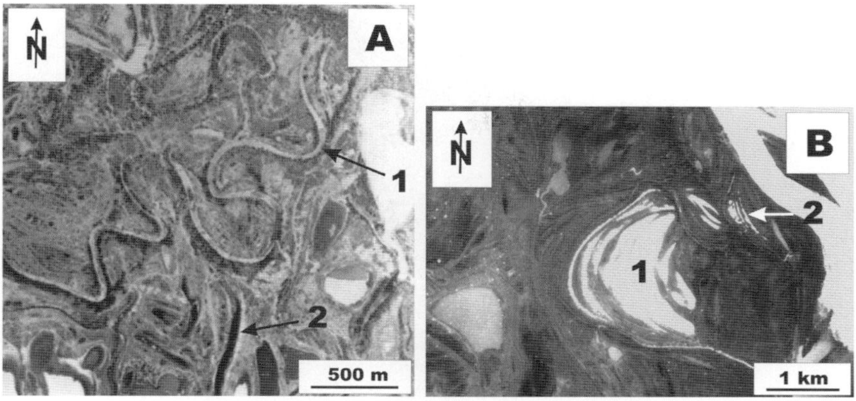

Fig. 3.8 A Lakes in abandoned river channels (*1* and *2*); **B** Lunate lake (*1*) and fishbone lake shapes (*2*)

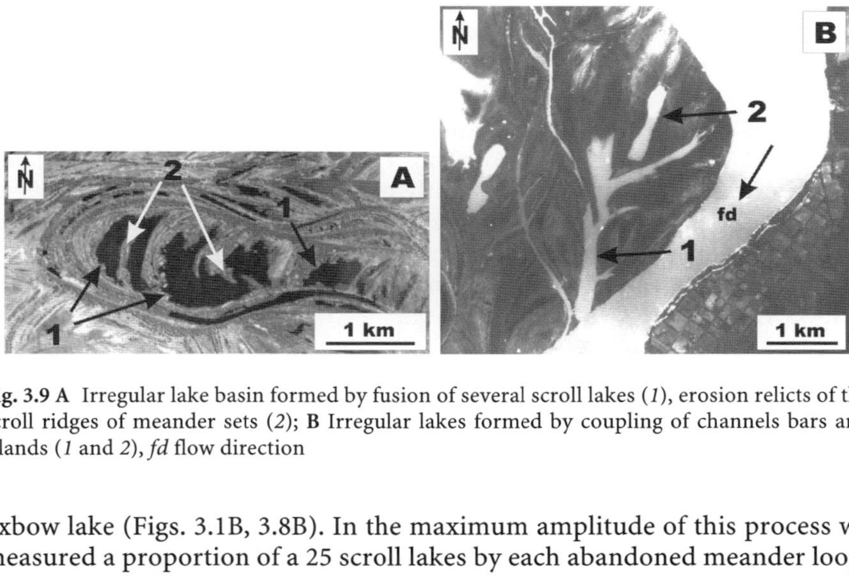

Fig. 3.9 A Irregular lake basin formed by fusion of several scroll lakes (*1*), erosion relicts of the scroll ridges of meander sets (*2*); **B** Irregular lakes formed by coupling of channels bars and islands (*1* and *2*), *fd* flow direction

oxbow lake (Figs. 3.1B, 3.8B). In the maximum amplitude of this process we measured a proportion of a 25 scroll lakes by each abandoned meander loop.

Irregular lakes are formed by the fusion of two or more lake basins. Typical examples are the large irregular lakes formed by the fusion of several scroll lakes as well as by the coupling process of large bars and islands of the main channel (Figs. 3.1A, 3.9A1, B1). They show higher densities in the floodplain areas affected by neotectonics (Iriondo 1988; Ramonell et al. 2000; Orfeo and Stevaux 2002), where there is a marked development of the anastomosing network (Figs. 3.1C, 3.2B). A striking example is the Laguna Coronda, the largest lake of the Middle Paraná floodplain, located 50 km downriver of Santa Fe (Fig. 3.10). This tectonic-fluvial lake has a surface area of 128 km^2 and irregular basin shape ($D_L = 4.5$), which is enhanced by numerous affluent and effluent

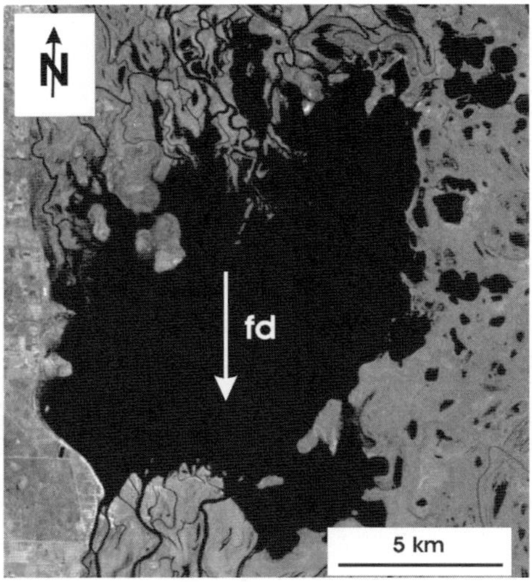

Fig. 3.10 A tectonic-alluvial lake "Laguna Coronda" (32°05′S–60°55′W), *fd* flow direction

floodplain channels, small bird's-foot deltas, wide bays and annexed lake basins (Paira and Drago 2006). This complex lake basin also contains drowned channels, discontinuous or flying levees, channels with anomalous curves or turns, which are also indicators of tectonic subsidence (Schumm et al. 2002). A less common basin type is the water body formed between a levee and the higher plain and cliff forming the right and left valley margins, respectively.

3.4 Lake Morphometry

The mean area of the computed lakes was 0.32 km^2, being the 78.7% the lakes with areas between 0.01 and 0.40 km^2 (Sect. 3.3, Table 3.6, Fig. 3.11A), while basins with areas from 0.40 to 0.78 km^2 showed a decrease to 12.7%. The 99.1% of the lakes not overcome 3.10 km^2 in surface area, however have the largest coefficient of variation (Cv) of the overall measured parameters (Cv = 217%). However, as was stressed by Sippel et al. (1992), the mean lake area is an inappropriate statistic to describe the lake population, mainly due to the large number of smallest lakes with areas lower than 0.01 km^2 that are not easy to measure from aerial photographs. Lake perimeters present a mean of 3.10 km and 39.6% of the lakes have perimeters ranging from 1.80 to 3.24 km, whereas 31% of these show perimeters between 0.36 and 1.8 km. Of the total lake perimeters 70.7% were smaller than or equal to 3.24 km, while 99.3% of the lakes display perimeters smaller or equal to 17.61 km (Table 3.6,

Table 3.6 Ranges of the morphometric parameters of the surveyed lakes

Parameter	Unit/categories	Range
Surface area (A)	km^2	0.01–14.68
Perimeter (P)	km	0.36–55.00
Maximum length	km	0.12–9.20
Maximum width	km	0.03–3.10
Mean width	km	0.01–2.04
Length-width ratio	Dimensionless	0.92–33.33
Shoreline development (D_L)	Dimensionless	1.00–7.61
Fetch	km	0.12–7.20
Maximum depth	m	0.20–5.30
	* Round-oval	
Basin shape	* Elongated (thin and wide basins)	
	* Irregular	
Distance to main channel	km	0.00–28.00
Distance to floodplain channel	km	0.00–28.00
Aquatic plant cover	%	0–100
Shoreline forest cover	%	0–100
Shoreline shrub cover	%	0–100

Fig. 3.11B). Fetch is closely related with the maximum effective length and is a very valuable parameter in the sedimentation-bottom sediment resuspension studies. Most of the Middle Paraná floodplain lakes have fetches between 0.12 and 1.91 km (95.7%) and about 99.5% are smaller than or equal to 3.47 km (Table 3.6, Fig. 3.11C; Paira 2003; Paira and Drago 2006). According to these values, the majority of lakes show high rates of sediment resuspension because the shallowness of the alluvial basins in relation to their large fetches. Lakes with fetches lower than 0.2 km and perimeters supporting dense gallery forests have high water transparency and minimum rates of bottom sediment resuspension (Drago and Vassallo 1980; Drago 1989). The maximum width average of lake basins is 0.23 km, and 63% of the water bodies have a mean width ranging from 0.01 to 0.38 km. These data indicate the abundance of elongated basins. This fact is also marked because 35% of the lakes show a length-width ratio over 4.50, being similar to the mean length-width ratio of 4.27 obtained for overall lakes population. However, it must be stressed that 81.6% of the lakes have a wide range of length-width ratio, which encompassed from 1.00 to 6.04 (Table 3.6, Fig. 3.11D; Paira 2003; Paira and Drago 2006).

The shoreline development (D_L) is a measure of the degree of irregularity of the shoreline. In the Paraná floodplain lakes, the mean D_L is 1.84 and 89.5%

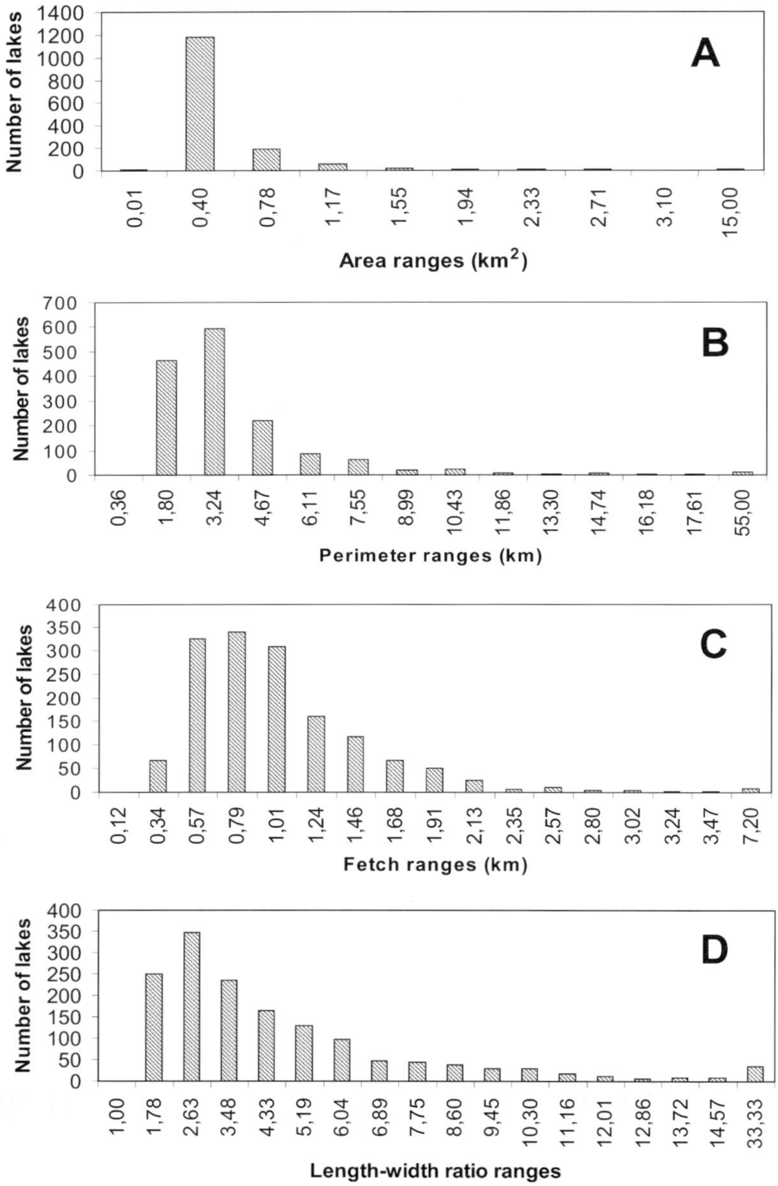

Fig. 3.11 Morphometric parameters of floodplain lakes. Number of lakes in relation with ranges in lake area (**A**), perimeter (**B**), fetch (**C**), and length-width ratio (**D**). Total number of lakes: 1,500

of the lakes range from 1.00 to 2.74. Despite the large diversity in lake shapes and areas D_L shows the lowermost coefficient of variation (Cv = 41.6%) of the overall analyzed morphological parameters (Table 3.6). This fact may be because lakes with different areas and shapes can have similar D_L values. There is a large group of lakes within low D_L ranges (1 to 3) and a very low number of basins with high D_L ranges (4 to 7). Thus, the most abundant lake basins (20.3%) have D_L ranges between 1.00 and 1.22 and 99.2% of lakes have D_L equal to or lower than 4.48 (Table 3.6, Fig. 3.12A; Paira 2003; Paira and Drago 2006). Despite D_L is largely applied in Limnology (Hutchinson 1957;

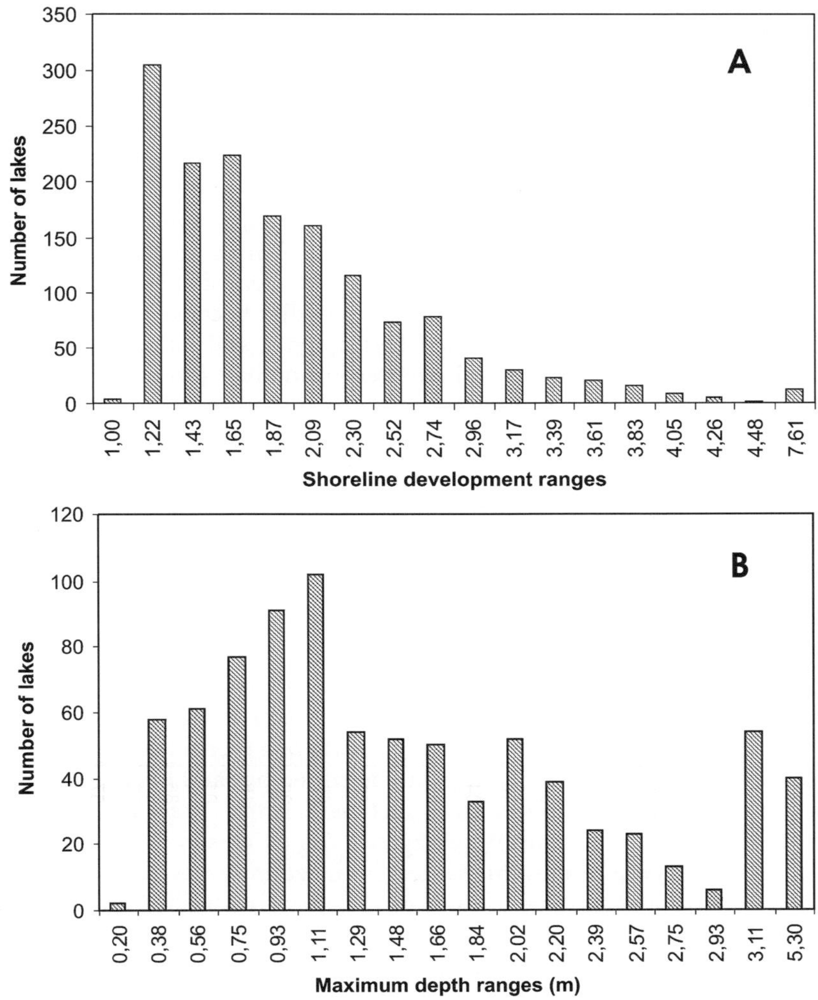

Fig. 3.12 Morphometric parameters in floodplain lakes. Number of lakes and shoreline development ranges relationship in 1,500 lakes (**A**) and maximum depth ranges in 830 lakes (**B**)

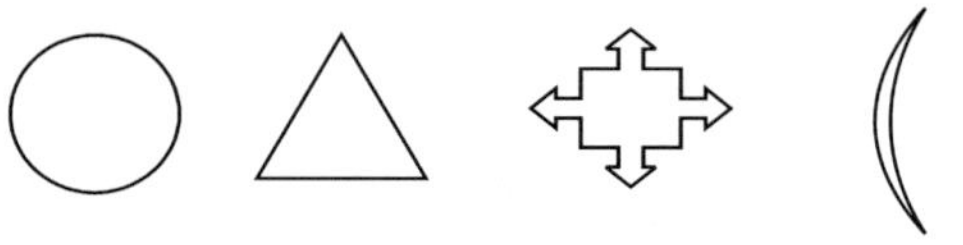

Fig. 3.13 Graphical representation of shoreline development for four lake shapes

Hakanson 1981; Timms 1992), from our field experience and data lake sets we have found that the D_L is not a reliable parameter to infer consistently the degree of irregularity of the lake shorelines in the Middle Paraná River floodplain (Fig. 3.13). This fact permit to assert that this conclusion can be applied to the other water bodies of large river-floodplain existing all over the world.

The mean maximum depth of the Middle Paraná floodplain lakes is about 1.46 m (Cv = 63.9%) and the maximum lake depth detected during mean river stages was 5.30 m and 95% of the basins show maximum depths equal to or lower than 3.11 m (Table 3.6, Fig. 3.12B). However, few lakes with maximum depths over 5 m can be found in areas where the annexation processes have recently occurred (e.g., the coupling of deep anabranches to the floodplain, Fig. 3.1A). Round-oval and irregular lakes tend to show flat bottoms with gentle slopes to the center of the water bodies because the higher sedimentation rates, overgrow of rooted hydrophytes and helophytes and the accumulation of autogenic organic matter are mainly produced along lake shorelines. These processes progress from the shoreline toward the center of the basins, causing the progressive elimination of the shoreline irregularities (e.g., inlets and bays) and increasing the circularity as well as the decreasing of the lake areas. Figure 3.9A clearly depicts the evolution of an irregular lake originated by fusion of scroll lakes to several rounded lakes. We call this type of morphological evolution the "roundness process" (Fig. 3.14; Paira and Drago 2006).

These processes are usually faster in round-oval lakes than in elongated basins, which are deeper and with steep slopes due to their origin (abandoned channels and meanders). In the channel-lakes, the terrestrialization process mainly begins from the alluvial plugs to downstream of dead channels. A very common fact is that the downriver open mouth of a recent lake, directly connected with an active channel, begins to be closed by sand bars mainly when they are overgrown by pioneer trees such as *Tessaria integrifolia* and *Salix humboldtiana*. On the contrary, elongated basins as scroll swale or "fishbone" lakes everywhere show higher terrestrialization rates due to their shallower and flatter basins. Due to their origin, the terrestrialization process in the scroll lakes usually begins from both ends. The narrowest channel lakes are commonly covered by dense floating meadows, which strongly contribute to the increase the sedimentation rates magnify by the coarse organic matter (see Chap. 4). These water bodies present layers up to 0.7 m of organic mud on their bottoms. In this case, the paramount factors are the area encom-

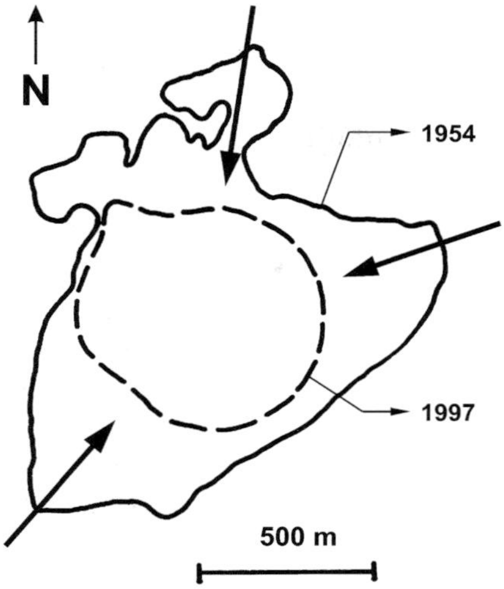

Fig. 3.14 Lake basin roundness process through the elimination of the shoreline irregularities by the aquatic and palustrine vegetation overgrown, siltation, etc., in Lake Cañas, Cañas Island, Santa Fe, Argentina (31°40′S–60°37′W). *Arrows* indicate the general directions of the shoreline progradation

passed by deep waters and the major or minor slope of the basin. However, the influence of these factors is overcome by the orientation of the lake basins in relation to the active channels. Thus, the lakes orientated parallel to the active channels are exposed to periodic rejuvenation processes (Paira and Drago 2006).

3.4.1 Analysis of Lake Shape

Table 3.7 depicts the mean morphometrical and associated parameters of the Middle Paraná floodplain. Elongated lakes are the most predominant shapes with 55.1%, whereas the round-oval and irregular lakes present 24.2 and 20.7%, respectively (Table 3.8). From the statistical analysis of the morphometrical parameters and the lake typology (Tables 3.5, 3.6, 3.8), we concluded that the Middle Paraná river network floodplain are mainly formed by meandering channels. Due to this characteristic, i.e., by the wandering of the floodplain channels, there are a large number of elongated basins as scroll and fishbone meander lakes. These types of lakes are narrower and shallower, their maximum width being five times lower than their length channels. They do not have tributaries or drainage channels and are usually densely (more than 50%) covered by aquatic vegetation. Moreover, they mostly present more than of the 60% of their perimeter with gallery forest and shrubs. Thus,

Table 3.7 Mean data of morphometrical parameters in lakes of the Middle Paraná floodplain

	Round-oval	Elongated	Irregular
Area (km^2)	0.27	0.20	0.74
Perimeter (km)	2.20	2.47	6.05
Depth (m)	1.56	1.26	1.89
Fetch (km)	0.78	0.82	1.34
Length (km)	0.85	0.94	1.64
Width (km)	0.34	0.29	0.61
Mean width (km)	0.27	0.19	0.33
Length/width radio (dimensionless)	3.43	4.54	3.55
Shoreline development (dimensionless)	1.43	1.76	2.44
Number of tributaries	1.00	1.03	1.77
Number of emissaries	0.95	1.04	1.77
Aquatic cover vegetation (%)	39.45	34.80	32.7
Distribution of surrounding forests (%)	37.50	34.95	35.2
Distribution of surrounding shrubs (%)	60.90	64.55	64.6
Distance from main channel (km)	7.79	9.34	6.36
Distance from secondary channel (km)	1.21	1.35	1.27

Table 3.8 Percentage and number of lakes for each basin-shape types in the Middle Paraná floodplain

Shape	%	Number of lakes
Round-oval	24.2	363
Elongated	55.1	827
Irregular	20.7	310
Total	100.0	1,500

the evolution of this type of lakes to a swamp condition is faster than other type of lakes (Paira 2003).

Irregular lakes commonly show the largest and deeper basins with extended perimeters and abundant tributaries and emissaries. The aquatic vegetation usually covers 33% of their surface and their perimeters present 33% covered by trees and more than 60% by shrubs. Irregular basins formed by annexation of meander scroll (e.g., fishbone and meander scroll lakes) are recognized because of the small, elongated and narrow islands scattered on their surfaces. These islands are erosion relicts of the scroll ridges belonging to the old meander sets and are principally overgrown by shrubs (Fig. 3.9A2).

On the contrary, irregular lakes formed by the coupling of channels bars and islands do not have islands and present twice the depth compared to the annexation scroll meander lakes (Fig. 3.9B1, 2; Paira 2003).

The meander scroll lakes as well as the irregular annexation scroll lakes are closely linked to oxbow lakes, due to the fact that the former was originated during the meander evolution. Comparing the oxbow lakes and the scroll basins we found that the width of the oxbows was 33% larger in length and show a larger area than the scroll lakes. Moreover, oxbow lakes present more tributaries and emissaries. Twenty-five percent of the oxbow area is usually covered with aquatic vegetation and 33% of the perimeter is overgrown by tall trees due to the high elevation and consolidation of the levees, which allows the development of dense gallery forests (Paira 2003).

3.5 Lake Evolution

The morphology and evolution of alluvial lakes is driven by river dynamics as a consequence of the annual inputs of water and sediment to the floodplain drainage system. Some floodplain sectors are also affected by neotectonics, which sometimes strongly change both the longitudinal continuum of the alluvial area and the morphological evolution of the lake basins.

Lake evolution also largely depends on three main factors: *spatial position* or the major or minor distance to the main channel and floodplain channels, *orientation* or the transversal or parallel position of lake basins in relation with the main flows and their *elevation* or topographic altitude on the active channels. Lake basins parallel to the largest active channels may show periodic rejuvenation processes and consequently, a decrease in their evolution progress to terrestrialization. In fact, the elongated basins that are transversal to the main flow of the active channels and with their perimeters covered by gallery forests, may overcome the strong wash out action during floods. Thus, these basins can maintain their aquatic plant covers for several years, which accelerates the siltation process. On the contrary, these types of channel shape basin usually completely or partially lose their aquatic vegetation cover in high floods (Drago et al. 2003). However, we observed during the highest flood periods similar processes in lakes with the same vegetation coverage located far away from the active channels and with a very low elevation within the general floodplain topography.

By means of measurements of lake sedimentation rates, the higher deposition values as well as the accumulation of coarse sediment and coarse organic particulate matter carried by floodwaters were usually detected on the shallow perimetral areas (Drago unpubl.). In the case of the studied lake of Fig. 3.14 (Cañas Island, 31°40′S–60°37′W), the roundness process, i.e., the basin area contraction, was ca. 61% in 43 years, while the migration of the shoreline towards the center of the basin reached a mean rate of 5 m year^{-1} (Paira and Drago 2006).

In the evolution of a meandering sector, the generation of one oxbow lake involved the formation of several scroll lakes, originating the typical ridge and swale topography. Moreover, in these areas, two evolution patterns can be detected. One of these is originated when very narrow and shallow scroll lakes show a rapid evolution to swamp conditions and terrestrialization. Thus, the majority of these basins are completely covered by recent stands of aquatic and palustrine vegetation, with the shallower areas beginning to be overgrown by dense stands of *Tessaria integrifolia* and *Salix humboldtiana*. The second evolution pattern is developed when the erosion of the ridges allows the annexation of two or more scroll swale lakes, thus forming an irregular lake basin. This kind of irregular lakes shows some morphological features that allow their genetic interpretation, as the narrow islands that are relicts of the erosioned scroll levees. Sometimes these islands maintain the alignment in the direction of the former scroll ridges (Fig. 3.9A2; Paira 2003; Paira and Drago 2006).

3.6 Aquatic and Riparian Vegetation as a Morphological Factor

A significant physical factor in floodplain building as well as in the evolution of their drainage channels is aquatic and riparian vegetation (Marston et al. 1995; Knighton 1998). In fact, as is well known, hydrophytes, helophytes, and inundation forests are closely related to aggradation processes and the stability of the stream banks, lake shorelines, bars, and islands. In many floodplain lakes during drought phases, a zonation of terrestrial and aquatic and semiaquatic vegetation along the flood gradient can be observed. These aquatic grasses are effective tools in the deposition of sediments during the flood phases, and large stands of these aquatic and semiaquatic plants can remain on the lake shorelines during drought phases, thus enhancing the deposition rates (Figs. 3.9A, 3.14). Nearly 25% of the Middle Paraná floodplain lakes have only 5% of their surfaces covered with free or rooted macrophytes (*Eichhornia crassipes*, *E. azurea*, *Pistia stratiotes*, *Panicum elephantipes*, *Echhinocloa polystachya*, *Ludwigia peploides*, *Nymphoides indica*, etc.). However, the cover of aquatic vegetation ranges from 0 to 100% in lakes with areas less than 2 km^2 (Fig. 3.15; Paira 2003; Paira and Drago 2006). This large amplitude in the coverage variation is due to the small areas shown by the large number of levee and scroll lakes at the beginning of their formation. The alluvial forest is mainly represented by the species developed on the levees, being the major species richness detected on the older levees. The typical landscape is the riparian forests covering the banks of the floodplain channels. Species as *Tessaria integrifolia* and *Salix humboldtiana* are the pioneer trees covering the levees as well as the alluvial plugs, bars, and islands. Twenty-five percent of the floodplain lakes have 10% of their shorelines

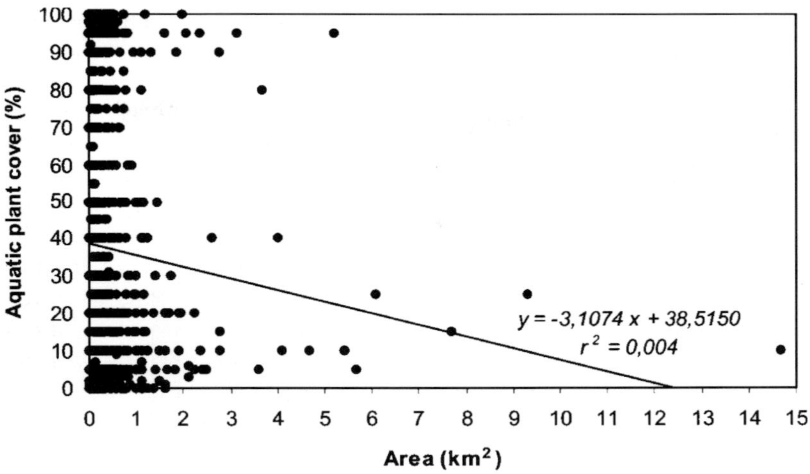

Fig. 3.15 Relationship between area an aquatic vegetation cover in 1,484 lakes

covered by alluvial forests, and more than 30% of the basins show their shorelines 90–100% covered by dense stands of *Schoenoplectus californicus*, *Polygonum* spp., *Panicum prionitis*, *Echinodorus grandiflora*, *Sagittaria montevidensis*, etc. (Fig. 3.16A, B; Paira 2003; Paira and Drago 2006). The gallery forests and shrubs also have an important role in the morphological evolution of the floodplain lakes because they act as physical barriers against the strong currents during flood phases. Long and narrow lakes, mainly those transversal to the major floodplain channels, are protected from the dragging of their aquatic and semiaquatic vegetation cover by the gallery forests and shrubs, even during larger floods. Thus, these lentic environments present a more continuous process of terrestrialization. On the contrary, in the channel-shaped lakes orientated parallel to the active channels, the free-floating and rooting hydrophytes as well as large portions of the thick organic matter deposited on their bottoms can be completely carried away during large floods, delaying the terrestrialization (Drago et al. 2003).

3.7 Concluding Remarks

River dynamics through scouring and filling processes together with long-term channel evolution at the riverscape scale generate and change lentic water bodies giving rise to a diverse morphology and hydrologic behavior. Thus, in a lake-rich terrain as river floodplains, the lakes are essentially the short- and long-term residual landforms of the river dynamics (Drago 1976, 1989). The origin of many kinds of floodplain lakes can be defined by their

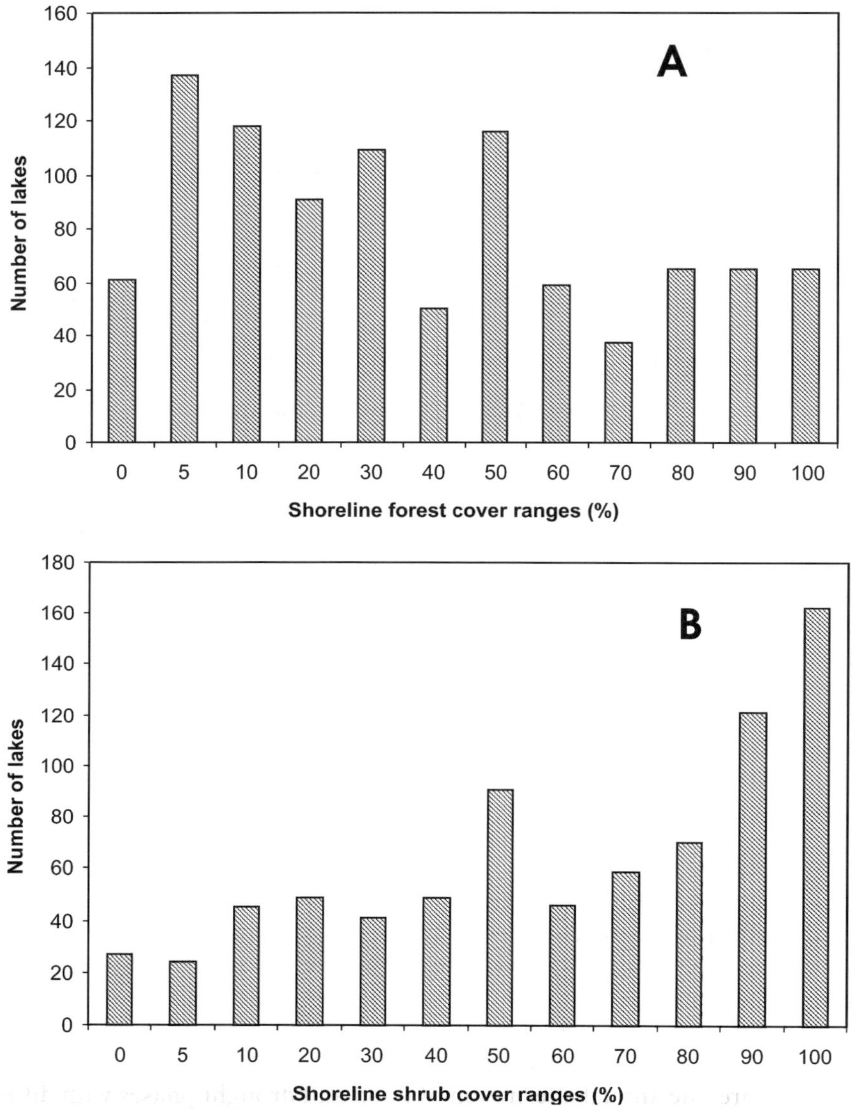

Fig. 3.16 Number of lakes relationship with shoreline forest cover ranges for 937 lakes (**A**) and with shoreline shrub cover ranges for 784 lakes (**B**)

basin size and shape, e.g., oxbow and island levee lakes, which are created by migrating channels and by the evolution of river bars and islands, respectively. However, due to the results of more than one fluvial process that gives rise to irregular and complex lake basins, the determination of their origin is sometime difficult to distinguish from the aerial photographs and satellite images. Our morphometric measurements of the Middle Paraná floodplain lakes suggest the following lotic-lentic hydrogeomorphological characteristics which may be applied to other similar large river floodplains. The majority of floodplain lentic environments have been formed from abandoned fluvial channels and therefore, elongated lake basins predominates on the floodplain. However, most round-oval lakes have also been old stream beds. Channel lakes on the Paraná main channel–floodplain border occur principally in association with scroll bars. Therefore, the morphology of the alluvial lakes generally allows one to determine their genetic processes. The floodplain is dominated by ridge and swale topography, which also contributes to the predominance of the narrow elongated lakes and therefore few channel lakes have dimensions that approach those of the main river channel or their anabranches. We concluded from this assessment that the wandering of the floodplain channels create most of the oxbows and scroll lakes of the Paraná alluvial plain.

The lake basin evolution is closely related to the distance to the main channel or active floodplain channels as well as with the connectivity degree between the lotic and lentic environments. The key factors in the morphological evolution are the parallel or perpendicular orientation and elevation of the lake basins in relation to the active channels. The evolution of the lakes through the "roundness process" tends to the generation of a large number of round-oval basins; therefore, higher densities of this shape basin are detected in the oldest areas of the floodplain. The largest and more irregular lake basins in the floodplain have been originated by neotectonic processes, fusion of several basins, or by the abandonment of the main channel anabranches by the islands and sand bars assemblages. Lakes with genetically shallow basins and that are isolated from active channels show the highest evolution rates to palustrine and terrestrial conditions. Due to their genesis, oxbow lakes and annexation lakes by island assemblages display the major maximum depths, and the scroll swale lakes and island levee lakes show the minor maximum depths. Elongated lakes transversal to the active channels and surrounded by gallery forests usually show dense floating meadows and a faster evolution to the terrestrialization. Lakes with large isolation phases usually hold higher biomass of aquatic and palustrine vegetation, which also contributes to a higher basin siltation.

The riverscape diversity is enhanced by active channel migration and flooding which creates, modifies, and destroys fluvial landforms. Furthermore, the annual alternation of flood and drought phases with different spatio-temporal magnitude, also contribute to enhance the biological diversity of these fluvialscapes. Thus, by preserving this mosaic of habitats and their hydrological connectivity we will preserve the integrity of the ecologic hydrodynamics of these important fluvial hydrosystems.

Acknowledgements. We are very grateful to Dr. Robert H. Meade, Dr. Steve Hamilton, Dr. Rip Sparks, and Dr. Mario Amsler for constant help and valuable discussions on several versions of this chapter. Technicians Ulises Molet, Walter Reutemann, Ramón Regner and Eduardo Lordi (Instituto Nacional de Limnología, CONICET-UNL) are thanked for their dedicated help in the field and laboratory.

References

Amsler ML, Prendes H (2000) Transporte de sedimentos y procesos fluviales asociados. In: Paoli C, Schreider M (eds) El Río Paraná en su tramo medio. Contribución al conocimiento y prácticas ingenieriles en un gran río de llanura. Centro de Publicaciones, Universidad Nacional del Litoral, Santa Fe, Argentina

Coleman JM (1969) Brahmaputra River: channel processes and sedimentation. Sediment Geol 3:129–239

Drago EC (1973) Caracterización de la llanura aluvial del Paraná Medio y de sus cuerpos de agua. Boletín Paranaense de Geociencias 31:31–44

Drago EC (1976) Origen y clasificación de ambientes leníticos en llanuras aluviales. Rev Asoc Cs Nat del Litoral 7:123–137

Drago EC (1977a) Erosión y sedimentación en un tramo de cauce del río Paraná medio (República Argentina). Asoc Geológica Argentina Revista 32:277–290

Drago EC (1977b) Campaña "Keratella I" a lo largo del río Paraná medio. II: Formas del lecho en su cauce principal. Rev Asoc Cs Nat del Litoral 8:57–62

Drago EC (1981) Grados de conexión y fases hidrológicas en ambientes leníticos de la llanura aluvial del río Paraná (Argentina). Ecología (Argentina) 6:27–33

Drago EC (1989) Morphological and hydrological characteristics of the floodplain ponds of the Middle Paraná River (Argentina). Rev Hydrobiol Trop 22:183–190

Drago EC (1990) Geomorphology of large alluvial rivers: lower Paraguay and Middle Paraná. Interciencia 15:378–387

Drago EC, Amsler ML (1988) Suspended sediment at a cross section of the Middle Paraná River: concentration, granulometry and influence of the main tributaries. Sediment Budgets IAHS 174:381–396

Drago EC, Paira AR (1987) Temperature and heat budget in a floodplain pond of the Middle Paraná River (Argentina). Rev Asoc Cs Nat Litoral 18(2):105–228

Drago EC, Vassallo M (1980) Campaña limnológica "Keratella I" en el río Paraná Medio: Características físicas y químicas del río y ambientes leníticos asociados. Ecología 4:45–54

Drago EC, Ezcurra de Drago I, Oliveros O, Paira A (2003) Aquatic habitats, fish and invertebrate assemblages of the Middle Paraná River. Amazoniana 17:291–341

Hakanson L (1981) A manual of lake morphometry. Springer, Berlin Heidelberg New York

Hutchinson GE (1957) A treatise on limnology: geography, physics and chemistry, vol 1. Wiley, New York

Iriondo MH (1972) Mapa geomorfológico de la llanura aluvial del río Paraná desde Helvecia hasta San Nicolás (República Argentina). Rev Asoc Geol Arg 27(2):155–160

Iriondo MH (1988) A comparison between the Amazon and Paraná River system. Mitt Geol-Paläont Inst Univ Hamburg SCOPE/UNEP Sonderband 66

Junk WJ, Bayley PB, Sparks RE (1989) The flood pulse concept in river-floodplain systems In: Dodge DP (ed) Proceedings of the International Large River Symposium (LARS). Can Spec Publ Fish Aquat Sci 106:110–127

Knighton D (1998) Fluvial forms and process: a new perspective. Arnold, London

Leopold L, Wolman G (1957) River channel patterns: braided, meandering and straight. Geol Surv Prof Paper 282B:45–62

Lima D, Campana N, Amsler M, Schreider M, Gaudin H (1990) Desplazamiento de dunas y carga de fondo en un tramo del río Paraná. Memorias del XIV Congreso latinoamericano de Hidráulica, Montevideo, Uruguay 3:1203–1214

Marchese MR, Ezcurra de Drago ID (1992) Benthos of the lotic environments in the Middle Paraná River system: transverse zonation. Hydrobiologia 234:1–13

Marchese M, Ezcurra de Drago I, Drago EC (2002) Benthic macroinvertebrates and physical habitat relationships in the Paraná River-floodplain system. In: McClain M (ed) The ecohydrology of South American rivers and wetlands. Publ N° 6 of the International Association of Hydrological Sciences. UNESCO, Venice

Marston RA, Girel J, Pautou G, Piegay H, Bravard JP, Arneson C (1995) Channel metamorphosis, floodplain disturbance, and vegetation development: Ain River, France. Geomorphology 13:121–131
Mertes LAK (1985) Floodplain development and sediment transport in the Solimoes-Amazon River, Brazil. MS Thesis, University of Washington, Seattle
Orfeo O, Stevaux J (2002) Hydraulic and morphological characteristics of middle and upper reaches of the Paraná River (Argentina and Brazil). Geomorphology 44:309–322
Paira AR (2003) Características morfométricas de lagunas de la llanura de inundación del río Paraná Medio. Tesis de Licenciatura en Cartografía, Universidad Nacional del Litoral, Santa Fe, Argentina
Paira AR, Drago EC (2006) Genetical, morphological and evolutional relationships of the floodplain lakes in the Middle Paraná River hydrosystem. Zeitschrift für Geomorphologie 145:207–228
Ramonell CG, Amsler ML, Toniolo H (2000) Geomorfología del cauce principal. In: Paoli C, Schreider M (eds) El río Paraná en su tramo medio. Contribución al conocimiento y prácticas ingenieriles en un gran río de llanura. Centro de Publicaciones de la Universidad Nacional del Litoral, Santa Fe, Argentina
Schumm SA (1971) Fluvial geomorphology: the historical perspective. In: Schumm SA (ed) River morphology. Dowden, Hutchinson & Ross, Stroudsburg, pp 365–395
Schumm SA, Dumont HF, Holbrook JM (2002) Active tectonics and alluvial rivers. Cambridge University Press, Cambridge
Sippel SJ, Hamilton SK, Melack J (1992) Inundation area and morphometry of lakes on the Amazon River floodplain, Brazil. Arch Hydrobiol 123:385–400
Timms BV (1992) Lake geomorphology. Gleneagles Publishing, Adelaide
Tockner K, Malard F, Ward JV (2000) An extension of the flood pulse concept. Hydrol Process 14:2861–2883

4 The Physical Dynamics of the River–Lake Floodplain System

Edmundo C. Drago

4.1 Introduction

The Paraná River is the second largest river in South America in terms of catchment area (1.51 million km^2), the second longest (4,400 km from the headwaters of Grande River in Brazil to the Río de la Plata estuary), and the third in terms of discharge (about 470 km^3 of freshwater carried to the sea annually). Within this fluvial hydrosystem, the segment of the Middle Paraná encompasses 700 km upstream from the Paraguay outlet to the apex of the delta upstream from Rosario city (Santa Fe Province, Argentina) (Chap. 1, Fig. 1.5). Along its main channel, the Middle Paraná has built a relatively wide fringing floodplain with a surface of ca. 20,000 km^2. This floodplain resembles other large tropical and temperate floodplains in its amplitude and duration of flooding, and in many of its ecological characteristics. The annual phases of flood and drought and the complex hydrological connectivity between lotic and lentic waters are the major variables driving the lateral and longitudinal exchanges of sediments, organic matter and organisms at the riverscape scale. Moreover, the hydrology of the floodplain lakes, as well as the sedimentation-resuspension processes, the thermal behavior and the heat content variations are closely linked with the flood and drought phases. The aim of this chapter is to extend the knowledge on the hydro-sedimentological dynamics, as well as the thermal structures and mixing processes that occur in the river–floodplain lake systems of the Middle Paraná during the annual connection and disconnection phases.

4.2 Water Regime of the Floodplain Lakes

4.2.1 Degrees and Types of Connection

The geomorphology and the evolutional dynamics in different areas of the floodplain drive the origin, evolution and connectivity degree of the floodplain lakes. Figure 4.1 depicts the interrelationships between hydrological,

Instituto Nacional de Limnología (CONICET-UNL), José Maciá 1933, 3016 Santo Tomé, Santa Fe, Argentina, e-mail: edmundodrago@arnet.com.ar

M.H. Iriondo, J.C. Paggi, and M.J. Parma (Eds.)
The Middle Paraná River: Limnology of a Subtropical Wetland

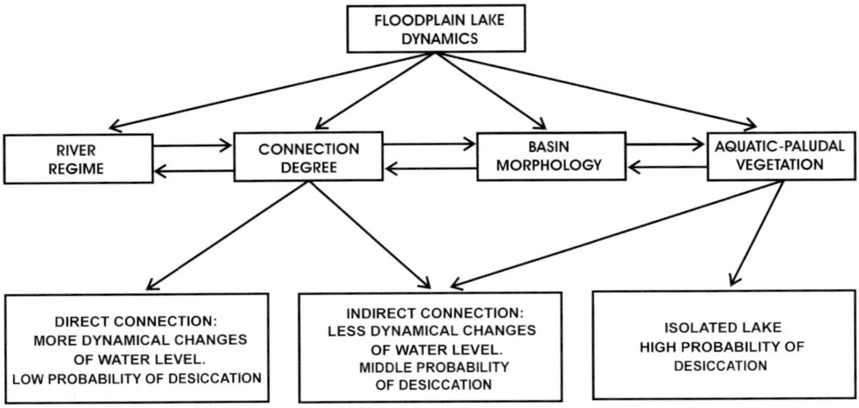

Fig. 4.1 Diagram illustrating the interactions between river regime and connection degrees, morphological, and vegetational lake components and their incidence on the in-lake hydrodynamics

morphological and vegetational components that drive the in-lake hydraulic dynamics of the Middle Paraná floodplain lakes. Three main degrees of hydrological connectivity between lotic and lentic waters can be distinguished according to the morphological connection type (Drago 1981): (a) *Direct connection*: permanent connection through a mouth, an erosion ditch levee or a short channel less than one kilometer in length. This length is adopted because a major channel longitude commonly implicates a stronger influence of the aquatic/terrestrial transition zone (ATTZ; Junk et al. 1989) or other water bodies, which sometimes markedly change the characteristics of the river water. Mouth and channel crevasse are detected in a major number in the less evolutioned floodplain areas, i.e., in the modern sand bar and meander areas located along the border of the main stem and the floodplain. The levee island lakes (Paira and Drago 2006) usually present a connection mouth at the downstream ends of the levees which allows the immediate response of the lake levels to the river stage variations. This connection mouth is progressively filled by sediments due to the formation of sand bars. Aquatic macrophytes and large woody debris stranded at the mouth during the river falling stages and overgrowth of helophytes accelerate the deposition process. Crevasse chutes are channel-side ridges ranging from less than 0.1 to 1 km in length, usually with sediments of sand size, fining away from the active channels. Some erosion levee ditches are strongly scoured during the falling water phases and vertical cutting. For example, in a levee break connecting a lake with a floodplain channel, 0.8 m of vertical cutting was measured in 15 days (Drago 1981). The back-and-forth flow can maintain an open connection for several years between the lotic and lentic environments. Direct connections are commonly observed in the levee island lakes and obstruction lakes. (b) *Indirect connection*: occurring from the rising phase up to the bankfull river stage, i.e., the lotic water has not overflowed the levees yet.

Lakes are connected with the river through a mouth or erosion ditch levee with intervening lake(s) or swamp(s), temporary floodplain channels of more than 1 km in length with intervening lake(s) or through a silted channel. It is typical of the more evolutioned floodplain areas and it can be permanent or temporary (Drago 1976, 1981, 1989). In tectonic sunken floodplain areas (Iriondo 1988), the indirect connectivity increases in density and pattern complexity. Large and irregular lakes can be fed and drained by several channels, some of them flowing from other interconnected lakes, while channels draining the same lake can join downstream and also feed other lakes (Chap. 3, Figs. 3.1C, 3.10), and (c) *Overbank connection*: overspill connection that occurs during the overbank phases. Lakes do not present any morphological connection, i.e., inflow of river water occurs only during the peaks of inundation phases through the floodplain surface.

The hydrological response to the rising channeled phase of the lakes with indirect connection is markedly slower than in those with direct communication, even between adjacent lakes (Chap. 3, Fig. 3.7A). Thus, the inflow of flood water is delayed for several weeks in relation to the lakes with direct a connection (Fig. 4.2). Completely isolated lakes, surrounded by swamps or marshes, only receive flood waters during the inundation phases (Figs. 4.2, 4.3). Therefore, the hydrology of the isolated lakes during prolonged drought periods is driven by the rainfall-evapotranspiration budget. As the rainfall is not important in the lentic hydrology, compared with the high evapotranspiration rates, some shallow isolated lakes (mean depth ca. 1 m) usually dry during the longest drought periods. There are also two levels of lateral-vertical subsurficial connectivity: (i) water river infiltration through levees and lake bottoms (aquifer-lake), that may be important during the first time of the river rising phase, originating small elevations of the lake levels, when the river is still under the bankfull stage, and (ii) the groundwater contribution from terrestrial aquifers.

The communication systems between lotic and lentic environments are usually more complex with increasing distance to the river main channel. This complexity is enhanced in the areas with tectonic influence where there are several large and irregular lakes interconnected by a dense anastomosed drainage network (Chap. 3).

4.2.2 Hydrological Lake Phases

The annual water regime in the floodplain lakes can be divided into four phases (Drago 1980, 1981): (1) The *rising-channeled water phase*, embraced from the inflow of channeled flood waters into the lakes until the river reaches the bankfull stage. The annual supply of allochthonous material to the lakes begins with this phase. During this phase, the diffuse inundation of the ATTZ from the internal part of the floodplain also starts (Drago et al. 2003). When the river inundates the floodplain lakes in the onset of this phase, mixing between the local lentic waters and the regional lotic waters begins. Afterwards, by the time the river floods its levees, the

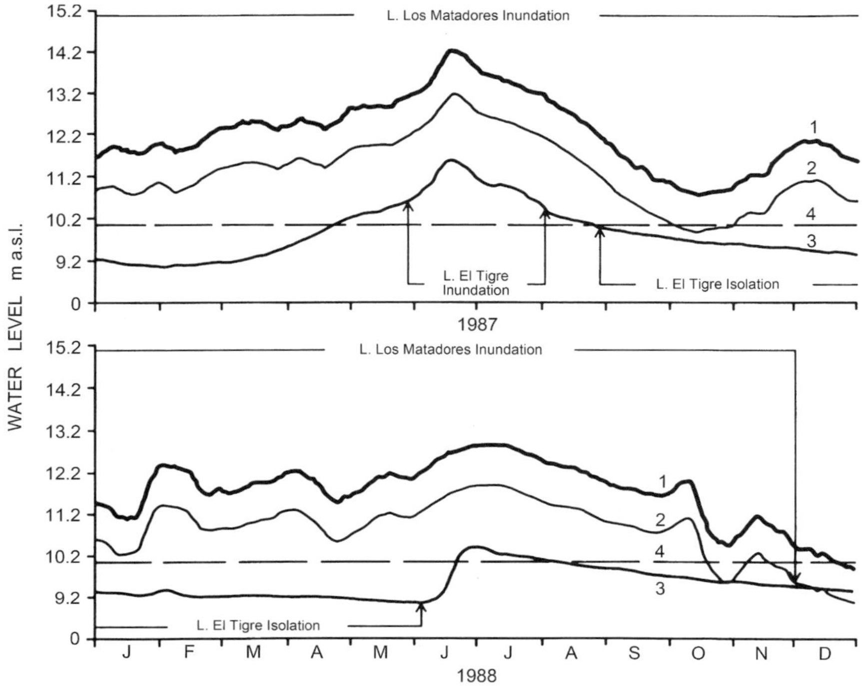

Fig. 4.2 Hydrographs of the height stage for the Paraná River at Santa Fe (*1*) and Los Matadores (*2*) and El Tigre (*3*) lakes, showing the duration of inundation and isolation phases in relation to the river level and the mean annual low water level of the Paraná River (*4*). The lakes are 500 m apart

floodplain is only partially flooded by river water, and several patterns of mixing occur. This mixing of the surface waters on the floodplain was defined as the "perirheic zone" (Mertes 1997). Several and irregular areas of mixing are usually formed according to the degree of connectivity between the channels and the "lentic water bodies-flatlands-forests" complex. In large and irregular lakes with direct connection it is possible to detect parallel bands of mixing, located progressively away from the connection point, which show sharp borders of mixing between local and regional waters (mainly in turbidity), and others nearest the lotic water inflow, which show a lesser contrasting physical border condition. (2) The *inundation phase* begins when the river overcomes the bankfull level and flood water spreads over the levees, flats and water bodies, and the majority of the lakes lose their lentic identity, attaining their maximum depths. Organic and inorganic materials of the floodplain flats become suspended and are carried into the lakes. Mixing between lotic and lentic waters reaches its maximum intensity during this phase. The shape, spatial position, and orientation of the lake basins in relation to the main water

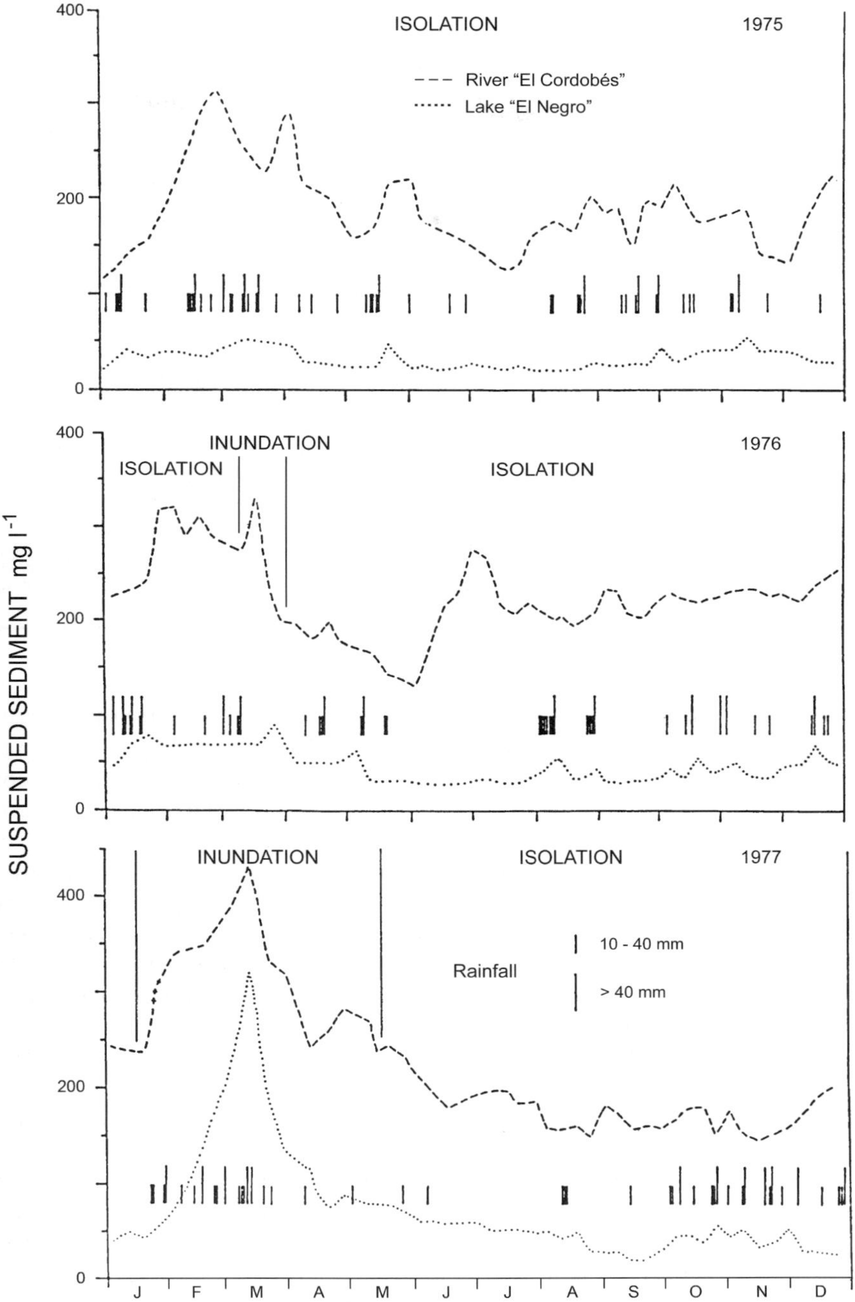

Fig. 4.3 Variation of suspended sediment concentrations during isolation and inundation phases in the Cordobés River–El Negro Lake system. Note the variations of suspended sediment concentrations due to the indirect influence of wind by the movements of the floating meadows during isolation phases. The winds are also strongly linked with the rainfall

courses promote completely or partially the temporary acquisition of some lotic characteristics, as current velocities of ca. 1 m s^{-1}. Figure 4.2 depicts the duration of the inundation and isolation phases in lakes with different connection degrees. (3) The *falling-drainage water phase* starts when the river stage falls below the bankfull level, originating a flow from the lentic water bodies and flatlands to the main stem and floodplain channels. The lotic waters will again be confined progressively to the channel banklines, and the lentic water bodies will return to their lacustrine conditions. (4) The *isolation phase* means a lack of communication between lotic and lentic water bodies, and the running waters are confined in their channels well down the bankfull level. Connections as mouths, levee ditches and short channels usually dry up completely if the river level is very low. If this phase is largely extended, the shallowest lentic water bodies may dry up wholly, sometimes producing massive fish mortalities. Furthermore, limnological conditions in the lakes may be influenced by meteorological factors, such as air temperature, wind, and rainfall. In El Negro Lake, for example, some daily rains, recorded during 1975–1977, ranged between 51 and 94 mm, giving rise to an average increment of 16 mg l^{-1} in the suspended sediment concentrations (Fig. 4.3). The rising-channeled and drainage water phases are important components of the expansion-contraction cycles occurring below and near the bankfull level (Tockner et al. 2000).

The river fluctuations and the communication systems between lotic and lentic components will determine the amplitude, duration, frequency, predictability, and the time of the year of each water phase. Clearly, the stage limits will not be the same for lakes with different types and degrees of connection (Tables 4.1, 4.2). Hence, it is possible to find isolated, as well as connected lakes, with the same river level. Furthermore, the persistence of a specific gauge height will determine the duration of the corresponding phase. The lake isolation phases are usually from July to December, while the flooding phases are between January and June. However, particularly low or high water level periods can strongly deflect this regime pattern (e.g., the ENSO events). Table 4.2 shows different annual water phases in Los

Table 4.1 Water stage heights of Paraná River at Santa Fe and for hydrological phases in different lakes (in meters a.s.l.) according to direct (*DC*) and indirect connections (*IC*) and isolated conditions (*I*) with stream channels

Lake phase	Los Matadores Lake (D.C.)	El Negro Lake (I.C.)	El Tigre Lake (I)
Isolation phase	<10.5	<12.3	<12.3
Rising-water phase	10.5–12.3	12.3–12.7	12.3–13.2
Inundation phase	>12.3	>12.7	>13.2
Falling-water phase	12.3–10.5	12.7–12.3	13.2–12.7

Table 4.2 Annual isolation and connection phases in three floodplain lakes (time in days)

Year	1975	1976	1977
Los Matadores Lake			
Isolated	19	0	17
Connected	346	366	348
El Negro Lake			
Isolated	354	243	224
Connected	11	123	141
El Tigre Lake			
Isolated	356	300	260
Connected	9	66	105

Matadores and El Negro lakes, 7 km apart; one of them was isolated about 61% of the year and the other one only 5% of the year. The hydrographs of the Paraná River sometimes vary considerably from year to year. The amplitude of these annual variations drives the frequency and degree of mixing between lotic and lentic waters. The minimum river level and its duration specify the magnitude of the contraction or expansion of the lakes, and the maximum river level determines the intensity of mixing of local and regional waters.

4.2.3 Lake Level Variations

During the rising and inundation phases, the flood water firstly enters into the floodplain through the floodplain channels, and then by overflowing the levees (Figs. 4.2, 4.4). Due to the flow across the levees, flatlands and lentic environments, flood waters usually reach the lakes with their physical and chemical conditions strongly modified. Furthermore, with the increasing river stage, the groundwater level rises, feeding the lakes (Fig. 4.4A, B). This fact is better detected in lakes with indirect connection (Fig. 4.4B), since groundwater seepage is masked by the inflow of river flood water in lakes with direct connection. Thus, lake levels are regulated by the river behavior which, in turn, controls the velocity of the underground seepage.

The stage hydrographs of Los Matadores Lake (direct connection) and El Tigre Lake (indirect connection) show the typical different influence of the Paraná River level changes (Santa Fe harbor staff gauge) on the selected lakes (Fig. 4.2). Both lakes are fed by the Correntoso River and are only 500 m apart. Los Matadores Lake was connected with the Correntoso River when the Paraná River overcame the level of 10.4 m a.s.l., while El Tigre Lake was connected at a river stage of 12.2 m a.s.l. Due to the close connection with the

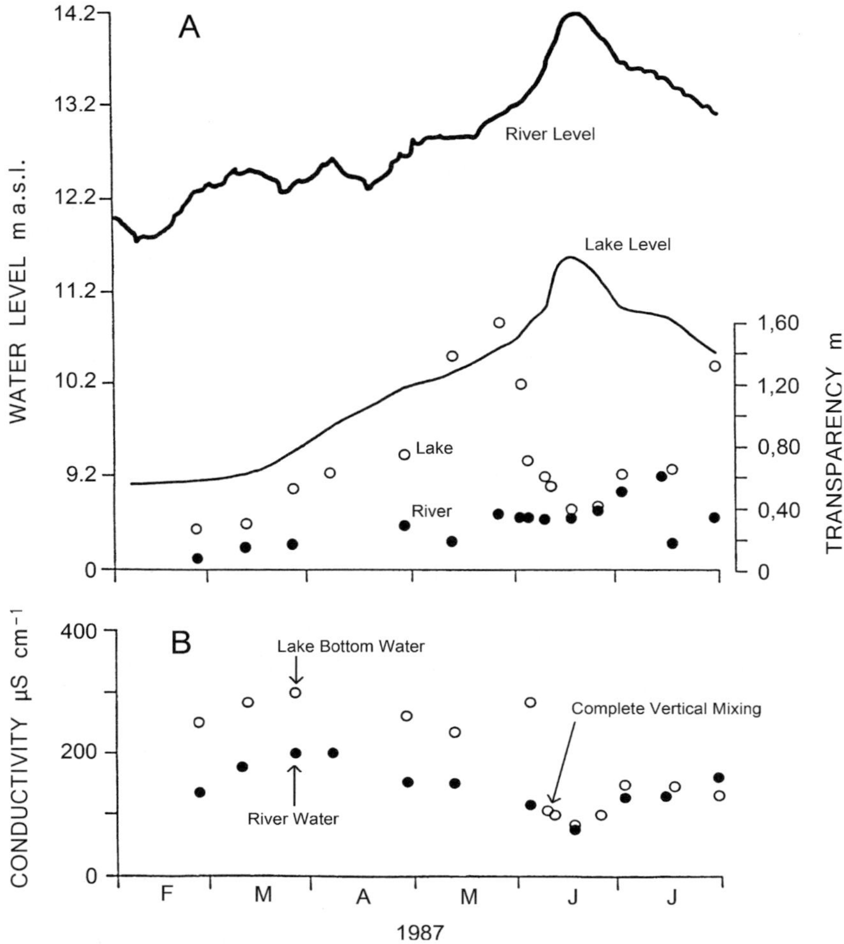

Fig. 4.4 **A** Hydrographs of the height stage for the Paraná River at Santa Fe and El Tigre Lake and the variation of the river (*filled circles*) and lake (*empty circles*) transparencies. **B** River water (*filled circles*) and lake bottom water (*empty circles*) conductivities during rising water and inundation phases. Note during the peak stage levels the complete mixing of the lake water column detected through the transparency and conductivity data

river, the water level in Los Matadores Lake follows closely the stage fluctuations of the Paraná River ($r = 0.99$, $P < 0.001$). Some lakes, such as El Tigre, were isolated from the river for 437 days, i.e., 60% of the time (1987–1988). At the same time, other lakes, such as Los Matadores, were isolated from the river for only 75 days and connected during 674 days, i.e., 8% and 92% of the time, respectively, for the considered period (Fig. 4.2). Several lakes of this last type are connected with the rivers for more than 50% of the year. Owing to their indirect connection, some lakes, as El Tigre, present diverse water regimes.

Nevertheless, the correlation between river and lake levels is good (r = 0.70, $P < 0.001$), being the lower figure the result of the incidence of the isolation periods. In turn, when lakes and river levels are correlated over the connection stage of 12.2 m a.s.l., the correlation rises up to r = 0.93 ($P < 0.001$). Conversely, the correlation is negative and non-significant (r = –0.31) under the river stage of 12.2 m a.s.l., showing a complete independence of the lake from the river. Changes in the lake level are due to seepage and evapotranspiration. River fluctuations over 10.2 m a.s.l. but under the level connection of 12.2 m a.s.l. have no effects on these types of isolated lakes. During a long isolation period from August 1987 to June 1988 (278 days), some lakes, as El Tigre, decreased their level in about 1 m, i.e., a daily average falling of 3.6 mm. In a second isolation period from August to December 1988 (131 days), the decreasing lake-level was 0.5 m, with an average of 3.8 mm d^{-1}. Data about lake levels changes during rising, inundation and drainage phases were obtained in El Tigre Lake during the 1987 flood (February–August) over the connection level of 12.2 m a.s.l. During the rising curve, from 12.2 to 14.2 m a.s.l. (maximum flood peak), the river level increased 17.5 mm d^{-1} in average, and the lake increased at a rate of 21.5 mm d^{-1}. For the same river level range, the curve of the drainage phase showed a falling rate of river and lake levels of 29.3 mm d^{-1} and 23.5 mm d^{-1}, respectively (Fig. 4.2).

The highest rising rate in the lake levels when they are compared to the increase in the river level occurs because the river continues "injecting" water to the aquifer, even if the river level is falling, but always over 12.2 m a.s.l. Thus, the major supply of the groundwater to the lakes accelerates the rising of lake levels in relation to the river stage (Figs. 4.2, 4.4). On the other hand, the decrease in the river stage presents higher rates, because its level falls almost without interruptions while the lakes drain into the river channels firstly over the levees, then across other water bodies and flats, and finally by seepage. The maximum level amplitude in El Tigre Lake was 2.55 m, with maximum and minimum depths during the inundation and isolation phases of 3.4 m and 0.85 m, respectively. During extended drought periods, the maximum depth decreases to around 0.40 m at the center of the shallower permanent lakes.

El Tigre Lake and the Correntoso River showed three annual variable periods of hydrological connectivity: (i) very indirect connection, when the river was at the rising phase (river stage over 12.2 m a.s.l.), raising the lake level by the entrance of groundwater through the lake bottom, with a rate of 15 mm d^{-1} (Figs. 4.2, 4.4A); (ii) indirect connection, when the river stage overcame 12.7 m a.s.l., originating the inflow of lotic waters through small swamps, increasing the lake level in about 20 mm d^{-1}; (iii) direct connection, when the river stage was over 13.2 m a.s.l. and lotic waters overflowed the levees. At this river stage (13.2 m a.s.l.), the lake increased its level at a rate of 45 mm d^{-1} (Figs. 4.2, 4.4A). When the floodplain became completely inundated, the maximum river and lake level peaks were coincident. Figure 4.4A, B depicts the three periods of connection levels and their influence on the changes in water stages, conductivity and transparency. It is possible to observe that from a river stage of 12.2 m a.s.l., the lake level begins to increase and, at the same time, as a result of the groundwater seepage, there is an increase in the bottom water conductivity. Meanwhile, the river conductivity is in average 100 $\mu S\ cm^{-1}$ lower than the lentic bottom water. Figure 4.2 illustrates that the level of the El Tigre Lake decreases very slowly when the river level falls under 12.2 m a.s.l. Strong fluctuations of the river level during its decrease from 12.2 m to 10.2 m a.s.l. are not detected at the lake level. In turn, during the same period, some lakes with direct connection, as Los Matadores Lake, show the same level fluctuations than the river. In spite of the decrease in the river level from 12.2 m to 10.7 m a.s.l., the lake loses water by seepage and evapotranspiration, with an average level falling of 7.5 mm d^{-1}. From October (spring), in spite of the increase in the river stage but without overcoming the level of 12.2 m a.s.l., the lake continues losing water, mainly by evapotranspiration, decreasing about 3.2 mm d^{-1}. This date coincides with the evapotranspiration rates (2.7 mm d^{-1}) measured experimentally by Lallana et al. (1987) during a similar seasonal period. The isolation period coincides with the maximum air temperatures (spring-summer), increasing the water loss by evapotranspiration (Figs. 4.2, 4.3).

4.2.4 Residence Time

The current velocities in the floodplain lakes are driven by the shape and bottom morphology of the basins (e.g., shore irregularity and depth of the old thalwegs), the river level, the type and degree of connection with the main and floodplain channels, the spatial location of the lakes into the floodplain, their topographic altitude, the orientation of the basin axes in relation to the nearest major lotic flow and the aquatic vegetation (Table 4.3). Drainage density of the floodplain network developed in different areas of the valley is also a relevant factor (Table 4.4). The basin shape will have a strong influence on the in-lake water velocities and current patterns. Less variation will be found in round or elongated lakes than in irregular lakes, both during high and low river levels.

During the lowest river stages, the lakes remain isolated from the running waters. Therefore, only the strong winds originate currents, mainly in the larger lakes, so fringing forests become important as shelters. At high river level, the entrance of flood waters promotes strong currents, mainly near the connection points, or when the channel-shaped lakes are located near, and parallel to, the active channels (Fig. 4.5). A channel-shaped El Negro Lake, for example, with a maximum length of 790 m, was isolated from the river during a long period of 2 years, lasting with 100 % of its surface covered predominantly by *Eichhornia crassipes* and aquatic grasses, forming a dense floating meadow where it was possible to walk on its surface. During the next 77 days of inundation, the aquatic plants were completely carried out of the lake and the thick mantle of organic matter and fine sediment deposited on its bottom was also washed away, allowing the exposure of the former sand bed channel. The maximum surficial current velocity reached was 0.95 m s^{-1} and the minimum retention

Table 4.3 Residence time at different sites of La Cuarentena Lake and other lakes according to connection degree and distance from the main stem and floodplain channels (*m* minutes; *h* hours; *d* days)

Lake	Distance from mail channel (m)	Distance from secondary channel (m)	Connection degree		Residence time	
			Direct	Indirect	Connect.	Isolat.
La Cuarentena (main basin)	2,500	-	X		7 d	50 d
La Cuarentena (secondary basin)	3,600	-	X		15 d	130 d
La Cuarentena (small bay)	1,000	-	X		30 d	180 d
La Cuarentena (large bay)	1,500	-	X		50 d	230 d
La Torre	200	-		X	3 h	150 d
Los Matadores	10,000	20	X		8 d	19 d
El Tigre	10,000	30		X	23 d	365 d
El Negro	15,000	40		X	20 m	354 d
El Alemán	11,000	50		X	10 d	310 d
Victoria	10,000	20	X		20 d	280 d
La Tina	11,000	100		X	15 d	306 d

Table 4.4 Interrelationships between lotic and lentic components at the Middle Paraná river-floodplain showing the increasing hydrographical-connectivity complexity from a young sector near of main channel (**A**) to an old sector away from the main channel (**B**). Surveyed area: 100 km^2. Tested area: 5,000 km^2

	A	B
Number of lakes by km^2	0.11	2.53
Number of swamps by km^2	0.75	3.55
Number of channels by km^2	1.02	2.41
Drainage density (km/km^2)	0.95	3.20
Number of lakes with tributaries or emissaries	3	18
Number or tributaries or emissaries of lakes	10	22
Number of marshes and swamps with tributaries or emissaries	2	12
Number of tributaries or emissaries of marshes and swamps	2	14
Number of lakes with associated marshes and swamps	4	29
Number of marshes and swamps associated with lakes	8	32
Number of junctions between secondary channels	13	29

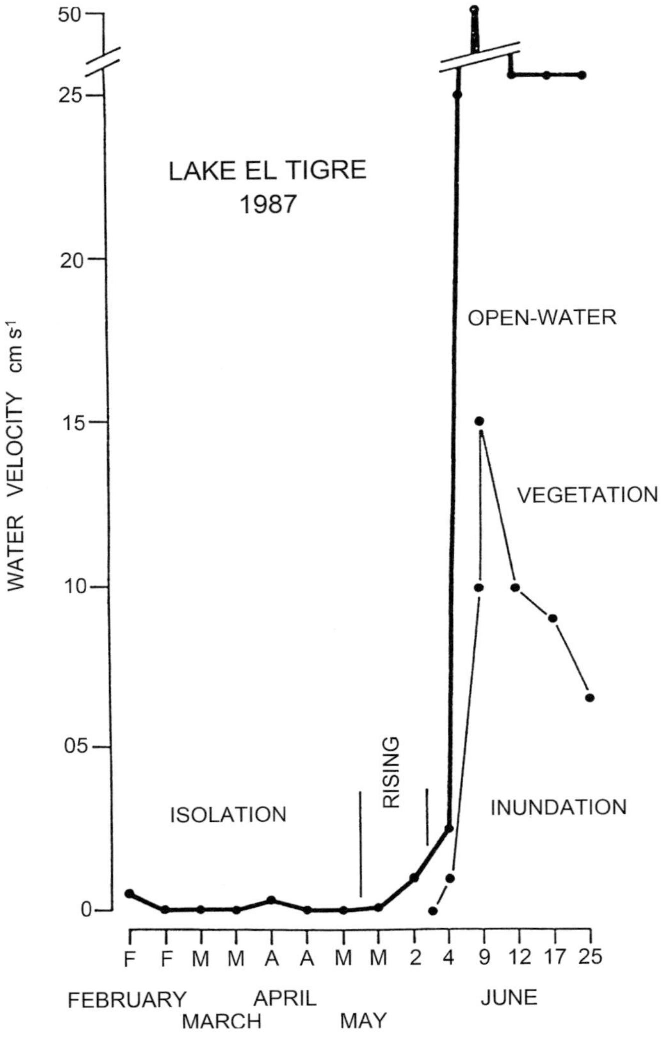

Fig. 4.5 Spatial variations of current directions and velocities during the start of an inundation phase in El Tigre Lake (June 9, 1987)

time of the lake water during this inundation phase was 23 min (Santa Fe stage: 14.1 m a.s.l.; March 1977). On the other hand, large irregular lakes, as La Cuarentena (Table 4.3), have their main basins and larger embayments free of aquatic vegetation, and the smaller narrow bays are usually 60–100% covered by dense floating meadows of *Paspalum repens*, *Panicum elephantipes*, *Echinochloa polystachya*, *Polygonum* spp., *Eichhornia crassipes* and *E. azurea*. Thus, currents that do not exceed 0.10–0.15 m s^{-1} are developed only in the largest and open basins during the isolation phase, whereas the water velocity is negligible in the bays. At high waters, the main basin of La Cuarentena Lake is directly connected with the Paraná main stem through a short connecting channel (Table 4.3). In-lake strong differences in current velocities and in the water renewal times occurred in that period. Thus, the embayments that were far away from the river connection, received flood waters from the main basin during the channeled rising phase, and then from other lakes and inundated flatlands. Therefore, they had very slow currents, reaching ca. 0.10 m s^{-1}. In these cases, the residence time of water is about one month or more. On the other hand, in the main basin, flood waters flow firstly along the old thalweg, reaching maximum current velocities of 0.5 m s^{-1} and the retention time is between 5 and 7 days. During the rising and falling phases, the current velocities of these water bodies do not exceed 0.4 m s^{-1}, and the retention times vary between 7–10 days (Table 4.3).

In isolated lakes or with indirect lotic connection, the current velocities and retention times change sharply when the river waters rise above the levees. Small shallow lakes such as El Tigre have wind-driven currents, with maximum velocities of 0.02–0.03 m s^{-1} during the isolation phase. Five days after the start of the inundation phase, the current velocities reached 0.5 m s^{-1} (Fig. 4.5). During this period, the residence time in El Tigre Lake was 23 days, and its volume increased approximately tenfold. In contrast to non-floodplain lakes, the water residence time in floodplain lakes varies greatly according to the magnitude of the inundation or isolation phases. Thus, in extreme cases, some isolated lakes, as El Negro, have a residence time of 354 days (isolation phase) or 23 min (inundation phase). Between these extremes, the retention times vary according to the river level, the degree of connection with the river, the spatial location, and the orientation of the lake within the floodplain drainage network (Table 4.3).

The lakes with a direct connection have the shortest water retention times during flood peaks, and rapid changes in lake volumes occur during inundation phases. Conversely, the volume changes are slow in the rising and drainage phases. The volume variations during isolation periods are very slow, being determined mainly by the evapotranspiration rates. During the falling phases, the lake boundaries retract from the forested area. Drainage flow decreases gradually throughout the falling phase, until it ceases, usually 1 or 2 months later. The lake volumes decrease by about 50% or more during the isolation phase because evaporative water loss exceeds rainfall on the lakes. In this process, the rains are unimportant.

4.3 Suspended Sediments

The spatial distribution of the suspended sediment concentrations in the floodplain lakes varies according to external factors and to an in-lake set of factors. The external factors are the duration and magnitude of the hydrological phases, the sediment amounts introduced into the lakes by the inflowing waters, the riparian vegetation, and the wind speed. The in-lake factors are the shore irregularities, depth, fetch, type of connection, current speed, hydrophyte coverage, and bottom composition. During the inundation, the inflow of river waters with high concentrations of suspended sediments originates temporal and spatial variations (Figs. 4.3, 4.6, 4.7). In large irregular lakes, the highest concentrations are usually detected toward the center of the open-water areas, because flood waters with the highest velocities follow the old thalwegs. The lowest concentrations are recorded in the narrow channel-shaped embayments and sheltered inlets covered by aquatic plants. These areas maintain fairly high transparency during inundation, quite similar to that measured during the isolation periods. Noticeable spatial variations in the sediment concentrations also occur during the developing and evolution of the perirheic zones. Rounded lakes with fetches over 300 m develop hydrophytes and helophytes only along the shorelines, owing to the following wind actions: (a) free-floating plants are destroyed or carried away by the mechanical action of waves and currents, and (b) the bottom sediment resuspension strongly increases turbidity, which has negative consequences on growth of rooted aquatic macrophytes (Jupp and Spence 1977; Spence 1982; Chambers and Kalff 1985; Chambers 1987). In irregular lakes, the narrow channel-form embayments are densely encroached by aquatic and palustrine vegetation, which covers up to 100% of their surfaces, showing high transparencies and, consequently, similar conditions to lakes with smaller areas and fetches lower than 300 m.

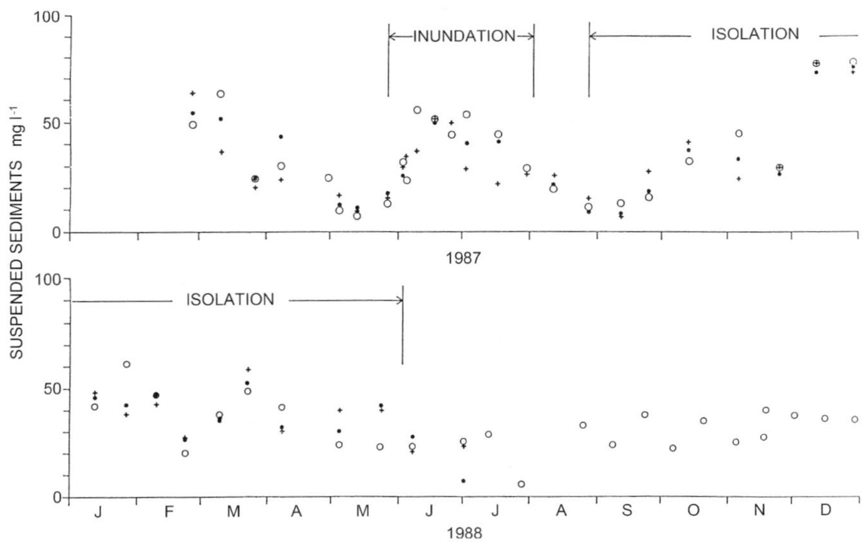

Fig. 4.6 Annual variation of the suspended sediment concentration in the surficial water across the main axis of El Tigre Lake basin during isolation and inundation phases. Northern (*crosses*), central (*empty circles*) and southern (filled circles) basin areas

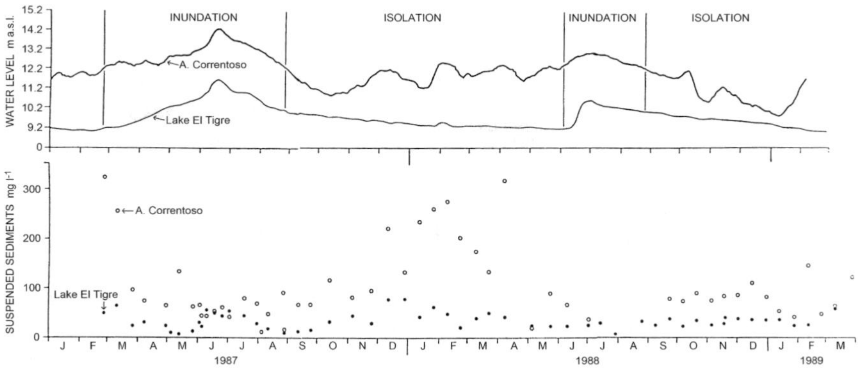

Fig. 4.7 Relationships between hydrological phases and the variation of suspended sediment concentrations in lotic and lentic waters in the Correntoso River-El Tigre Lake system

During isolation, the direct or indirect action of the wind is the main factor producing temporal and spatial variations in the suspended sediment concentrations. In lakes with large areas and fetches, the wind-generated turbulence causes sediment resuspension. In smaller water bodies with vegetation cover, even with 100% of covering, the movement of the aquatic plants by high winds promotes the release of the sediment attached to the roots and stems, increasing temporarily the sediment concentration. However, high levees, with trees and shrubs surrounding narrow channel-shaped basins, prevent bottom sediment resuspension.

In lakes with indirect connections, the communication between rivers and lakes starts when the Paraná level rises up to 12.2 m a.s.l. at the Santa Fe staff gauge (Figs. 4.2, 4.3, 4.7; Table 4.1). For example, in El Tigre Lake, the river water enters by direct seepage through the levee or by a rising water table when the river level is between 12.2 and 12.7 m a.s.l. (Figs. 4.4, 4.7). During this period, suspended sediment concentrations decrease by dilution. Some small fluctuations in the concentrations may be caused by the sediment released from the floating meadow movements. When the river level increases from 12.7 to 13.1 m a.s.l., the flood water enters into the lake through swamps, and is strongly filtered by the palustrine and aquatic vegetation, also decreasing its sediment load (Figs. 4.4, 4.6). Therefore, when the river level is between 12.2 and 13.1 m a.s.l., the increase in sediment concentration in the lake water is not very important (Figs. 4.4, 4.6, 4.7). When the river rises above 13.1 m a.s.l., the inundation phase begins, increasing consequently the suspended sediment concentrations (Figs. 4.4, 4.6, 4.7). During this short period (01–07 July, 1987), the northern part of the lake did not receive the direct influence of lotic waters because they flowed through dense vegetated flatlands and other lentic environments (Fig. 4.6). During isolation periods, variations in lake suspended sediments are mainly caused by bottom sediment resuspension because the surficial fine sediment layer is frequently in suspension. Wave action is usually effective over almost the entire lake bottom in the very shallow lakes. Nearshore areas, in particular, are often characterized by wide variations in suspended sediment concentrations due to localized belts of floating vegetation and shallow shores with large accumulation of fine materials. The important effect of the wind-induced currents in the resuspension process is to transport suspended materials over the lake.

4.3.1 Relationships between Suspended Sediment, Lake Morphology, Wind and Aquatic Vegetation

Temporal and spatial fluctuations of suspended sediment concentrations by wind action are linked with two main mechanisms: (1) in large lakes with effective fetch over 300 m and without surrounding forest, the wind generates wave-turbulence and bottom resuspension; (2) in lakes with effective fetch lower than 300 m and shorelines with forests and dense floating meadows, the turbulence caused by light winds is not sufficient to originate bottom sediment resuspension (wind speeds <30 km h^{-1}). The sporadic increase in suspended sediment concentrations originates in the supplies of materials from the aquatic vegetation due to the "shaking effect" of the wind. For example, in lakes areas 1–2 m in depth and covered by aquatic vegetation, wind speeds over 50 km h^{-1} increased the suspended sediments from 38 to 60 mg l^{-1}. However, it is possible to state that bottom resuspension is not a common feature in lakes with small areas, short effective fetch, large cover of aquatic and semiaquatic plants, and sheltered by forests. In small lakes with fetches lower than 300 m and surrounded by high levees, with trees and shrubs and large coverage of floating and rooted macrophytes, bottom sediment resuspension only occurs during strong rainstorms. Samplings and visual observations allowed us to verify these processes. Thus, samples obtained before and after a rainstorm with strong winds of 100 km h^{-1} and a shower of 39 mm (11:00 pm–8:00 am; 17/18-11-88) showed resuspension in a water column of 1.3 m depth. At the same time, the increase in suspended particulate and dissolved organic matter concentrations originated a sharp decrease in the dissolved oxygen concentration in the bottom water layer (Table 4.5). The resuspension processes are clearly observed during isolation periods. The inflow of sediment-laden river waters and the mixing processes generated during the overflow usually masked the above-mentioned features during the inundation phase.

The significance of the fetch is illustrated in Figs. 4.3 and 4.8. In small lakes, similar suspended sediment concentrations were detected between sampling stations. For example, in El Tigre Lake, where the effective fetch is 110 m, there are no important spatial variations in resuspended material concentrations (Fig. 4.6). In larger lakes, a common fact is to detect important spatial variations in the suspended sediment concentrations, even during inundation phases. During these

Table 4.5 Changes in some physical and chemical parameters measured in El Tigre Lake before and after a spring storm originated from 11:00 pm to 8:00 am (November 17/18, 1987). *Tw* water temperature; *K* conductivity; *O_2* dissolved oxygen; *SSC* suspended sediment concentration; *POM* suspended particulate organic matter concentration; *DOM* dissolved organic matter concentration

Date		Tw (°C)	K (µS cm^{-1})	O_2 (mg l^{-1})	pH	SSC (mg l^{-1})	POM (mg l^{-1})	TDS (mg l^{-1})	DOM (mg l^{-1})
17-Nov-87	Surface	25.4	263	4.8	6.2	28.0	1.1	132	12.6
	Bottom	24.8	259	2.8	6.3	50.0	2.1	130	12.0
18-Nov-87	Surface	26.7	258	6.7	6.3	40.0	6.3	129	17.8
	Bottom	25.4	253	0.4	5.7	69.0	6.2	124	16.6

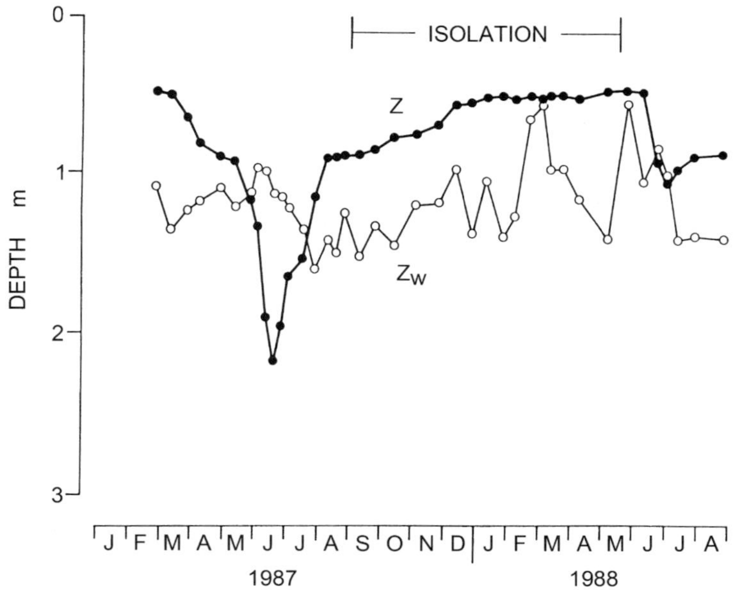

Fig. 4.8 Mean depth (z) and wave-mixed depth (z_w) during the 1987–1988 isolation-resuspension phase in El Tigre Lake

phases, differences in concentrations mainly depend on the location of the sampling sites in relation to the inflow points of the river waters. A sampling station near the direct input of the river water will show higher concentration and current velocity than other site located where the river water has previously crossed vegetated levees, flatlands or other lentic environments (Fig. 4.6).

4.4 Bottom Sediment Resuspension, Mixing Depth, and Sedimentation Rate

As seen in Sect. 4.3, the intensity of the resuspension-sedimentation process is a function of several variables, including wind velocity, effective fetch, depth of water column, aquatic-terrestrial vegetation cover, physical properties of bottom sediments and their distribution on the basin. Despite the complexity of the phenomenon, Carper and Bachmann (1984) demonstrated that sediment resuspension in an open small lake was predictable from wind velocity. In the Paraná floodplain lakes, sediment resuspension may be predictable if the effective fetch, depth and aquatic-terrestrial vegetation are considered simultaneously. Fetch and aquatic-terrestrial vegetation determine the degree of wind-driven turbulence at the lake surface, while depth gives the degree to which this turbulent energy contacts the bottom surface.

The relationship between wind action and basin morphometry was used to survey bottom resuspension. The wave-mixed depth (z_w) data during the sampling date were compared to the lake mean depths (z) (Fig. 4.8). Wind speed reached values for the generation of turbulence with sufficient strength to promote sediment resuspension (Figs. 4.3, 4.8). The small waves generated in these lakes are the so-called "deepwater waves", i.e., the waves that travel in water with a depth greater than one-half of their wavelength; the wave is said to "feel" the bottom (Carper and Bachmann 1984; Fig. 4.8). This means that there is an oscillatory horizontal motion of the water immediately over the bottom surface which may be sufficient to resuspend bottom sediments. In lakes with effective fetches over 300 m, these processes were detected during storms with wind speeds ranging from 30 to 50 km h^{-1}, while in small lakes, as El Tigre, this kind of waves are only generated during wind velocities over 50 km h^{-1}.

In small lakes, the deepwater wave exceeded the mean depth throughout the isolation phases, indicating that sediments could have been resuspended from much of the lake bottom (Fig. 4.8; Table 4.6). Figure 4.9 shows how sediment resuspension and sedimentation rate increased in intensity over the course of the isolation phase. As expected, the z_w/z ratio is directly related to the intensity of sediment resuspension because it reflects the increasing proportion of the lake bottom from which sediments could be resuspended (Häkanson 1981). At the same time, it is possible to detect a marked process of sedimentation that coincides with the highest suspended sediment concentrations and the lowest lake water transparencies (Fig. 4.9). Lakes with an indirect connection to the active channels show higher values of suspended sediment concentrations by resuspension, more than by the entrance of the sediment-laden river flood-waters. The "filtration" action of the levees, the aquatic-terrestrial vegetation and other lentic water bodies is very effective, decreasing the suspended sediment concentrations of the river water during its flow across the floodplain (Fig. 4.6; Drago 1989). During a sampling period from July 1987 to March 1989, the average sedimentation rate was 31.6 g m^2 d^{-1}, with maximum rates during the isolation periods (spring 87 and

Table 4.6 Composition of deposited material in a sediment trap located at the center of El Tigre Lake basin. *Cg%* grams carbon; *OM* organic matter; *IM* inorganic material

Date	Cg%	OM%	IM%
11-Sep-87	2.879	4.96	95.04
24-Apr-87	2.458	4.24	95.76
13-Oct-87	1.805	3.11	96.89
03-Nov-87	2.680	4.62	95.38
17-Nov-87	2.606	4.49	95.51
30-Nov-87	2.815	4.85	95.15
18-Dec-87	2.465	4.25	95.75
30-Dec-87	1.509	2.60	97.40
15-Jan-88	2.007	3.46	96.54
29-Jan-88	3.306	5.70	94.30
10-Feb-88	2.855	4.92	95.08
25-Feb-88	3.394	5.85	94.15
15-Mar-88	3.488	6.01	93.99
05-Apr-88	1.953	3.37	96.63
21-Apr-88	4.486	7.73	92.27
05-May-88	3.342	5.76	94.24
26-May-88	2.589	4.46	95.54
07-Sep-88	5.555	9.58	90.42
22-Sep-88	4.728	8.15	91.85
06-Oct-88	0.140	0.24	99.76
20-Oct-88	0.154	0.26	99.74
04-Nov-88	2.885	4.97	95.03
17-Nov-88	2.575	1.44	98.56
01-Dec-88	2.427	4.18	95.82
15-Nov-88	2.467	4.25	95.75
29-Dec-88	2.133	3.68	96.32
12-Jan-89	1.853	3.19	96.81
27-Jan-89	2.387	4.11	95.89
10-Feb-89	1.693	2.92	97.08
24-Feb-89	10.667	18.32	81.61
10-Mar-89	11.300	19.48	80.52

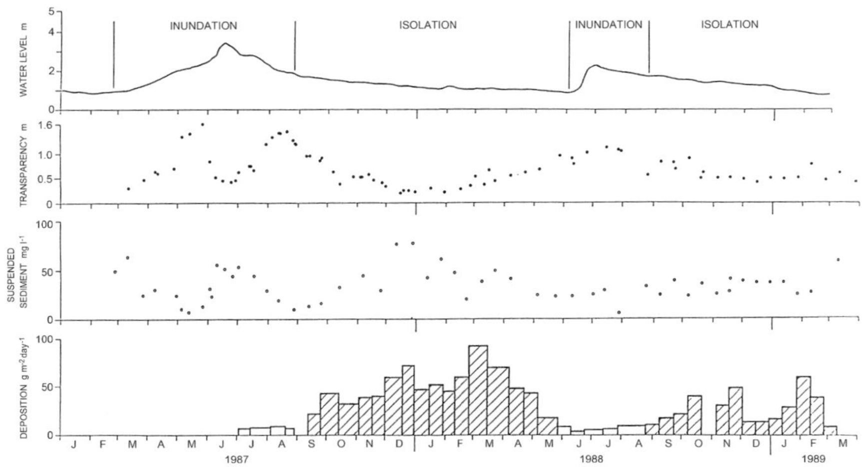

Fig. 4.9 Maximum depth (water level), transparency, suspended sediment concentration and sedimentation rate during isolation (*streaky bars*) and inundation (*empty bars*) phases in El Tigre Lake

summer-autumn 88) of 92 g m^2 d^{-1} (March 88; Fig. 4.9). Minimum sedimentation rates were measured during the inundation phases, reaching 4.4 g m^2 d^{-1} (June 88; Fig. 4.9). The sedimentated material was composed of 80–99% of inorganic fraction, and the organic material ranged from 0.14 to 11.3 C g % (Table 4.6).

During floods, the currents originated in the lakes are the main factor preventing the sedimentation of fluvial sediments, whereas the increase in lake depth decreases the probability of resuspension of lake bottom sediments. These processes were tested in the laboratory and in the field, during studies on sediment traps dynamics (Gardner 1980a, 1980b). Furthermore, floating macrophytes are another factor contributing to decrease resuspension of lake bottom sediments, as was demonstrated by Drago and Vassallo (1980), who found a high correlation between aquatic plant cover and Secchi disk transparency ($r = 0.82$, $P < 0.001$).

4.5 Water Temperature

4.5.1 Annual Variation

Water temperatures of the Middle Paraná River and its floodplain streams and lakes mainly follow the fluctuations imposed by solar radiation and air temperature (Drago 1984; Drago and Paira 1987; Figs. 4.10, 4.11A). Furthermore, the small depths of the lakes markedly enhance the immediate response of the entire water column to the atmospheric changes, even during maximum river levels. Figures 4.10 and 4.11A, B, show the strong dependency

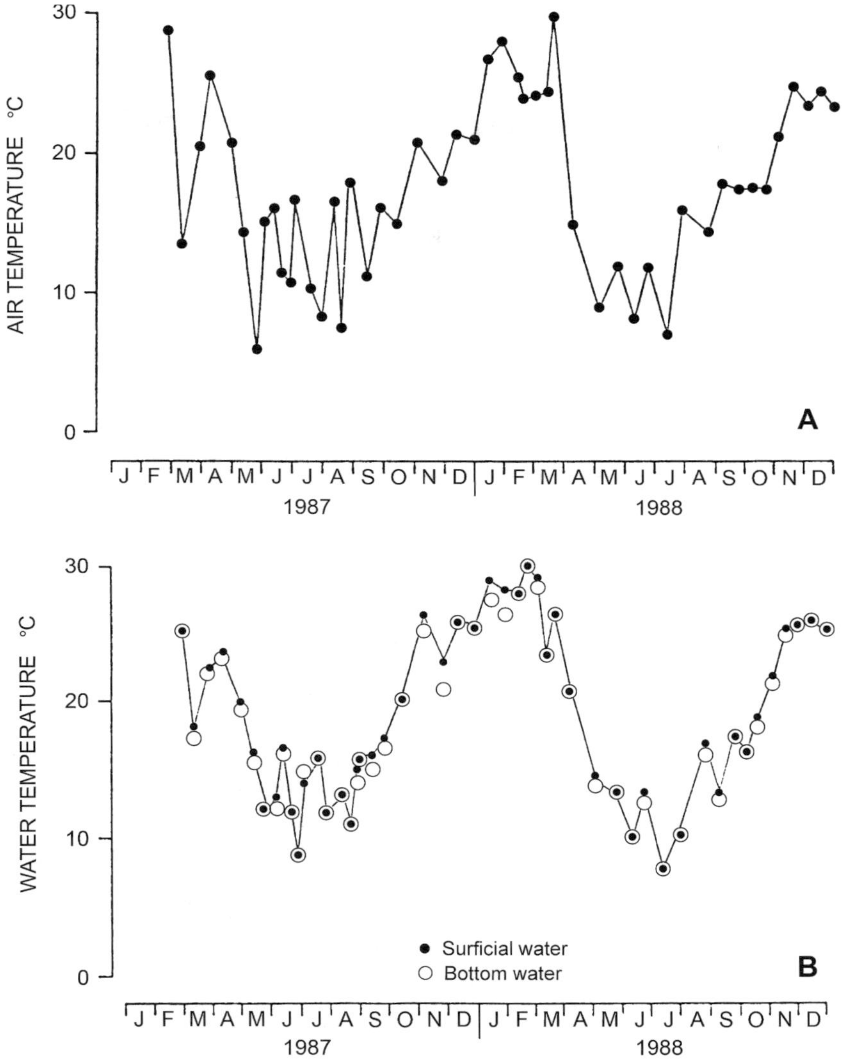

Fig. 4.10 Time diagrams for daily mean air temperature (**A**) and surficial (*full dots*) and bottom (*empty circles*) water temperatures (**B**) in El Tigre Lake

of surface and bottom water temperatures to air temperatures in a floodplain lake. Thus, the correlation between the mean daily air temperature and mean water column temperature for each sampling date was $r = 0.82$ ($P < 0.001$). Likewise, it was possible to observe that the difference between surficial and bottom water temperatures does not rise above 2°C (November 1987; Fig. 4. 10B), which demonstrates the polymictic character of these lakes. The annual amplitude in open water areas of the studied lake reached about

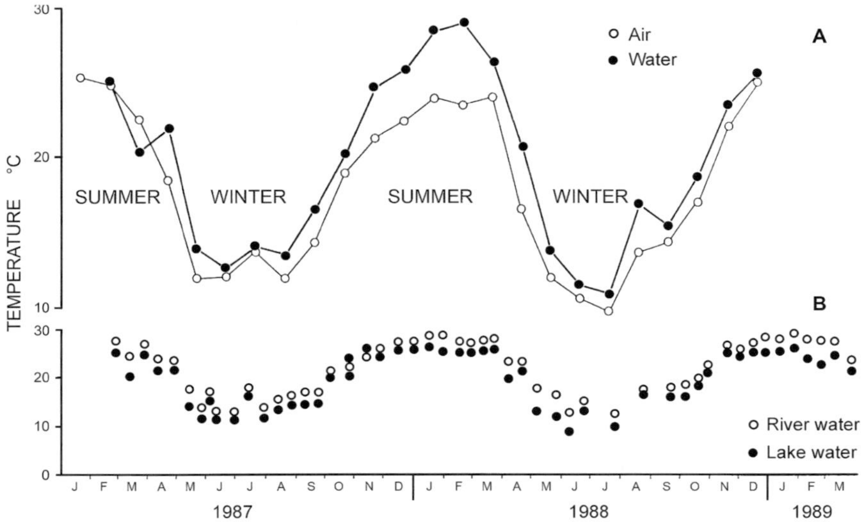

Fig. 4.11 A Monthly mean air (*empty circles*) and surficial water (*filled circles*) temperatures in El Tigre Lake. **B** Surficial water temperature in Correntoso River (*empty circles*) and El Tigre Lake (*filled circles*)

23°C, with maximum and minimum water temperatures of 30°C and 7°C, respectively (Figs. 4.10, 4. 11A, B).

Single linear correlations were estimated using daily mean water temperatures and daily mean air temperatures (air temperature values from 1 day to the mean of the 12 days before the sampling date were used). All the correlation coefficients were higher than $r = 0.86$ ($P < 0.001$), but the highest positive correlation was obtained by testing the relationship between the average of the daily mean air temperature of the 2 days previous to the sampling date and the mean temperature of the water column ($r = 0.91$, $P < 0.001$). Values of the coefficient of determination (r^2) indicated that the calculated regression coefficients accounted for a high proportion of the variance in water temperature (usually never lower than 74%). In this way, it is reasonable to stress that air temperature is the best single estimator of the floodplain lake water temperature (Table 4.7).

4.5.2 Seasonal Variation

Daily mean air and water temperature variations during the months with maximum (February 1987) and minimum (August 1987) air temperatures are illustrated in Fig. 4.12A, B. Air temperature is driven by active components as solar radiation and wind. In summer, the entrance of southern cold-dry polar air to the region originates strong fluctuations in insolation, air temperature, wind, and rainfall, sometimes producing marked changes in lake water

Table 4.7 Correlation coefficients showing the relationships between monthly mean lake water temperature (T_W) and maxima (Ta_{min}), mean (Ta_{me}), minima (Ta_{min}), absolute maxima (Ta_{max-a}) and absolute minima (Ta_{min-a}) air temperatures in El Tigre Lake

Tw - Ta_{max}	r = 0.95, $P < 0.001$
Tw - Ta_{me}	r = 0.95, $P < 0.001$
Tw - Ta_{min}	r = 0.93, $P < 0.001$
Tw - Ta_{max-a}	r = 0.92, $P < 0.001$
Tw - Ta_{min-a}	r = 0.94, $P < 0.001$

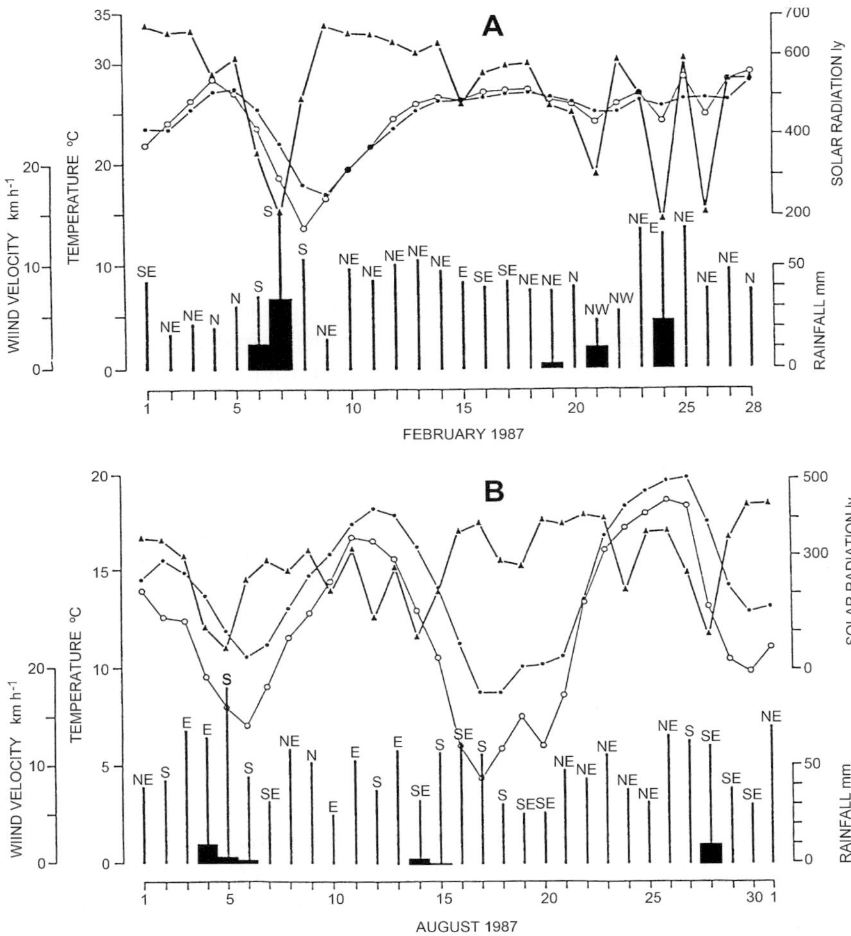

Fig. 4.12 Daily solar radiation (*filled triangles*), daily mean air temperature (*empty circles*), daily mean of the water column (*filled circles*), daily mean wind speed and direction (*sticks*) and rainfall (*bars*) during summer (**A**) and winter (**B**) in El Tigre Lake

temperatures. Storms coming from the north and northeast sectors of the country, with hot and humid winds, affect mainly the insolation and, to a lesser degree, the air temperatures. The low insolation, combined with cold winds and rains, originates sharp fallings in air temperatures (Fig. 4.12A). Thus, during the first days of February, the meteorological conditions presented a warm north wind and a mean air temperature of 28.1°C, but, subsequently, the temperature decreased to 23.5°C on February 6, and to 18.7°C on February 7, as a consequence of a squall with strong south cold wind and heavy showers. At the same time, there was a lowering in the isolation due to the high cloudiness. The cold and dry southwest wind "Pampero" lowered even more the air temperature, reaching 13.7°C on February 8. However, the decrease in the water column temperature was slower and of lesser magnitude, delaying 1 day in relation to the minimum air temperatures. During those days, the water temperature decreased 9.8°C (Fig. 4.12A). Turning to the normal summer weather conditions with high insolation originated the immediate response of air and water temperatures (Fig. 4.12A). After such a cooling episode, a rapid increase in air and water temperatures was measured in the middle of February, as well as some day-to-day variations at the end of the month, related to the unstable weather conditions and the decrease in the incident solar radiation. Water temperatures were always warmer than air temperatures during the winter months, as in August (Fig. 4.12B). South and southwest cold winds originate a strong lowering of air and water temperatures and north and northeast winds show the opposite effects. Despite the high insolation occurring from 16 to 21 August, for example, continuous cold-dry south–southwest winds lowered the water temperatures (Fig. 4.12B). These conditions indicate that solar radiation and air temperature are strongly associated with the type of winds which, in turn, generate different cloudiness and air temperature variations. Several authors stressed the importance of these active components on the thermal regimes of both lotic and lentic waters (Smith 1975; Timms 1975; Walker and Lawson 1977; Crisp et al. 1982; Lewis 1973, 1983, 1984; Melack 1984; McIntyre and Melack 1984; Ward 1985).

The interaction of maritime and continental influences under which the climate of the region is exposed turns it quite variable, with seasonal sharp changes, and also with changes within every season. Furthermore, the valley of the Middle Paraná River is located in a "corridor" through which the Atlantic anticyclonic warm-humid winds from the north-northeast, as well as the polar cold-dry winds from south-southwest zones of South America, are conveyed. Water temperatures during summer are usually the same or slightly lower than air temperatures, except during the installation of south-southwest cold air masses. In winter, water temperatures are usually higher than air temperatures (Fig. 4.12B). The strong climatic influence on the thermal behavior of floodplain lakes is also demonstrated by the similarity between air temperature values of the July winter isotherm (11°C) and the January summer isotherm (26°C) and the mean lake water temperatures for those months (Fig. 4.11A). Time series of stream and lake water temperatures, depicted in Fig. 4.11B, also demonstrated the similar thermal behavior between lotic and lentic floodplain water bodies. At the same time, these data showed a prevailing tendency of a slightly higher temperature of lotic water, which can be explained by the turbulence and continued vertical mixing showed by flowing water. Furthermore, temperature of lotic water is unlikely to be much affected by shading, as the rivers' size conveys considerable thermal inertia and virtually ensures that they are largely exposed to the sun. Lotic temperatures in the Middle Paraná River rarely exceed 30–32°C under normal weather conditions (Drago 1984). In the main channel, the maximum difference measured between surficial and near-bed waters (24 m depth) was 2.5°C. The average water temperature in the main channel was 21.84°C for a period of 5 years (1977–1981), the highest summer temperatures ranged from 28 to 30°C, and the minimum winter water temperature detected was 12.9°C (Drago 1984). Drago (1984) found the best linear correlation between the mean temperature of the water column at the channel center and the averages of the daily maximum air temperatures of the 15 days previous to the sampling date ($r = 0.95$, $P < 0.001$). The monthly average air temperature was also closely related to the mean water temperature ($r = 0.93$, $P < 0.001$). The maximum extent of diel river temperatures in the Middle Paraná main channel ranged from 0.5 to 1.5°C, which is according to the temperate climate of the region and its drainage basin hierarchy, greater than tenth order (sensu Strahler 1957). Thus, seasonal changes in water temperature of the river-floodplain water bodies follow seasonal trends in monthly air temperatures. The annual degree-day (above 0°C) calculated for the river is around 7,600, which is in agreement with values estimated for streams located at a latitude of 28°–32°S, and is also typical of rivers flowing through temperate and low altitude areas (Allan 1995; Vannote and Sweeney 1980). Furthermore, it must be stressed that the northern part of the Middle Paraná fluvial segment is under the subtropical climate influence.

4.5.3 Diel Variations

It is a well-known fact that stratification is more easily destroyed by nocturnal cooling in tropical lakes than in temperate lakes (Drago and Paira 1987; Lewis 1973, 1983a, 1983b, 1984; MacIntyre and Melack 1984, 1988). Usually, it is assumed that there is more or less a permanent stratification if the thermocline is detected under the surficial temperature gradient. For example, Froehlich et al. (1978) considered the thermocline as "a temperature discontinuity located below the level of diurnal warming and nocturnal cooling and sufficient to constitute an effective barrier to mixing". In the Middle Paraná floodplain lakes, a typical localized thermal discontinuity does not exist. According to a large number of vertical profiles of water temperature measured in these alluvial lakes (Drago and Vassallo 1980; Drago 1984; Bonetto et al. 1984; Drago and Paira 1987), the shallowness of the lakes prevents the formation of a permanent discontinuity layer. Therefore, the existence of a thermocline "sensu lato" in the lakes of the Paraná floodplain is discarded. However, it is very common to detect temporary thermal discontinuity layers that could be defined as "tropical thermoclines" (sensu Coche 1974). Furthermore, the effect of such tropical thermoclines is only temporary because they are mere products of diurnal heating and are periodically broken down by nightly cooling. In Amazonian floodplain lakes, McIntyre and Melack (1984, 1988) showed that the persistent stratification begins when maximum depths of 6 m or more are reached, and it ceases with depths lower than 3.5 m. The Orinoco lakes show a similar thermal behavior when they are compared to the Paraná lakes (Vásquez 1992).

Several features of the thermal structure show the variety of responses of the lakes to insolation, air temperature, wind, and type and cover degree of aquatic vegetation. Detailed diel measurements in open and vegetated lake areas for different seasonal periods show a diurnal stratification with a strong warming in the first 0.5 m depth, a parallel warming in the intermediate layers, and a slight variation in bottom waters (Figs. 4.13, 4.14).

The daily period of warming and stratification (if there is one) begins early in the morning (8:00 am), with maximum temperatures at the first hours in the afternoon (12:00 am–4:00 pm) and a later development of stratification. Maximum air temperature arises about 3:00 pm; therefore, the maximum water temperature delay is of about an hour. The layers lower than 1 m depth presented a lag of their maximum temperatures of 1 or 2 h in relation to the maximum surficial temperatures. From 8:00 am to 4:00 pm, the surface water temperature raises about 5°C. The preceding features are detected only during extended periods of high insolation and slight or calm winds. During these periods, the maximum water temperature within the first meter depth begins to decrease around 4:00 pm (Figs. 4.13, 4.14). If the weather shows no change, an isothermal condition will be established at all levels, which will be maintained until ca. 8:00 am. During the night, thermal values of the most surficial layers usually present the same or lower temperatures than those of bottom waters. This condition is driven by the heat flux from the lakes surface to the atmosphere by conduction. The amplitude of thermal fluctuations is similar both in surficial and bottom waters (Figs. 4.13, 4.14).

This kind of thermal stratification and the daily mixing patterns are usually observed in the Middle Paraná floodplain lakes. Particular meteorological conditions, as cloudiness or gusty winds, can decrease water temperature during the day, producing free convection and avoiding the formation of temporary stratification. Cloudiness associated to winds prevents or destroys the thermal stratification, while winds associated with unstable density gradients cause forced convections. Likewise, free convections can only be originated by unstable thermal stratification. In the Paraná floodplain lakes, the seasonality of the stratification and circulation processes is not very marked as in the stratified deeper lakes, but thermal differences are stronger in spring and

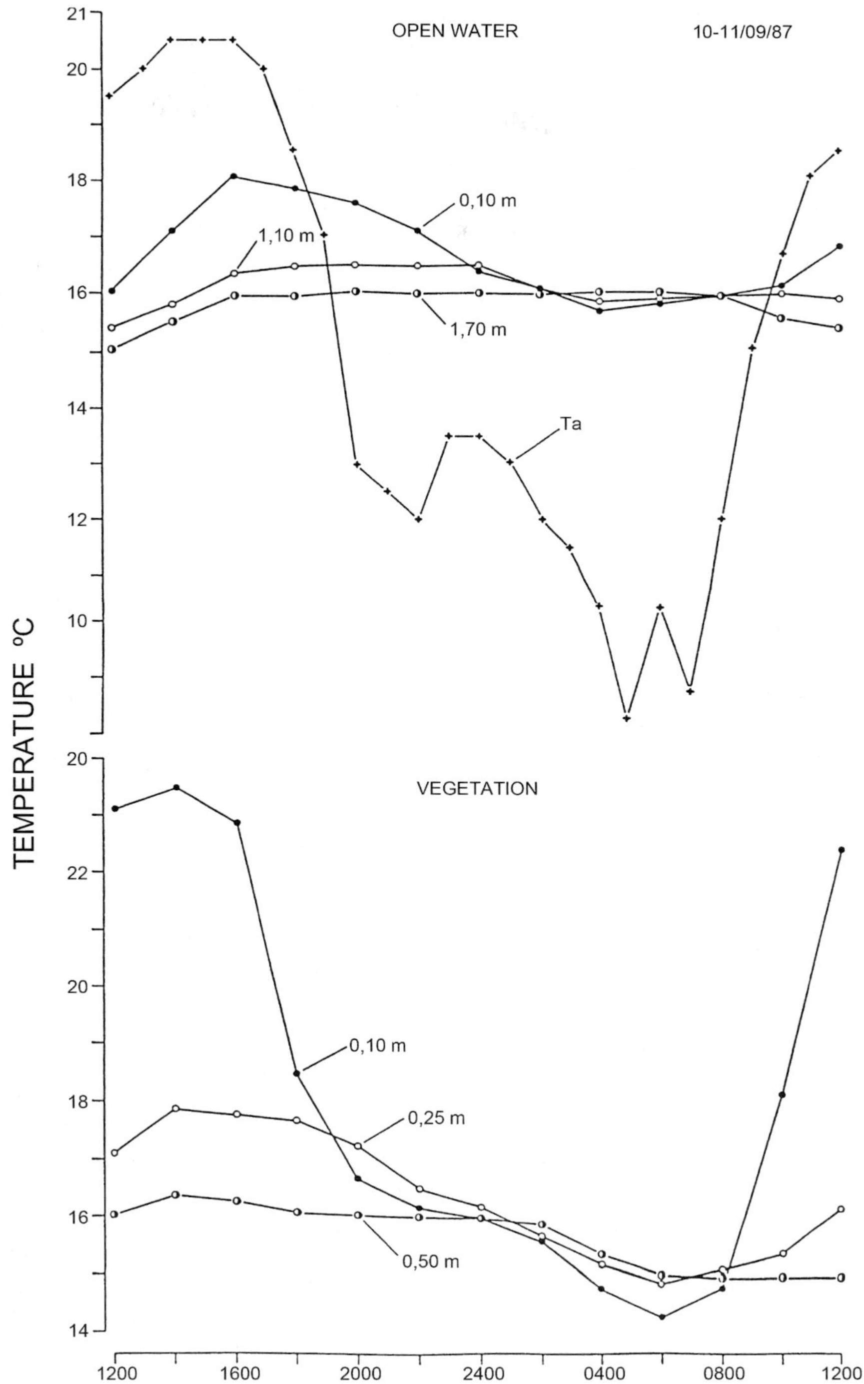

Fig. 4.13 End-winter diel variation of water temperature at different depths in open water and under an on-shoreline floating meadow in El Tigre Lake. *Ta* air temperature

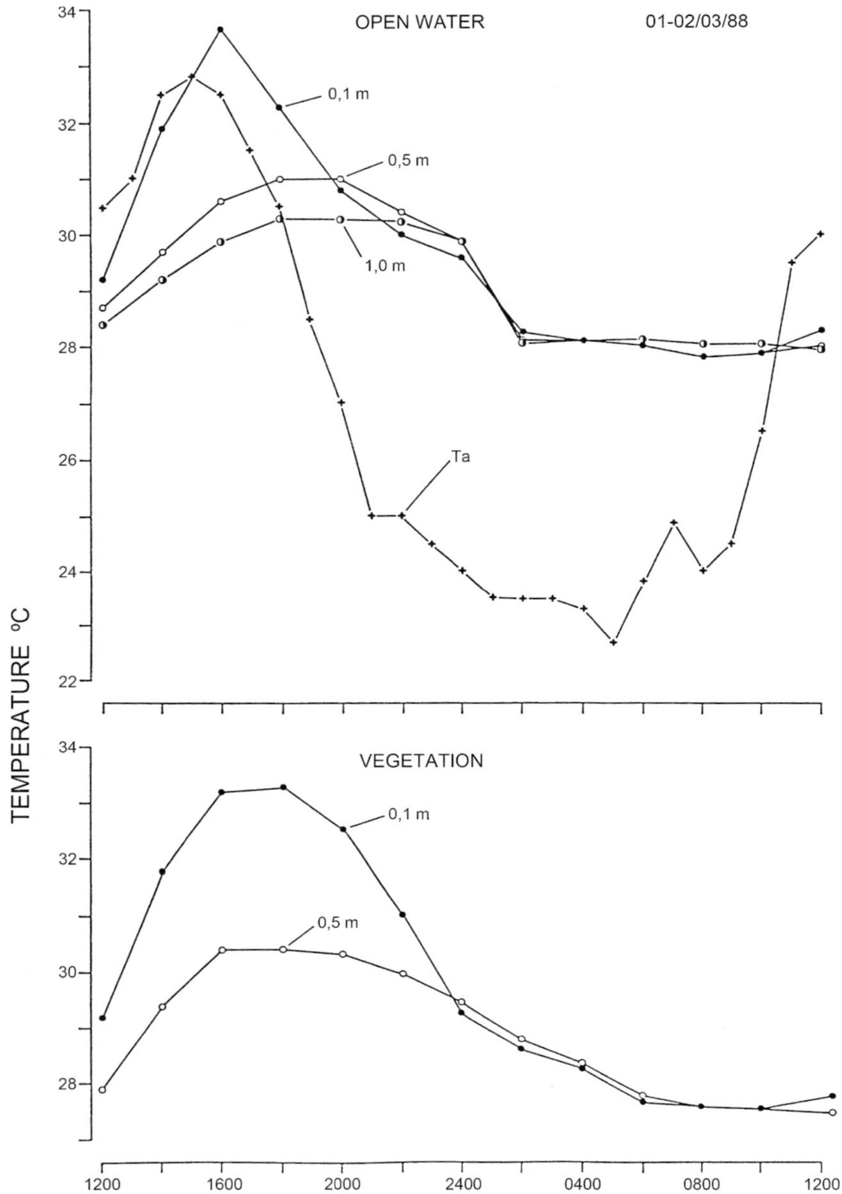

Fig. 4.14 End-summer diel variation of water temperature at different depths in open water and under a shoreline floating meadow in El Tigre Lake. *Ta* air temperature

summer during short stratification periods. Water temperature stratification in the Middle Paraná lakes is unstable and shows swift and strong changes according to the water level, periodical and rather short weather changes and the daily rate of insolation and air temperature. The impossibility of maintaining a direct stratification beyond 24 h, and the irregularity of the density changes caused by nocturnal cooling, rainfall and winds, are the characteristics of these continuous warm polymictic lakes (Drago and Paira 1987).

4.5.4 Lake Thermal Microstructures

The type and coverage density of aquatic vegetation in floodplain lakes are responsible for diverse thermal microstructures that are mainly detected during spring, summer, and at the beginning of autumn (from September to March), when hydrophytes are plentiful. The aquatic vegetation is one of the internal factors that influence the amplitude and rate of fluctuations in water temperature (Martin 1972). The longest stratification periods are observed in lakes with more than 80% of vegetation cover because there is a decrease in wind stress. Species as *Panicum elephantipes*, *Echinochloa polystachya* and *Polygonum* spp. also cover quite large percentages of water surface in some lakes during winter, and important thermal microstructures can be generated during periods with several sunny and calm days. During winter and spring, noticeable high water temperatures (20°C in sunny winter days) can be recorded in narrow-elongated lakes, with very shallow depths (<0.4 m) and covered by dense stands of *Paspalum repens*, *Panicum elephantipes*, *Eichhornia crassipes*, *Pistia stratiotes*, *Limnobium spongia*, *Enydra anagallis*, *Ludwigia peploides*, and *Azolla filiculoides* (Figs. 4.13, 4.14). Overgrowth of aquatic plants in these lakes produces the interception of light before it impinges on the bottom. A small fraction of this light is used in the photosynthetic process, but a much larger proportion is converted into heat at the leaf surface where it causes a localized rise in water temperature (Dale and Gillespie 1976, 1977; Drago and Paira 1987; Vásquez 1992). These authors have shown that floating hydrophytes may have a strong influence on the diurnal temperature fluctuation at, and close to, the air–water interface. Moreover, the daily cycle of increasing and decreasing radiant energy impinging on the aquatic plants results in changes in the temperature profile of the water column below them. Different thermal values are generated at the same time of the day according to the type of vegetation, surface cover and, especially, density. Spatial variations in lake water temperature under different stands of aquatic plants, from the shore to the center, during a summer day are very marked. These differences are usually reflected in the vertical thermal structure. Water temperatures ranging from 34–45°C were measured in the first 0.10 m under *Salvinia biloba*, 32–34°C in *Azolla filiculoides*, 30–31°C in *Panicum elephantipes* and *Echinochloa polystachya* and 28°C in open water along a cross section of 60 m (2:00 pm; 01/03/88).

The maximum temperatures reached by surficial waters covered by hydrophytes are detected from 2:00 pm to 6:00 pm (Figs. 4.13, 4.14). There is a clear tendency in the dense floating masses of *Salvinia biloba*, *Azolla filiculoides* and *Pistia stratiotes* that the maximum temperatures are reached around 2:00 pm, with the development of a strong vertical stratification. Floating carpets of the above-mentioned floating species show a daily fluctuation of the superficial water temperatures of 9.5°C. The vertical gradient is also noteworthy, with differences of 4.8°C at 5 cm depth, i.e., with gradients of 0.96°C cm^{-1}. Furthermore, a difference of 5.6°C at 25 cm depth was measured in *Azolla filiculoides* (Fig. 4.14). In the deeper layers, the daily amplitude reached 6.7 and 3.0°C under floating carpets of *Salvinia minima*, *S. biloba* and *Azolla filiculoides*, respectively (Figs. 4.13, 4.14). Below these floating carpets, thermal stratifications with profiles and daily cycles similar to the annual cycles showed by the temperate deep lakes were observed. The striking difference is that in spite of the strong density gradients, the stratification is destroyed during the night and there is a complete mixing of the entire water column (Figs. 4.13, 4.14). Furthermore, thermal inversions could be measured. For example, at the end of spring (December), the surficial temperature was 1°C lower than at the 5 cm deeper layer. In floodplain lakes that support dense aquatic plant populations, vertical profiles of light and water temperature are roughly similar one to another. This feature has been demonstrated by Dale and Gillespie (1977), Bowes et al. (1979), Barko et al. (1982), Drago and Paira (1987), Hamilton et al. (1990) and Vásquez (1992).

4.6 Lake Heat Content

The heat content of the floodplain lakes displays a complex behavior due to the coupling of active (climatologic-hydrological) and passive (morphological) factors. The Paraná regime, one of the active factors, has and indirect influence on the caloric content fluctuations through the driving of the lake level regimes. A noteworthy example is the heat content evolution during the riverine flood inflows to the lakes. Thus, the entrance of flood waters to the El Tigre Lake changed the heat pattern, delaying the maximum heat peaks to the winter and spring months (Fig. 4.15A; June–July 1987 and August, November–December 1988).

During the two studied years at that lake, the heat content ranged from 3,154 cal cm^{-2} (June 1987) to 480 cal cm^{-2} (May 1988), i.e., a heat budget for the complete period of 2,674 cal cm^{-2}. A value of 3,000 cal cm^{-2} was also calculated by Drago and Paira (1987) in another lake of the Paraná River floodplain. The peak of maximum heat content in El Tigre Lake was in June 1987 (Fig. 4.15A), as a consequence of the relatively high temperature of the water column to the end of autumn and the beginning of winter (16.6°C).

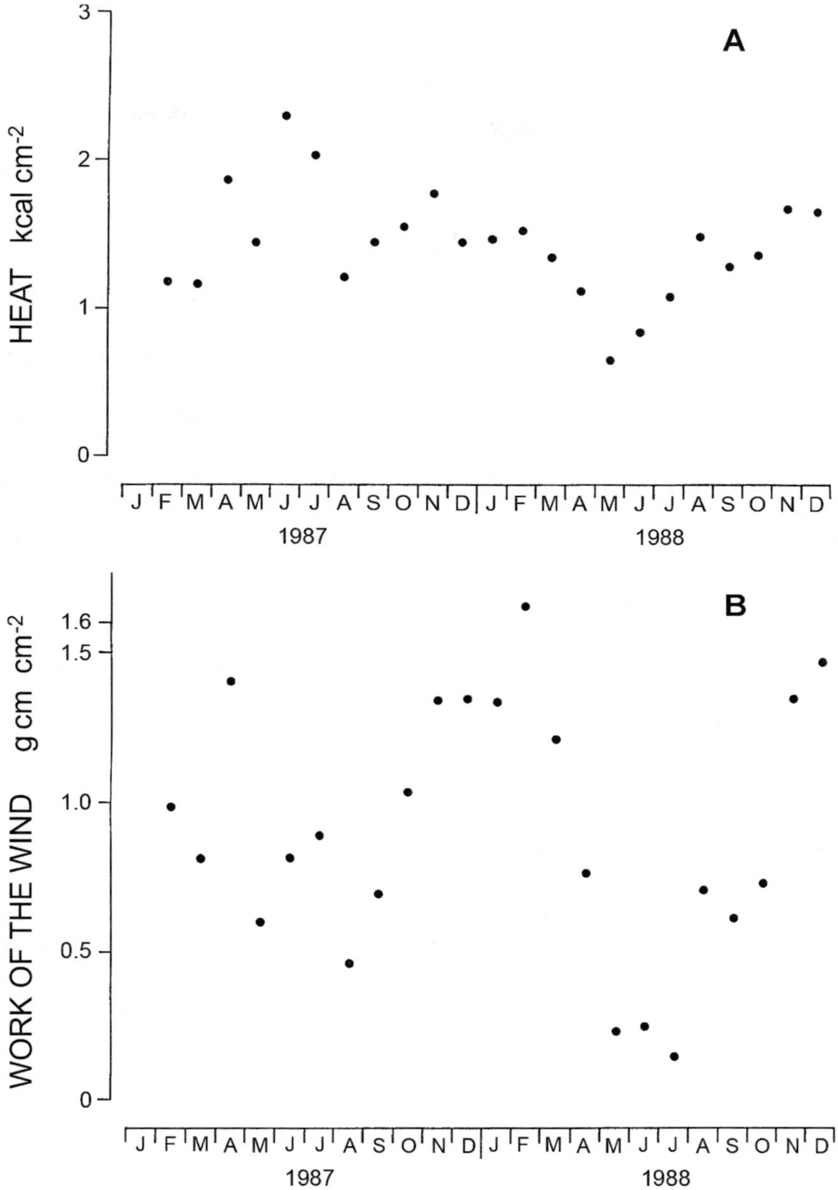

Fig. 4.15 Annual evolution of heat content (**A**) and the work of the wind (**B**) in El Tigre Lake

Other important factors were the high mean depth and volume of the lake when the maximum values were reached during the 2 years of lake monitoring, as a consequence of uncommon higher lake water levels imposed by the river regime. The Paraná River usually shows low water stages in June, and lakes having indirect connections with the lotic environments also show low water levels (Drago 1980, 1989). For example, in 1987, the water stage of the river-lake systems was quite unusual (Fig. 4.2). As it has been shown by Gorham (1964) and Timms (1975), volume and mean depth are the main regulators of heat content in large lakes. Timms (1975) stated that the morphometric parameters should be passive components, while active components would include the climatologic factors and the protection from the wind. Timms (1975) also stressed that passive components should be more important in the heat budgets of lakes, because the efficacy of heat distribution by the wind in a lake depends on its basin morphology. For floodplain lakes of the Paraná River, the above conclusions are valid, although the river regime must also be considered as a main factor because it can originate noticeable changes on the physical dynamics of these shallow lakes (Drago 1980, 1981; Drago and Paira 1987). Similar features were demonstrated for lakes of the Orinoco floodplain (Hamilton et al. 1990; Hamilton and Lewis 1987, 1990a; Vásquez 1992). Therefore, in these water bodies, the air temperature and the river stage would be the active factors that determine the heat content fluctuations. Considering mean depth, volume and lake surface as independent variables, they account for 55% of the variance in heat content. The regression coefficients of heat content with mean depth and volume are $r = 0.67$ ($P < 0.001$) and $r = 0.60$ ($P < 0.001$), respectively. Insolation and air temperature would be completing the explanation for such variations. Thus, it is possible to observe that with similar mean depth, an increase in air temperature and, consequently, an increase in water temperature, account for an increment of the heat content of the lakes (Figs. 4.10, 4.11A, B, 4.15A).

The calculated tropicality index (TI) (Coche 1974) was 545.5 cal cm^{-2} m^{-1}, lower than those recorded for tropical lakes (Hutchinson 1957; Coche 1974; Wood et al. 1976; Heide van der 1982; Lewis 1983, 1984; Henry and Barbosa 1989). This TI value would place the Paraná floodplain lakes closer to the subtropical lakes than the temperate ones.

4.7 Work of the Wind and Stability of Stratification

The work of the wind (B) in heat distribution (in the Birgean concept) is the work to heat the lake from an isothermal condition (determined by the thermal profile in winter) to the stratified condition. In the Paraná lakes, B attains the highest values during the periods of highest water temperatures, and usually coincides with the heat content fluctuations (Fig. 4.15A, B). During spring and summer, it is evident that wind must overcome higher density gradients and, therefore, a greater mechanical work. A very good correlation between water temperatures and the work of the wind is seen when Figs. 4.12 and 4.15A are compared. The greater B value (1.62 g cm cm^{-2}) was coincident with the highest water temperature (29.9°C), measured in February 1988, and the lower B value (0.08 g cm cm^{-2}) was coincident with the lowest detected water temperature (7.9°C), in July 1988. Stability (S) variations follow the same trend as heat content, with a maximum in spring and summer and zero or lightly negative values in winter (Figs. 4.15A, 4.16).

Similar stability fluctuations had been found in other floodplain lakes (Melack 1984; MacIntyre and Melack 1984). Quite large modifications of stability occur from week to week, and this has been attributed to variations in wind stress (Lewis 1983). Nevertheless, a rather complicate behavior of S, deflected from the common trend, has sometimes been observed. Thus, a lack of S was detected in summer (February 1987; Fig. 4.16) as a consequence of an isothermal condition, where the whole water column ($z_m = 0.95$ m) showed 25°C. During summer, S accounts for the amount of work necessary to prevent the entire water column from developing a thermal stratification. The maximum S was found in spring (November 1987), reaching 0.06 g cm cm^{-2} (Fig. 4.16). These variations in B and S may be explained by the morphometry and hydrology of the lakes, and also by their continuous warm polymictic condition (sensu Lewis 1983). In these water bodies, the work of the wind is a very important factor, whether to develop or to destroy a thermal stratification.

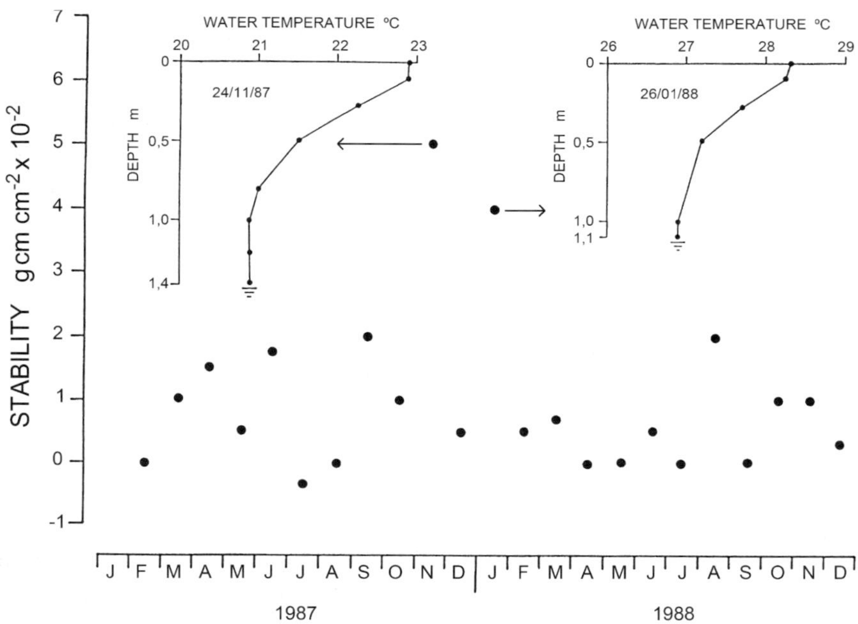

Fig. 4.16 Annual evolution of the stability in El Tigre Lake. The vertical profiles of water temperatures belong to the highest lake stability conditions

4.8 Conclusions

The Paraná fringing floodplain is annually inundated either partially or completely by the overflow of the main channel and the floodplain drainage network. The seasonal water regime behavior has changed over the last three decades of the last century, as well as at the beginning of the present one. Thus, the increase in discharges since about 1970 was very noticeable, with regard to the historical mean discharges, and they were even more evident if the 1944–1970 period is taken as a reference (García and Vargas 1998). The 1944–1970 period was characterized as the driest one, and the period that began in 1970 presented hyperhumid characteristics, a tendency that according to García and Vargas (1998), does not seem to revert. The data presented by these authors show an increase in the mean discharges of the Paraná River and its main tributary, the Paraguay River, that is 50% on average. Moreover, Paoli and Cacik (2000) stressed that during the last 25 years of the last century the highest discharges (>30,000 $m^3\ s^{-1}$) occurred not only in the high-water phase (February–March) but also during some months of the low-water phase (May, June, July). These changes in the magnitude and distribution of discharges have been attributed to a climatic change in a regional scale (García and Vargas 1998). For example, as a consequence of this variation in the Paraná River regime, the mean water stage in Santa Fe increased ca. 0.45 m. The above-mentioned facts indicate an increase in time and magnitude of the connectivity between the lotic and lentic components of the Middle Paraná floodplain. However, from 1999 up to the present, the mean water levels of the river at Paraná Harbor began to show a decreasing trend (Cristina 2006), showing a mean value for the 1999–2005 period of 13.4 m a.s.l., similar to that presented by the river in the 1970s (12.5 m a.s.l) (Fig. 4.17). This fact demonstrates the complexity and variability of the global climatic change.

The most active boundaries of the channels are usually located where the floodplain is being built or eroded, i.e., at the floodplain-main channel border, although during the largest floods this activity can be extended across most of the floodplain surface. During the highest floods, the entire floodplain is covered by the river water and we can therefore define it as a geomorphologic active floodplain. In this way, it is necessary to recognize the genetic association between rivers and the floodplain they construct. Because of this, we stressed to differentiate the river floodplain from other areas inundated by overflow of lakes, reservoirs, direct precipitation, or groundwater that had been defined as inundation plains (Junk et al. 1989). Thus, the Paraná floodplain, as well as other river floodplains, has a typical river-lake system, which shows particular physical, chemical and biological behaviors that markedly differentiate them from other periodically inundated areas (Drago et al. 2003).

The relationship between the floodplain drainage network and its lakes with different connection degrees is very important for the floodplain

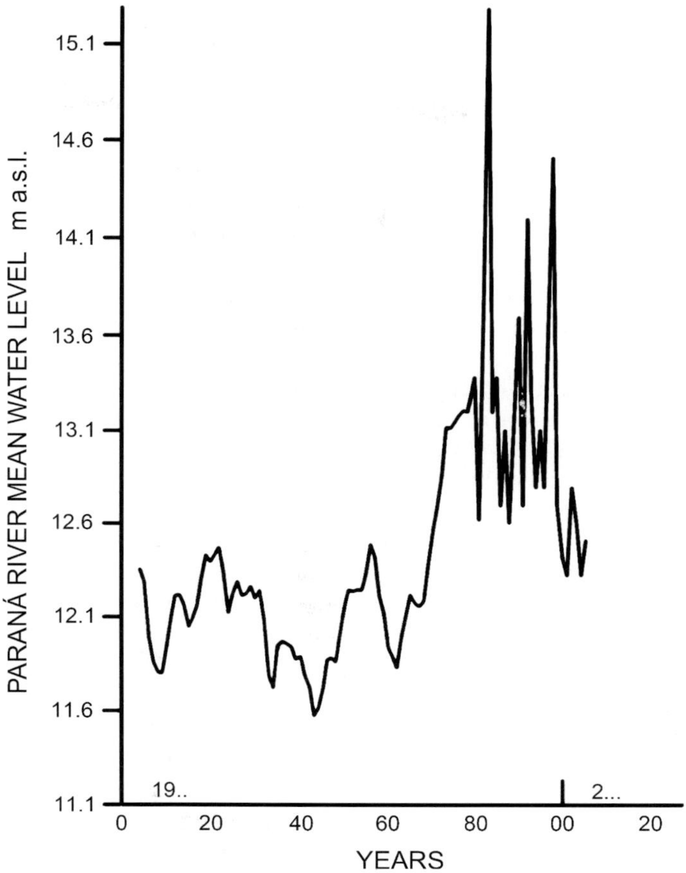

Fig. 4.17 Historical evolution of water stage heights at Paraná Harbour (Entre Ríos Province, Argentina)

dynamics. Furthermore, this means that the main river and its secondary channels, through erosion and deposition processes, can modify those hydrological relationships between floodplain lotic and lentic environments. Therefore, lakes with direct connection may lose that condition by deposition, or lakes with indirect connection may be reconnected with the river by fluvial erosion.

The different type and degree of lotic-lentic connections will give a different time span for each hydrological phase, even during the same river stage. Because of this, in the Paraná floodplain lakes, as well as in the majority of these types of water bodies, it is necessary to differentiate the water renewal time during the isolation and inundation phases. The extremes were ranged from minutes to about 1 year or more. The amplitude, duration, frequency, regularity, and period of the year of the high, middle, and low water phases drive the hydro-sedimentological behavior of the lakes. These changes affect the floodplain habitats. Thus, in the Paraná floodplain, the expansion and contraction of the lake basin are different according to the lake morphology and its connection degree with the channels. In large water bodies with a direct connection, this process shows a strong dynamics but, at the same time, presents the shortest lifespan. On the other hand, small and shallow lakes with an indirect connection are less dynamic, but show a longer period of existence. Some of these may temporarily disappear by desiccation during pronounced isolation phases.

The banks of the main stem and floodplain channels change continuously by erosion or deposition, sometimes showing a strong shifting according to the river level and the movements of the thalweg (Drago 1990). The main stem and the floodplain channels of the highest hierarchies show mobile sand beds and the formation of sand bars and islands. Some of the secondary channels present well-developed meander sets and their migrations originate new lotic and lentic habitats. Small secondary channels have silted-clayed beds with sand patches and deep and shoal sequences. All floodplain channels present meander pools with different sizes and depths at the bends and scour holes at the confluence with other secondary channels. The bank slumping is a very common feature and it also causes changes, not only in the secondary channels, but also in the levees and flatlands. Owing to these processes, lakes with indirect connection change to a direct connection with the active channels or vice versa. This means a noticeable change of the physical, chemical, and biological lake dynamics.

The inflow of running water with diverse suspended sediment concentrations affects light and sediment patterns of the lakes. The amounts of suspended solid inputs received by lakes are variable, depending on their shape, spatial position, and orientation into the floodplain (Chap. 3). Strong differences of suspended sediment concentrations and water transparency are detected in large and irregular lakes or in those lakes far away from the rivers. In small lakes with a direct connection, small spatial variations in suspended solids were usually observed across the lake surface. The indirect wind action, causing the movement of the floating meadows or rooted hydrophytes, sometimes promotes large fluctuations of suspended sediment concentrations. This material is released from roots and stems of aquatic plants.

During low waters, the largest floodplain channels follow the main river suspended sediment pattern variations. Smaller secondary channels show a noteworthy decrease in current velocity until the flow ceases in some of them, achieving temporal lentic characteristics, with a high decrease in suspended sediment concentration. Thus, in the stream confluences between major and minor floodplain channels, we measured strong gradients of suspended sediment concentrations, transparency, and temperature. Sedimentation-rate data allow us to know only a part of their dynamics. Sedimentation rates measured in a small lake with indirect connection with the river showed larger increments during isolation periods.

This fact supports the conclusion that when the lakes are not disturbed by the currents generated by flood waters or high speed winds, a strong period of sedimentation begins. Moreover, the highest rates of deposition are presented immediately after the connection with the river ceases. The intensity of sediment-resuspension in the Paraná floodplain lakes may be predictable if the wind, effective fetch, depth, and aquatic-terrestrial vegetation are considered simultaneously. Riparian vegetation, helophytes and hydrophytes prevent the wind action on the lake surfaces and the development of deep waves. Nevertheless, small- and medium-sized lakes could present bottom sediment resuspension during periods of strong winds. In large lakes with effective fetches over 300 m, it is possible to detect resuspension during isolation phases, and mainly during periods with wind velocities over 30–40 km h^{-1}. Fetch determines the degree of wind-driven turbulence at the lake surface, while depth determines the degree in which this turbulent energy contacts the bottom sediment. Aquatic and riparian vegetation acting as shelter is another main factor to avoid resuspension from the bottom. However, some of the largest lakes may be too deep for sediment resuspension in some areas and the fetch of the shallowest lakes may be too small for large waves to develop. The Paraná River floodplain lakes are continuous warm polymictic lakes without permanent thermal stratification. Under floating meadows, thermal microstructures were detected, although they are usually daily broken by wind action or nocturnal cooling. The heat budget ranges from 2,000 to 3,000 cal cm^{-2}, being the air temperature and the river regime the main factors that regulate their temporal fluctuations. B and S variations may be explained by the morphometry and hydrology of the lakes and mainly by their thermal condition of continuous warm polymixis. Both parameters usually follow the heat budget trends. The tropicality index shows values around 545.5 cal cm^{-2} m^{-1}, which would place them closer to the subtropical lakes than to the temperate ones.

Acknowledgments. I am very grateful to Dr. Rip Sparks, Dr. Steve Hamilton, and Dr. Wolfgang Junk for constant help and valuable discussions on several versions of this chapter. B.Sc. Aldo Paira and Technicians Ulises Molet, Walter Reutemann, Ramón Regner and Eduardo Lordi of the Instituto Nacional de Limnología-CONICET are thanked for their dedicated help in the field and laboratory.

References

Allan JD (1995) Stream ecology: structure and function of running waters. Chapman & Hall, London

Barko JW, Hardin DG, Matthews MS (1982) Growth and morphology of submersed freshwater macrophytes in relation to light and temperature. Can J Bot 60:877–887

Beach Erosion Board (1972) Waves in inland reservoirs. US Army Corps Eng, Beach Erosion Board Tech Memo 132

Birge EA (1915) The heat budgets of American and European lakes. Trans Wis Acad Sci Arts Lett 18:166–213
Bloesch J, Burns NM (1980) A critical review of sedimentation trap technique. Schweiz Z Hydrol 42:15–55
Bonetto C, Zalocar Y, Lancelle H (1984) A limnological study of an oxbow lake covered by *Eichhornia crassipes* in the Paraná River. Verh Internat Verein Limnol 22:1315–1318
Bowes G, Van T, Garrard L, Haller W (1977) Adaptation to low light levels by Hydrilla. J Aquat Plant Manage 15:32–35
Carper GL, Bachmann RW (1984) Wind resuspension of sediments in a prairie lake. Can J Fish Aquat Sci 41:1763–1767
Chambers PA, Kalff J (1985) Depth distribution and biomass of submersed aquatic macrophyte communities in relation to Secchi depth. Can J Fish Aquat Sci 42:701–709
Coche AG (1974) Limnological study of a tropical reservoir. In: Balon EK, Coche AG (eds) Lake Kariba: a man-made tropical ecosystem in Central Africa. Junk Publishers, The Hague, pp 1–247
Crisp DT, Matthews AM, Westeake DF (1982) The temperatures of nine flowing waters in southern England. Hydrobiologia 89:193–204
Cristina I (2006) Evolución de los niveles del río Paraná. Centro de Informaciones Meteorológicas, Facultad de Hidrología y Ciencias Hídricas, Universidad Nacional del Litoral, Santa Fe, Argentina
Dale HM, Gillespie T (1976) The influence of floating vascular plants on the diurnal fluctuations of temperature near the water surface in early spring. Hydrobiologia 49:245–256
Dale HM, Gillespie T (1977) Diurnal fluctuations of temperature near the bottom of shallow water bodies as affected by solar radiation, bottom color and water circulation. Hydrobiologia 55:87–92
Drago EC (1976) Origen y clasificación de ambientes leníticos en llanuras aluviales. Rev Asoc Cienc Nat Litoral 7:123–137
Drago EC (1980) Aspectos limnológicos en ambientes próximos a la ciudad de Santa Fe (Paraná Medio): Comportamiento hidrológico y sólidos suspendidos. Ecología (Argentina) 5:31–48
Drago EC (1981) Grados de conexión y fases hidrológicas en ambientes leníticos de la llanura aluvial del río Paraná (Argentina). Ecología (Argentina) 6:27–33
Drago EC (1984) Estudios limnológicos en una sección transversal del tramo medio del río Paraná. VI: Temperatura del agua. Rev Asoc Cienc Nat Litoral 15:79–82
Drago EC (1989) Morphological and hydrological characteristics of the floodplain ponds of the Middle Paraná River (Argentina). Rev Hydrobiol Trop 22:183–190
Drago EC (1990) Geomorphology of large alluvial rivers: Lower Paraguay and Middle Paraná. Interciencia 15:378–387
Drago EC, Vassallo M (1980) Campaña limnológica "Keratella I" en el río Paraná Medio: Características físicas y químicas del río y ambientes leníticos asociados. Ecología Argentina 4:45–54
Drago EC, Paira AR (1987) Temperature and heat budget in a floodplain pond of the Middle Paraná River (Argentina). Rev Asoc Cienc Nat Litoral 12:193–201
Drago EC, Amsler ML (1988) Suspended sediment at a cross section of the Middle Paraná River: concentration, granulometry and influence of the main tributaries. IAHS Publ 174:381–396
Drago EC, Amsler ML (1998) Bed sediment characteristics in the Paraná and Paraguay rivers. Water Int 23:174–183
Drago EC, Ezcurra de Drago I, Oliveros OB, Paira AR (2003) Aquatic habitats, fishes and invertebrate assemblages of the Middle Paraná River. Amazoniana 17:291–341
FICH (Facultad de Ingeniería y Ciencias Hídricas) 1991a. Informe de diagnóstico sobre navegación en la ruta Rosario–Océano. El río Paraná Inferior, Convenio: Universidad Nacional del Litoral – Instituto Nacional de Ciencia y Técnicas Hídricas, Santa Fe, Argentina

Froehlich CG, Arcifa-Zago MS, de Carvalho MA (1978) Temperature and oxygen stratification in Americana Reservoir, State of Sao Paulo, Brazil. Verh Internat Verein Limnol 20:1710–1719

Gardner WD (1980a) Sediment trap dynamics and calibration: a laboratory evaluation. J Mar Res 38:17–39

Gardner WD (1980b) Field assessment of sediment traps. J Mar Res 38:41–52

Gorham E (1964) Morphometric control of annual heat budgets in temperate lakes. Limnol Oceanogr 9:525–529

Häkanson L (1981) A manual of lake morphometry. Springer, Berlin Heidelberg New York

Hamilton SK, Lewis WM Jr (1987) Causes of seasonality in the chemistry of a lake on the Orinoco River floodplain, Venezuela. Limnol Oceanogr 32:1277–1290

Hamilton SK, Lewis WM Jr (1990a) Basin morphology in relation to chemical and ecological characteristics of lakes on the Orinoco River floodplain, Venezuela. Arch Hydrobiol 119:393–425

Hamilton SK, Lewis WM Jr (1990b) Physical characteristics of the fringing floodplain of the Orinoco River, Venezuela. Interciencia 15:491–500

Hamilton SK, Sippel SJ, Lewis WM Jr, Saunders JF III (1990) Zooplankton abundance and evidence for its reduction by macrophytes mats in two Orinoco floodplain lakes. J Plankton Res 12:345–363

Heide van der J (1982) Lake Brokopondo. Filling phase limnology of a man-made lake in the humid tropics. Offsetdrukkerij Kanters B.V., Alblasserdam

Henry R, Barbosa FA (1989) Thermal structure, heat content and stability of two lakes on the National Park of Rio Doce Valley (Minas Gerais, Brazil). Hydrobiologia 171:189–199

Hutchinson E (1957) A treatise on Limnology. I. Geography, physics and chemistry. Wiley, New York

Idso SB (1973) On the concept of lake stability. Limnol Oceanogr 18:681–683

Iriondo MH (1988) A comparison between the Amazon and Paraná River systems. Mitt Geol Paläont Inst Univ Hamburg SCOPE/UNEP Sonderband 66:77–92

Junk WJ, Bayley PB, Sparks RE (1989) The flood pulse concept in river–floodplain systems. In: Dodge DP (ed) Proceedings of the International Large River Symposium. Can Fish Aquat Sci Spec Publ 106:110–127

Jupp BP, Spence DH (1977) Limitations on macrophytes in an eutrophic lake, Loch Leven. II. Wave action, sediments and water fowl grazing. J Ecol 65:431–446

Köeppen W (1948) Climatología. Fondo de Cultura Económica, México

Lallana V, Sabattini R, Lallana M (1987) Evapotranspiration from *Eichhornia crassipes, Pistia stratiotes, Salvinia hersoggi* and *Azolla caroliniana* during summer in Argentina. J Aquat Plant Manage 25:48–50

Lewis WM Jr (1973) The thermal regime of lake Lanao (Philippines) and its theoretical implications for tropical lakes. Limnol Oceanogr 18:200–217

Lewis WM Jr (1983a) Temperature, heat and mixing in lake Valencia, Venezuela. Limnol Oceanogr 28:273–286

Lewis WM Jr (1983b) A revised classification of lakes based on mixing. Can J Fish Aquat Sci 40:1779–1787

Lewis WM Jr (1984) A five-year record of temperature, mixing, and stability for a tropical lakes (Lake Valencia, Venezuela). Arch Hydrobiol 99:340–346

Lewis WM Jr, Weibezahn FH, Saunders III JF, Hamilton SK (1990) The Orinoco River as an ecological system. Interciencia 15:346–357

MacIntyre S, Melack JM (1984) Vertical mixing in Amazon floodplain lakes. Verh Internat Verein Limnol 22:1283–1287

MacIntyre S, Melack JM (1988) Frequency and depth of vertical mixing in an Amazon floodplain lake (L. Calado, Brazil). Verh Internat Verein Limnol 23:80–85

Martin NA (1972) Temperature fluctuations within English lowland ponds. Hydrobiologia 40:455–469

Melack JM (1984) Amazon floodplain lakes: shape, fetch, and stratification. Verh Internat Verein Limnol 22:1278–1282
Melack JM, Fisher TR (1983) Diel oxygen variation and their ecological implications in Amazon floodplain lakes. Arch Hydrobiol 98:422–442
Melack JM, Fisher TR (1990) Comparative limnology of tropical floodplain lakes with an emphasis on the central Amazon. Acta Limnol Brasil 3:1–48
Mertes LA (1997) Documentation and significance of the perirheic zone on inundated floodplains. Water Resour Res 33:1749–1762
Paira AR, Drago EC (2006) Genetical, morphological and evolutional relationships of the floodplain lakes in the Middle Paraná River Hydrosystem. Zeitschrift für Geomorphologie 145:207–228
Paoli C, Cacik P (2000) Régimen de crecidas y análisis de caudales máximos. In: Paoli C, Schreider M (eds) El Río Paraná, tomo 1, chapter 3. Universidad Nacional del Litoral, Centro de Publicaciones, Santa Fe, Argentina, pp 105–171
Sioli H (1984) Former and recent utilizations of Amazonia and their impact on the environment. In: Sioli H (ed) The Amazon: limnology and landscape ecology of a mighty tropical river and its basin. Junk Publishers, Dordrecht
Sippel SJ, Hamilton SK, Melack JM (1992) Inundation area and morphometry of lakes on the Amazon River floodplain, Brazil. Arch Hydrobiol 123:385–400
Smith IR (1975) Water temperature variations within a major river system. Nordic Hydrol 6:155–169
Smith IR (1979) Hydraulic conditions in isothermal lakes. Freshwater Biol 9:119–145
Smith IR, Sinclair IJ (1972) Deep water waves in lakes. Freshwater Biol 2:387–399
Spence DH (1982) The zonation of plants in freshwater lakes. Adv Ecol Res 12:37–125
Strahler AN (1957) Quantitative analysis of watershed geomorphology. Trans Am Geophys Union 38:913–920
Timms BV (1975) Morphometric control of variations in annual heat budgets. Limol Oceanogr 20:110–112
Tockner K, Malard F, Ward JV (2000) An extension of the flood pulse concept. Hydrol Process 14:2861–2883
Vannote RL, Sweeney BW (1980) Geographic analysis of thermal equilibria: a conceptual model for evaluating the effect of natural and modified thermal regimes on aquatic insect communities. Am Nat 112:667–695
Vásquez E (1992) Temperature and dissolved oxygen in lakes of the Lower Orinoco River floodplain (Venezuela). Rev Hydrobiol trop 25:23–33
Walker JH, Lawson JD (1977) Natural stream temperature variations in a catchment. Water Res 11:373–377
Ward JV (1985) Thermal characteristics of running waters. Hydrobiologia 125:31–46
Wood RB, Prosser MV, Baxter RM (1976) The seasonal pattern of thermal characteristics of four of the Bishoftu crater lakes, Ethiopia. Freshwater Biol 6:519–530

5 Fluvial Sediments: Main Channel and Floodplain Interrelationships

Mario L. Amsler[1,2], Edmundo C. Drago[2] and Aldo R. Paira[2]

5.1 Introduction

The quantity and quality of sediments supplied to such a large and complex system as the middle Paraná River, its differential transport through the drainage network, and the eventual sedimentation rates on the adjacent alluvial plain, are all variables that need to be known due to their crucial role on the type and survival of communities characterizing certain environments and on the possible interrelationships among them. Since the water transparency, the transfer of some nutrients and salts, the periodical renovation of alluvial plain surface soils, and the substratum features of lotic and lentic environments are all factors (processes) closely related to sediment, its pondering as a key variable for the state regulation of complex fluvial ecosystems has a sound justification.

This chapter will address the topic of sediment. The amount of the total sediment transport in the main channel and the principal secondary streams of the alluvial plain is focused distinguishing between the two fractions composing the solid discharge: the bed sediment (fine and medium sands in this case), and the finer sizes, silts and clays. These small particles predominate in the sediment transported through the system. They are essentially supplied by the Bermejo River from the Andean range, on the west side of the Paraná River basin. The silts and clays transport is considered with certain detail due to its quantity as well as the reductions verified in the amounts supplied by the upper Paraná River owing to anthropogenic factors occurred in the Brazilian areas of the basin. Those reductions reflected on the periodical sedimentation along the alluvial valley of the middle reach at several hydrological circumstances. This fact is stressed within the chapter conclusions together with the necessity of preserving the Bermejo River sediment supply, in order to assure the large wetland that involves the middle Paraná River reach.

[1]Facultad de Ingeniería y Ciencias Hídricas (UNL), CC 217, 3000 Santa Fe, Argentina, e-mail: mamsler@fich1.unl.edu.ar
[2]Instituto Nacional de Limnología (CONICET-UNL), J. Maciá 1933, 3016 Santo Tomé, Argentina

M.H. Iriondo, J.C. Paggi, and M.J. Parma (Eds.)
The Middle Paraná River: Limnology of a Subtropical Wetland

5.2 Basic Concepts

Some basic concepts used throughout this chapter are covered for any readers unfamiliar with the symbols and topics commonly cited in civil engineering literature that refer to fluvial hydraulics and sediment transport. The list below (or similar ones) should be found in any of the well-known textbooks and/or manuals of the specialty (e.g., Vanoni 1975):

Bed material transport. This term refers to that transport made up of particles found in appreciable quantities in the shifting portions of the bed. A certain part of this is transported as bed load (G_{sf}), which is defined as material moving on or near the bed jumping, rolling, and sliding. The rest moves as suspended bed material (G_{ss}). Summing G_{sf} and G_{ss}, gives the total bed material load (G_s).

Wash load (G_w). The moving particle sizes found in very small quantities in the active areas of the bed constitute the wash load. Obviously, both the bed sediment load and the wash load may move partially as bed load and partially as suspended load, although by definition, practically all the wash load is carried in suspension. Moreover, in many alluvial streams with sand bed channels the limiting size between the wash load and bed sediment is taken at the lower limit of the sand size, i.e., $d_s = 0.062$ mm, which means that G_w is made up of the finer particles, in the order of silt and clay sizes, transported normally in suspension and with nearly uniform concentrations at a given cross section. Finally, by adding G_s and G_w, the total sediment load (G_T) of an alluvial river is obtained.

5.3 The Main Channel

5.3.1 The Bed Sediment

5.3.1.1 Bed Material Size

The Paraná River in its middle reach is a large alluvial sand bed channel as is shown by Drago and Amsler (1998). These authors reported that the bed sediment is composed mainly of strongly unimodal and moderate to well-sorted medium sands (Table 5.1; Fig. 5.1).

Though the available samples that allowed these features to be known were taken along the river thalweg, they represent the sand of the entire active channel bed. Close to the banks, the percentages of silts and clays increase markedly since the stream velocity reduces significantly allowing deposits of the finer suspended sizes. Drago and Amsler (1998) also report the results of a study of 62 bed material samples obtained systematically in the Aguas Corrientes cross section (Fig. 5.2) over 5 years from 1976 to 1981, and discharges varying between 10,000 and 24,900 $m^3\ s^{-1}$.

Table 5.1 Granulometric characteristics of the bed sediment at the Paraná River thalweg in its middle reach (from: Drago and Amsler 1998)

Km from mouth	Site	d_g[a] (mm)	σ_g[b]	Size percentages of sand: Coarse	Medium	Fine	V. Fine
605	T. Subfluvial	0.316	1.29	6.8	63	28	0.5
680	-	0.260	1.21	1.4	77	20	0.2
745	La Paz	0.371	1.99	22	49	22	2.2
867	Esquina	0.316	1.27	10	82	6.2	0.2
1,008	-	0.363	1.51	14	69	13	0.5
1,060	-	0.323	1.65	18	38	40	1.6
1,140	Empedrado	0.281	1.44	7	46	45	2
1,170	-	0.224	1.69	3	47	32	15
1,208	Corrientes	0.306	1.57	66	27	2.3	1.0
1,244	P. de la Patria	0.444	2.04	29	52	12.2	0.7

[a]Mean diameter
[b]Geometric standard deviation of the size distribution

The analysis, carried out with the median sizes of the bed material, did not reveal any correlation with the river stage, something that relates with the characteristics of the upper sandy strata where the main channel is carved (Chap. 2).

5.3.1.2 *Bed Material Transport*

In spite of the importance of reliable information concerning the sand transported in the Paraná River for scientific as well as engineering purposes, the available studies and specific data on this topic were rather scarce or covered only limited river stages (Amsler and Prendes 2000). Only recently did Alarcon et al. (2003) report a comprehensive analysis concerning the bottom sediment transport at a reach ≈15 km long immediately upstream of Paraná City (Fig. 5.2), based on detailed data sets gathered over the last 19 years at several cross sections of the reach. These data showed that the Engelund-Hansen (1967) and Engelund-Fredsøe (1976) formulas yield reliable predictions of the total sand transport (G_s) and bed load transport (G_{sf}) in the Paraná River (Prendes et al. 1994; Amsler and Prendes 2000; see also Pujol et al. 1985).

The average values of G_{sf}, G_{ss}, and G_s in certain periods between 1976 and 1998, together with those for particular years, are given next (Table 5.2).

Note that the total sand transport in the main channel averaged 25.24×10^6 t/year in front of Paraná city, nearly at the end of the middle reach and 9% of

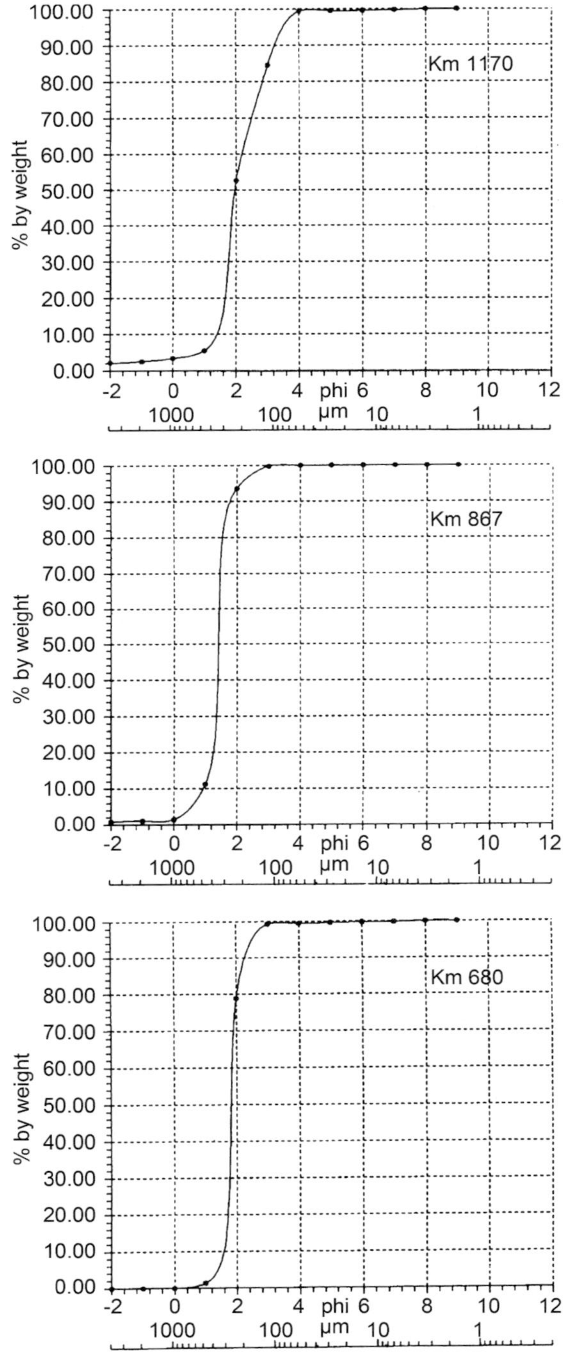

Fig. 5.1 Granulometric distributions of channel bed sands at several locations in the Paraná River thalweg (from: Drago and Amsler 1998)

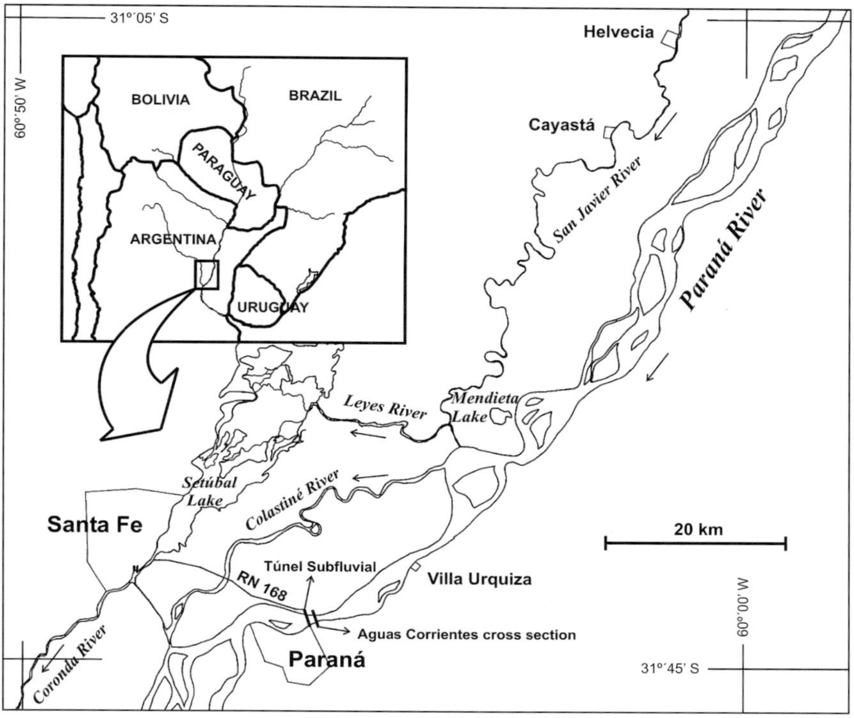

Fig. 5.2 Map with locations of the main secondary channels in the Santa Fe and Paraná cities area

Table 5.2 Annual averages of bed sediment transport in the Paraná River. Reach: Villa Urquiza-Paraná city. Period: 1976–1998 (from: Alarcón et al. 2003)

Period	G_{sf} (×10⁶) (t year⁻¹)	G_{ss} (×10⁶) (t year⁻¹)	G_s (×10⁶) (t year⁻¹)
1976–1980	1.995	19.945	21.940
1981–1990	2.568	25.866	28.434
1991–1998	1.933	21.374	23.307
1976–1998	2.223	23.016	25.239
Years of minimum and maximum bed sediment transport			
1978	1.534	15.153	16.687
1983	4.931	53.106	58.037

this is transported as bed load and the rest as suspended load. In 1983, when the largest flood of the last 100 years occurred (discharges larger than 30,000 $m^3\ s^{-1}$ during 9 months and maximums up to 60,000 $m^3\ s^{-1}$), the total bed sediment transport was more than twice the average.

5.3.2 The Wash Load

5.3.2.1 Origin

The finer fractions transported in suspension by the middle Paraná are supplied by the Bermejo River (Fig. 1.5) as was reported by several authors a long time ago (Soldano 1947; Cotta 1963). The Bermejo discharges are only ca. 5% of those of the Paraná in its middle reach but the sediment that supplies may be more than 80% of the total volumes transported downstream (Sect. 5.3.3). A large amount of very fine particles are eroded annually by the summer heavy precipitations between October and April in the upper basin of the Bermejo River on the east side of the sub-Andean range (Chap. 1, Sects. 1.2, 1.3). The largest sediment yield occurs from December to February when 3–6 monthly storms give rise to important flows with sediment concentrations sometimes higher than 100 g l^{-1}. This fine sediment is transported along the middle and lower Bermejo River reaching its mouth at the Paraguay River. Then, they continue until the confluence of this stream with the Paraná River. As shown in Fig. 5.3, the main stream's distinct colors clearly indicate the way followed by the finer particles from the lower Bermejo River. Also note the difference in water color between the right and left banks of the Paraná River downstream from the Paraguay confluence. Prendes et al. (1996) reported concentrations of 1.15 g l^{-1} and less than 0.05 g l^{-1} near both banks, respectively, from measurements performed in front of Corrientes on 28/03/95 and 04/04/95. Bonetto and Orfeo (1984) found similar values at the same site during March and April from samples taken monthly in 1981–1982. These authors also showed that as far as 200 km downstream a complete mixing of water with uniform wash load concentrations across the channel had not been reached.

5.3.2.2 Granulometric and Mineralogic Features

Although in sand bed rivers the wash load is defined broadly as the part of suspended load composed mainly by silts and clays, Drago and Amsler (1988) determined the separation between bed material and wash load specifically for the Paraná River. Based on bottom sediment samples taken along the middle reach and at the cross section of Aguas Corrientes (5.3.1.1), they concluded that the 31.2 μm diameter is the limit between wash load and the suspended coarser fractions. This size is quite similar to that specified by Lelievre and Navntoft (1980) (37 μm) in a study conducted in front of Corrientes City. Drago and Amsler (1988) also analyzed the size distribution of suspended sediment samples obtained monthly (and twice a month in the periods of maximum concentrations) during 1977 and 1978 at the Aguas Corrientes section. These authors showed that 60.8% of the mean diameter values of the single granulometric distributions fall within the fine silt range (8–16 μm). Moreover, their results (depicted in Fig. 5.4a, b, c) reveal that the

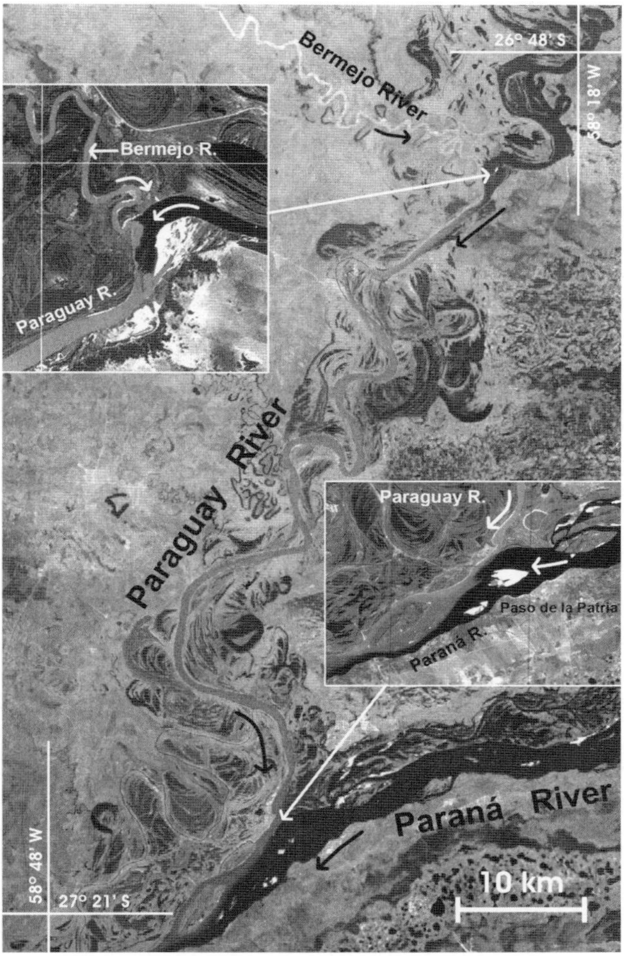

Fig. 5.3 Confluences of the Bermejo-Paraguay rivers and Paraguay-Paraná rivers during a period of maximum wash load supply. See the difference in the water color downstream the confluence of the Bermejo River

finest and best sorted suspended sediment is transported when the concentration peaks flow through the Aguas Corrientes section. Simultaneously, the maximum washload and clay percentages were also recorded (Table 5.3).

Amsler and Prendes (2000) reported also similar values related with the wash load particle sizes from frequent measurements performed between February of 1980 and April of 1981 at a cross section 30 km upstream of Paraná city. During this year, 92% of the total suspended transport was wash load with the following composition:

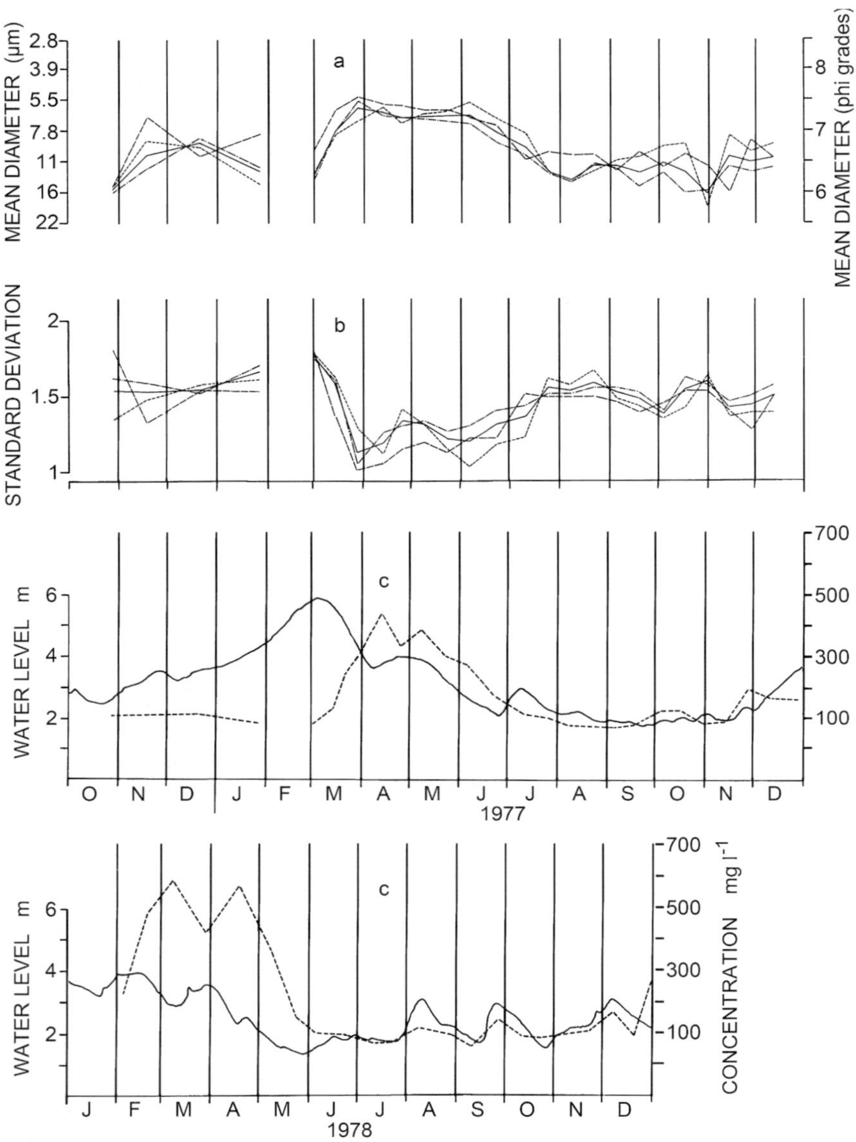

Fig. 5.4 Fluctuations of (a) mean diameters and (b) standard deviations at mid-channel (-.-.-), right bank (—-), left bank (– – – –), cross section (____); (c) concentrations (– – –) and water level (____). Section: Aguas Corrientes (from: Drago and Amsler 1988)

50 μm>d>10 μm: 32%
d<10 μm: 60%
(*d* is the particle diameter. Note that particles lower than 50 μm were considered as wash load in this case).

Table 5.3 Maximum and minimum percentages of wash load and clay fraction with the corresponding mean diameters and standard deviations (averages at the sampling vertical in the mid channel) (from: Drago and Amsler 1988)

	Wash load parameters			
Date	d_g[a] (μm)	σ_g[b]	Wash load (%)	Clay fraction (%)
28/10/76	16	3.07	66	22
27/01/77	12	3.25	73	36
01/03/77	14	3.34	66	31
28/03/77	6	2.08	97	55
13/04/77	7	2.39	93	51
25/04/77	7	2.46	92	51
09/05/77	7	2.51	92	49
23/05/77	7	2.41	94	47
06/06/77	7	2.48	92	47
25/07/77	12	2.87	77	31
08/08/77	14	2.87	73	23
31/10/77	15	2.99	67	22
14/11/77	11	2.77	81	26

[a]Mean diameter
[b]Geometric standard deviation of the size distribution

Again, maximum concentrations and very fine (d <10 μm) suspended sediment proportions were observed in April (see Fig. 5.4a, b, c).

Though there is limited information regarding the mineralogic composition of wash load, it is valuable to throw light on this topic and on its relationship with the river behavior. Depetris and Griffin (1968), reported qualitatively that the suspended silts of the Paraná River near Santa Fe city were composed of abundant quartz, mica, plagioclase and K-feldspar. Referring the minerals of the suspended clays at the same site, these authors presented the following results of two samples taken on June and July of 1967:

	Illite	Montmorillonite	Kaolinite	Chlorite
June 1967	38%	14%	33%	15%
July 1967	34%	18%	31%	17%

Mangini et al. (2003) studied the composition of clay minerals of the wash load sampled in the inflow of a lake directly connected to the main channel in the alluvial plain of the Paraná River (see 5.4.3). The samples were taken in eight field trips performed between March 15 and May 15 of 1999. The average percentages were:

Illite	Montmorillonite	Kaolinite	Chlorite	Other minerals
50%	24%	10%	10%	6%

Note the differences in the amounts of illite, montmorillonite, and kaolinite between the two data sources. According to Depetris and Griffin (1968), the illite is provided mainly by the Bermejo River when it drains the Andean range environments, which would tentatively be an explanation of the larger percentages detected by Mangini et al. (2003), since their sampling period coincides with the maximum clay supplies coming from that site. In relation to kaolinite, Depetris and Griffin (1968) suggest that the Paraguay and upper Paraná rivers basins could be the principal furnishing areas. In the last case, the numerous reservoirs (some of them very large) built since 1965 until now retain large amounts of the wash load annual supplies to the middle reach (see 5.4.2), which would account for the kaolinite percentages reduction at present. Possible reasons for the differences in the montmorillonite quantities cannot yet be proposed.

5.3.2.3 Wash Load Discharges

Amsler and Drago (2007) updated the sediment budget (essentially wash load) at the entrance of the Paraná River middle reach based on concentration data obtained during 1993–2003. Their results are shown in Fig. 5.5.

Two facts are clearly noticeable in Fig. 5.5: first, the decisive, already stated, role of the Bermejo River supplies on the wash load concentrations in the Paraná River, downstream from the confluence with the Paraguay River, for the 1970s data as well as for the new ones and, secondly, the smaller wash load concentrations coming from the upper Paraná River at present, cited above and further discussed in 5.4.2.

Figure 5.4c shows another distinctive feature of the wash load peak concentrations in the middle reach that stands out: they normally lag behind the highest streamflows, since the sediment wave coming from the Bermejo River arrives at the Paraguay-Paraná confluence after the annual flood of the latter river and is detected downstream with delays of 10 days on average (Drago and Amsler 1988; Alarcón et al. 2003). This lack of coincidence gives rise to the known hysteretic effects when concentrations and discharges are related (Walling 1974), which may prevent the application of the classic sediment rating curve method to compute the annual fine sediment transport in a given river. Alarcón et al. (2003) solved this problem by designing a procedure based on a combination of 8 years of concentration data obtained near Paraná City and at a section of the Bermejo River. They finally successfully appraised the wash load transport (G_w) in the Paraná River at the end of its middle reach between 1991 and 1998 (Table 5.4).

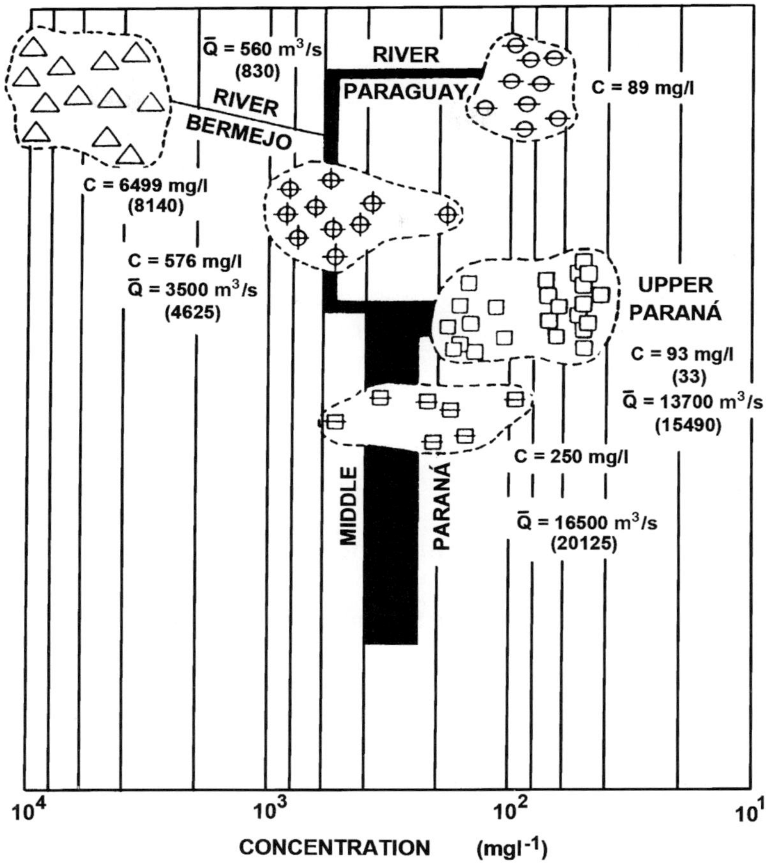

Fig. 5.5 Suspended sediment concentration changes along the Bermejo-Paraguay-Paraná hydrosystem during the period of maximum solid discharges. C: mean value for each group of total concentrations for a seventies' set of data and the new (between brackets) set of 1993–2003 (from: Amsler and Drago 2007)

5.3.3 Total Sediment Transport

In Table 5.4, the bed material transport and, thus, the total sediment transport during the same period at the Túnel Subfluvial cross section, have also been included (see Fig. 5.2). It is seen that the Paraná River had been transporting an average of 130–135 $\times$ 10^6 t year^{-1} of sediment through that section. The wash load represented 80% of the total amount. These figures may change slightly since sand, derived through the Colastiné and Leyes Rivers (Fig. 5.2), is not included in the computation of the annual G_s values. Some information about the sand transported in these secondary channels of the alluvial plain is given in 5.4.1.

Table 5.4 Bed material (sand), wash load and total sediment transport in the Paraná River in its middle reach (Section: Túnel subfluvial) (from: Alarcón et al. 2003)

Year	G_s (t year^{-1})	G_w (t year^{-1})	G_T (t year^{-1})	G_w/G_T
1991	18.2×10^6	220.8×10^6	239.0×10^6	0.92
1992	29.1	120.9	150.0	0.81
1993	21.4	85.7	107.1	0.80
1994	18.9	78.7	97.6	0.81
1995	21.1	88.6	109.7	0.81
1996	19.1	78.7	97.8	0.80
1997	24.8	130.4	155.2	0.84
1998	34.0	78.0	112.0	0.70
Averages	23.3×10^6	110.2×10^6	133.5×10^6	0.81

5.4 Sediment Transport and Processes in the Alluvial Plain

5.4.1 Sediment Transport in the Secondary Channels

The available information concerning the sediment transport along the secondary channels of diverse hierarchy which drains the Paraná River alluvial plain is extremely scarce compared to that for the main channel. Notwithstanding, based on suspended sediment concentrations and instantaneous values of sediment transport measured (and/or computed) principally in the main secondary channels over the last 30 years, it is possible to obtain an approach of the sediments amounts derived through them and of their proportions respecting the main channel transport. This topic is treated next for rivers ordered from north to south according to its general position in the alluvial plain (Fig. 5.2).

San Javier River. This stream flows along the west border of the Middle Paraná floodplain and shows a planform pattern that varies between anastomosing and meandering. The channel has a very small slope (1–1.5 cm/km for mean stages) and divides into minor courses that derive discharges towards a system of lakes and swamps particularly complex near its mouth. These features reduce the sediment transport capacity of the river preventing the sand to be carried to the mouth where the channel perimeter is completely cohesive (Ramonell and Amsler 2005).

The total sand transport, G_s, was computed by means of the Engelund-Hansen (1967) formula at sand bed reaches near Helvecia and Cayastá Cities with geometric, sedimentologic and hydraulic parameters measured at those sites (Schreider et al. 2004; Ramonell and Amsler 2005). For a mean discharge of 616 $m^3\ s^{-1}$ and a median diameter of 0.250 mm for the bed sediment, resulted:

$G_s = 53{,}700\text{–}73{,}800$ t year^{-1} (slope: 0.9 cm/km and 1.1 cm/km, respectively)

Regarding the wash load, 11 values of total mean concentrations measured during 1971 and 1972 (LHA 1974) at a cross section near the mouth, revealed:

$C_{max} = 230$ mg l^{-1} $\quad C_{min} = 80$ mg l^{-1} $\quad C_{av} = 130$ mg l^{-1}

where C_{max}, C_{min}, and C_{av} are the maximum, minimum, and average values of the concentration data.

Considering the same mean discharge and the C_{av} value, an annual wash load transport may be estimated:

$$G_w = 2.5 \times 10^6 \text{ t year}^{-1}$$

These figures show that G_s would represent less than 5% of the total sediment transport in the San Javier River.

Leyes River. An old branch of the Colastiné River together with the San Javier River supply the water and sediment discharges of the Leyes River, an important sand bedded channel flowing across the Paraná alluvial valley nearly perpendicular to its general direction. At the end of the Leyes channel, those discharges have built an enormous internal delta that increases its surface continuously into a vast lake system surrounding the east side of Santa Fe City. The largest recorded flood of the Paraná River in 1982–1983 changed the Leyes River morphology very much by increasing its cross section in such a way that the mean discharges varied from ≈1,000 m^3 s^{-1} before the flood to 1,550 m^3 s^{-1} after it.

In spite of recent key studies concerning the hydrology, geology, and geomorphology of the Leyes area (Ceirano et al. 2000; Ramonell 2005; Montagnini et al. 2005), little is known about its sediment transport. Montagnini et al. (2005) report only a rough estimation of the relative average increase in the total bed material load due to the changes resulting from the 1982–1983 flood. The lack of proper hydraulic information prevents any reliable computation of G_s. Perhaps the total mean concentrations obtained from seven suspended sediment measurements performed during 1971 and 1972 reported in LHA (1974) may be the single data set available to approach a fair estimation of the total sediment transport. The extreme and average values of those concentrations are:

$C_{max} = 190$ mg l^{-1} $\quad C_{min} = 80$ mg l^{-1} $\quad C_{av} = 135$ mg l^{-1}

Combining C_{av} with the mean discharges referred to above, G_T in the Leyes River amounts to:

$$G_T \approx 4.3 \times 10^6 \text{ t year}^{-1} \text{ (before 1982–1983)}$$
$$6.6 \times 10^6 \text{ t year}^{-1} \text{ (after 1982–1983)}$$

Note that a certain (yet for the moment unknown) percentage of these quantities is suspended bed material transport and the rest is wash load.

Colastiné River. This is one of the most important secondary channels draining the Paraná River floodplain. It shows a meandering planform pattern, a sandy bottom throughout its extension, and, due its higher slope (2–4.5 cm/km) and mean discharge (~1,700 $m^3 s^{-1}$ for the series 1904–2000), a bed sediment transport capacity markedly larger than that of the San Javier River. The median of the bottom sand size distribution varies between 0.150 and 0.400 mm, according to measurements performed at a reach immediately upstream from the RN 168 (National Road 168; Fig. 5.2). The most frequent sizes, on the order of 0.200 mm, correspond to the river thalweg.

The annual transports G_s and G_{sf} in the Colastiné River for its mean discharge were computed with the Engelund-Hansen (1967) and Engelund-Fredsøe (1976) formulas applied at the RN 168 reach. Hydraulic and sedimentologic measurements performed during 2001 and 2005 were used in the computation. The results were as follows:

$$G_s = 1.4\text{–}1.6 \times 10^6 \text{ t year}^{-1}$$
$$G_{sf} = 0.19\text{–}0.22 \times 10^6 \text{ t year}^{-1}$$

In relation to the mean total concentrations in this reach, LHA (1974) and Szupiany et al. (2005) reported results of measurements performed between 1971–1973 and 2004, respectively. From this information, the following results were obtained:

$$C_{max} = 215 \text{ mg l}^{-1} \qquad C_{min} = 100 \text{ mg l}^{-1} \qquad C_{av} = 155 \text{ mg l}^{-1}$$

Szupiany et al. (2005), also showed that an average of ~15% of the total concentrations is suspended bed material, i.e., ~130 mg l^{-1} would be wash load for a total C_{av}=155 mg l^{-1}. Thus the annual transport of wash load for the mean discharge in the Colastiné River is:

$$G_w = 7.0 \times 10^6 \text{ t year}^{-1}$$

This amount involves 82% of the total sediment transported annually on average by the Colastiné River, a percentage similar to that reported for the main channel (Table 5.4).

Coronda River. The sand bed channel of the Coronda River, which mainly drains the discharges flowing out of the lake system on the east side of the city (see "Leyes River"), and those supplied by the canal connecting the Paraná River with the city harbor, begins south of Santa Fe city. This stream also flows along the west border of the Paraná alluvial plain but on the south end of its middle reach. The mean discharges of the Coronda River increased after the 1982–1983 large flood due to the morphologic changes (described above) that occurred in the Leyes River, the entrance to the system. If the 1950–2000 series is considered, a mean discharge of 1,062 $m^3\ s^{-1}$ is obtained. But comparisons of some recent stream flow measurements with similar information taken during the seventies (LHA 1974) suggest a mean discharge of 1,950 $m^3\ s^{-1}$ at present.

Regarding the channel bed sediment, fine and medium sands with median diameters between 0.200 and 0.300 mm prevail, at least, along the first 10 km of the river. The knowledge concerning the bed material transport is also nearly nonexistent in the Coronda River. Applying the Engelund-Hansen (1967) formula with recent hydraulic measurements made at a straight reach near the river beginning, a total bed sediment transport (G_s) of 1.56 kg s^{-1} was computed (discharge: 1,130 $m^3\ s^{-1}$). In principle this value suggests a rather small bed sediment transport capacity of the Coronda River in spite of its relatively high mean water discharges i.e., in the same order than those of the Colastiné River. However, that capacity would be compatible with its lower water surface slope (≈1 cm/km).

The wash load transport may also be estimated in this case through the LHAs set of total concentration data (LHA 1974) measured during 1971–1972. The extreme and average values of the 11 suspended sediment measurements performed in the Coronda River, were:

$C_{max} = 195$ mg l^{-1} $C_{min} = 90$ mg l^{-1} $C_{av} = 140$ mg l^{-1}

Thus:

$$G_w \approx 4.7 \times 10^6 \text{ t year}^{-1} \text{ (mean discharge: 1,062 m}^3\text{ s}^{-1}\text{)}$$
$$8.6 \times 10^6 \text{ t year}^{-1} \text{ (mean discharge: 1,950 m}^3\text{ s}^{-1}\text{)}$$

Note that G_w would be comparable to G_T viewing the negligible amount of bed sediment transport for mean stages in the Coronda River.

5.4.2 Suspended Sediment Budget in the Middle Paraná River and its Alluvial Plain

The results presented in Fig. 5.4 show the decisive role of the Bermejo River on the wash load concentrations in the Middle Paraná River as well as the notable supply reduction occurring in the Upper Paraná River over the last ~10 years. This study was expanded recently by Amsler (2006), in order to quantify the general wash load sedimentation in the flood plain of the Middle Paraná River and the influence on the sedimentation rates

Table 5.5 Sedimentation of wash load in the alluvial plain of the middle Paraná River between Paso de la Patria and Paraná city during the months of maximum sediment discharges (from: Amsler 2006)

Period	Bermejo[a] G_w ($\times10^6$) [t]	Paraguay[b]	Upper Paraná[c]	Middle Paraná[d]	Deposition ($\times10^6$) [t]
Averages in the period					
1968–1974	52.69	1.44	14.60	53.55	15.18
1979–1982	142.20	2.69	12.50	96.69	60.70
1993–2004	95.27	2.63	2.67	82.89	17.70
Years with similar sediment discharges in the Bermejo River					
1972–1973	73.72	1.51	22.71	73.58	24.36
1981–1982	138.35	2.90	14.19	96.59	58.85
1995–1996	74.59	2.76	2.27	75.32	4.30
1998–1999	134.99	3.54	3.51	110.70	31.34
1997–1998 large flood					
1997–1998	51.62	3.62	3.58	68.62	–9.80

[a]El Colorado Section
[b]Puerto Pilcomayo Section
[c]Itatí Section
[d]Túnel Subfluvial Section (Fig. 5.2)

of the referred reduction at the entrance of the middle alluvial valley. The main results are presented in Table 5.5.

The reduction in the wash load supply from the Upper Paraná River would have begun in the early 1990s. Amsler and Drago (in press) and Amsler (2006), show that the large number of dams built in the Paraná River upper basin since the late 1960s and, particularly during the 1970s and 1980s, in the past century would account for this reduction. When comparing the average values of 1968–1974, 1979–1982 and 1993–2004, note that the sedimentation on the alluvial plain does not decrease due to the larger wash load supplies of the Bermejo River, which compensates the remarkable decrement in the Upper Paraná. But when specific years with similar wash load discharges in the Bermejo River are considered, the sedimentation reduction becomes clearly depicted: six times when comparing 1972–1973 with 1995–1996 and 50% if 1981–1982 and 1998–1999 are compared. An extreme event verified in 1997–1998 when the fourth largest flood in 100 years occurred and coincided with a minimum wash load supply from the Bermejo River. The result was a degradation in the order of 10 million t of the alluvial plain surface soil in the middle reach.

5.4.3 Additional Remarks about the Sedimentation Processes on the Alluvial Plain

Amsler et al. (unpublished) quantified the ~300-km sedimentation area of the alluvial plain north of Paraná City when the sediment is transferred from the main channel and its principal secondary streams through the mechanisms described in Chap. 4. These authors measured the surface of the colored incoming water on satellite images taken on February 1994 and March 1997 when the maximum wash load discharges supplied from the Bermejo River flowed through the system with a stage near to bankfull level. From their measurements, they concluded that approximately 28% of the investigated floodplain area would be the most affected annually by sedimentation, at mean rates of 4 to 8 mm $year^{-1}$ over the last 15 years.

Considering the prevailing type of sediment transported by the Paraná River and the explanations given above, the deposits on the alluvial plain contain large proportions of silts and clays, i.e., the particles transported as wash load. This is particularly true for the bottom sediments of the floodplain lentic environments (their lakes), as stated in Chap. 4. Mangini et al. (2003) made detailed studies of the mechanisms intervening in the sediment deposition transported into one of these lakes, the Lake Mendieta (Fig. 5.2), which is connected directly to the main channel. As was stated in 5.3.2.2, suspended and bed material samples were taken during field trips performed during maximum wash load discharges in 1999. Some important conclusions of this investigation are as follows:

(1) It is necessary to divide the Paraná River wash load into two fractions in order to study its deposition. One fraction is composed by the coarser particles that deposit individually without interacting with each other; the other, involving the finer sizes, is affected by flocculation and aggregates formation, which account for their sedimentation. The limit diameter between the two fractions would be 13–18 μm (15 μm in average).
(2) An average of 68% of the total wash load deposited into the lake pertained to the fraction lower than 15 μm, i.e., fine silts and clays, proving that flocculation is an important mechanism for the finest particle sedimentation in the lentic environments of the Paraná River.
(3) Montmorillonite content found in the lake bed sediment was nearly 38% larger than the one observed in the suspended sediment (see 5.3.2.2). This fact agrees with the properties of this type of clays that show a strong tendency to form flocs even in waters with low salt concentrations, e.g., the Paraná River.
(4) Nearly all the particles lower than 11–15 μm remain in suspension till the flow velocity decreases under a critical value of 0.15–0.20 m s^{-1}. When this condition is reached a sudden, massive and intense sedimentation occurs due to flocculation.

5.5 Conclusions

The Paraná River middle reach transports annually an average of 130–135 million tons of sediment. Near 80% of this amount is "wash load" made up by the finest particles with sizes in the order of silts and clays. The rest are fine and medium sands supplied by its own bed and carried as "bed load" (2.2 million tons) and "suspended load" (23 million tons).

In the main secondary channels of its alluvial plain (San Javier, Leyes, Colastiné and Coronda rivers), fed by the own main channel directly or indirectly, the proportion of transported wash load is also absolutely predominant (between 2.5 and 9 millions of tons per year). In these cases, the capacity of bed sand transport varies according to the respective drainage slopes and flows. The few available estimates of sand annual discharges range between ca. 50 thousands tons per year in the San Javier River and 1.5 million tons per year in the Colastiné River.

The Bermejo River, through the Paraguay River, gives the middle reach the major proportion of wash load during the December–May period of each hydrological year. Maximum concentrations of this type of sediment in all the system of the Middle Paraná River occur usually in March and April, lagging behind the peak flows, since these peaks, originated in the upper basin of the Paraná River, reach the confluence with the Paraguay River before the high concentrations coming from the Bermejo River.

The clay content (with illite and montmorillonite as predominant minerals) increases notably in the wash load during maximum concentrations, so during that period the minimum transparency of waters is recorded in all channels of the system and lakes connected directly to them. In Fig. 5.6, this phenomenon can be observed for the main channel during 1977.

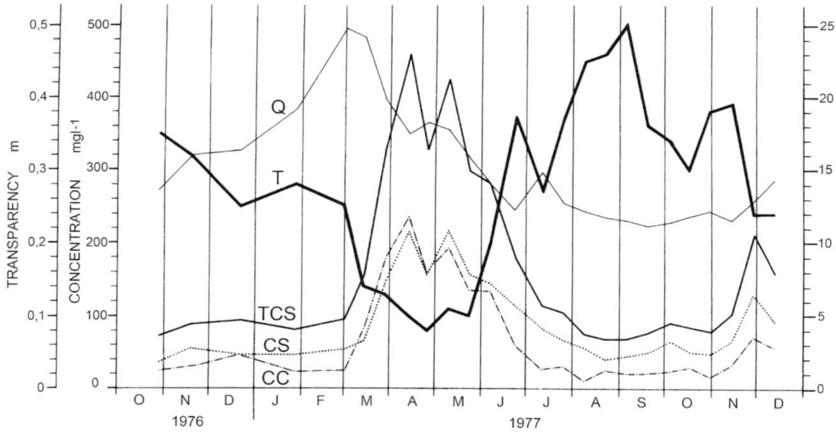

Fig. 5.6 Temporal variation of the water transparency (T), discharge (Q), total concentration of suspended sediments (TCS), concentration of clay fraction (CC) and silt fraction (CS). Concentrations represent surface water values at mid channel (from: Drago 1984).

The anthropic action shown in the construction of numerous reservoirs in the upper basin of the Paraná River during the last 30 years has decreased drastically the wash load supplies coming from that area. Though the Bermejo River supplies have counterbalanced the consequences of this fact in the alluvial valley of the middle Paraná River, a decrease in the deposits of fine sediments was detected given certain circumstances. Surface soils degradation during an extraordinary flood was the extreme result derived from the upper Paraná wash load deliveries reduction. Considering the role of the Bermejo River as the main source supplying silt and clay downstream, the eventual management of its basin should be conceived very carefully to avoid the irreversible alteration of the transference of that sediment to the Middle Paraná River system.

It is inferred that the wash load, the predominant sediment transported by the Paraná River, implies a renovation of surface soils of its alluvial plain through its already mentioned periodical deposition. Moreover, strongly positive correlations have been recently informed between sediments of the Bermejo River and concentrations of calcium and phosphorus in the Lower Paraná River (De Cabo and Seoane 2005). These facts imply key factors to preserve if this huge wetland is intended to be sustained.

References

Alarcón JJ, Szupiany R, Montagnini MD, Gaudin HE, Prendes HH, Amsler ML (2003) Evaluación del transporte de sedimentos en el tramo medio del río Paraná. Primer Simposio Regional sobre Hidráulica de Ríos, Buenos Aires, Argentina

Amsler ML (2006) Evolución de la carga de lavado en el alto Paraná (1968–2004). Incidencia sobre las sedimentaciones en la planicie aluvial del Paraná Medio. Actas III Congreso Iberoamericano sobre Control de la Erosión y los Sedimentos, Buenos Aires, Argentina

Amsler ML, Drago ECE (2007) A review of suspended sediment budget at the confluence of the Paraná and Paraguay Rivers. Hydrolog Process (in press)

Amsler ML, Prendes HH (2000) Transporte de sedimentos y procesos fluviales asociados. In: Paoli C, Schreider M (eds) El Río Paraná en su Tramo Medio. Contribución a su conocimiento y prácticas ingenieriles en un gran río de llanura, tomo 1. Centro de Publicaciones de la Universidad Nacional del Litoral, Santa Fe, Argentina, pp 233–306

Bonetto AA, Orfeo O (1984) Caracteres sedimentológicos de la carga en suspensión del río Paraná entre Confluencia y Esquina (provincia de Corrientes, República Argentina). Revista de la Asociación Argentina de Mineralogía, Petrología y Sedimentología 15:51–61

Ceirano EB, Paoli C, Schreider MI (2000) Las inundaciones en el área de Santa Fe. Interpretación de sus efectos y simulación de los subsistemas Leyes-Setúbal y Ruta Nacional 168-Alto Verde. In: Paoli C, Schreider M (eds) El Río Paraná en su Tramo Medio. Contribución a su conocimiento y prácticas ingenieriles en un gran río de llanura, tomo 2. Centro de Publicaciones de la Universidad Nacional del Litoral, Santa Fe, Argentina, pp 239–297

Cotta R (1963) Influencia sobre el río Paraná del material sólido transportado por el río Bermejo. Comisión Nacional del Río Bermejo. Publ N°92 EH, Buenos Aires, Argentina

De Cabo L, Seoane R (2005) Suspended sediments in the Bermejo River and their impact on the hydrochemistry of the Lower Paraná River. In: Dynamics and biogeochemistry of river corridors and wetlands. IAHS Publ N°294, pp 151–158

Depetris PJ, Griffin JJ (1968) Suspended load in the Río de la Plata drainage basin. Sedimentology 11:53–60

Drago ECE (1984) Estudios limnológicos en una sección transversal del tramo medio del río Paraná. IV: Influencia de la composición granométrica de los sedimentos suspendidos sobre la transparencia del agua. Rev de la Asoc de Cs Nat del Litoral 15:47–55
Drago ECE, Amsler ML (1988) Suspended sediment at a cross section of the Middle Paraná River: concentration, granulometry and influence of the main tributaries. In: Sediment budgets. IAHS Publ N°174, pp 381–396
Drago ECE, Amsler ML (1998) Bed sediment characteristics in the Paraná and Paraguay Rivers. Water Int (IWRA) 23:74–183
Engelund F, Fredsøe J (1976) A sediment transport model for straight alluvial channels. Nordic Hydrol 7:293–306
Engelund F, Hansen E (1967) A monograph on sediment transport in alluvial streams. Teknisk Vorlag, Copenhagen, Denmark
Lelievre J, Navntoft E (1980) Measuring sediment load in the Paraná River. Water Power & Dam Construction April:21–24
LHA (Laboratorio de Hidráulica Aplicada) (1974) Estudio del sistema fluvial Paraná-Santa Fe. PNUD-UNESCO Prog SF/ARG/66/521. Laboratorio de Hidráulica Aplicada, vols. I y II. Ezeiza, Argentina
Mangini S, Prendes HH, Amsler ML, Huespe J (2003) Importancia de la floculación en la sedimentación de la carga de lavado en ambientes del río Paraná. Revista Ingeniería Hidráulica en México XVIII:55–69
Montagnini MD, Basualdo A, Alarcon JJ, Ramonell CG, Amsler ML (2005) Evolución morfológica de la embocadura de los ríos Colastiné y Leyes, planicie aluvial del Paraná Medio. Segundo Simposio Regional sobre Hidráulica de Ríos, Neuquén, Argentina
Prendes HH, Huespe J, Schreider MI, Amsler ML, Zanardi L (1994) Pronóstico de evolución de una trinchera dragada en un cauce aluvial. Influencia de las distintas modalidades de transporte. Proc XVI Congreso Latinoamericano de Hidráulica, vol 2. Santiago, Chile, pp 393–404
Prendes HH, Torres GR, Zapata CR (1996) Aspectos sedimentológicos del riacho Barranqueras. Proc XIV Congreso Nacional del Agua, Neuquén, Argentina
Pujol A, Amsler ML, Sabatier MA, Gaudin HE (1985) Verificación de tres predictores Q/H en el río Paraná. Anales XII Congreso Nacional del Agua, Mendoza, Argentina
Ramonell CG (2005) Geología y geomorfología de la laguna Setúbal y su entorno (Santa Fe, Argentina). Trabajo Final de Licenciatura en Ciencias Geológicas. Dep de Geologia, Fac de Cs Fisico-Matemáticas y Naturales. Universidad Nacional de San Luis, San Luis, Argentina
Ramonell CG, Amsler ML (2005) Avulsión y rectificación de meandros: consideraciones para su predicción. Ingeniería del Agua 12:231–248
Schreider MI, Scacchi G, Amsler ML, Ramonell CG, Franco F, Romano C (2004) Determinación de Parámetros Hidráulicos en un Tramo de Cauce Secundario de la Llanura de Inundación del Río Paraná (Vuelta del Dorado-Río San Javier). XXI Congreso Latinoamericano de Hidráulica. (IAHR), Sao Pedro (Sao Paulo), Brasil
Szupiany RN, Amsler ML, Fedele JJ (2005) Estimation of Suspended Sand Concentrations with an ADP, Colastiné River-Argentina. Proc 4th IAHR Symposium on River, Coastal and Estuarine Morphodynamics: RCEM 2005. Illinois, EUA
Soldano F (1947) Régimen y aprovechamiento de la red fluvial argentina. Cimera, Buenos Aires, Argentina
Vanoni VA (ed) (1975) Sedimentation engineering. ASCE-manuals and reports on engineering practice N°54, New York
Walling D (1974) Suspended sediment and solute yields from a small catchment prior to urbanization. Inst Brit Geogr Spec Publ 6:169–192

6 The Geochemistry of the Paraná River: An Overview

PEDRO J. DEPETRIS[1] AND ANDREA I. PASQUINI

6.1 Introduction

River basins are not inert continental features. Moreover, it can be said that rivers have a life, and their evolution is usually predictable. In biological terms, for example, rivers are processors of materials as the biota they contain take up, convert, use, and release resources that come to them. In other words, rivers are active biological systems that metabolize the organic matter they transport. From a geological point of view, rivers transport sediments and solutes whose dynamics is also determined by a set of complex interacting variables, such as lithology, climate, and relief. Consequently, the water that reaches a river's mouth is far different, both qualitatively and quantitatively, from the water that entered the system as rain or snowfall. Summarizing, the chemical signatures of rivers are reflections of complex natural and interdependent relationships involving the chemistry of precipitation, the weathering of minerals, the cycling of vegetation, and the evolution or history of its water. The recently published geochemical treatise (Drever 2005) is a major step towards the elucidation of such complexities.

Not only natural factors intervene in the functioning of river basins. Human activities also affect rivers in many ways: directly, through dams, pollution, or eutrophication; indirectly, through the use of the land; and in much more subtle ways, by means of global warming and acid rain. Built to foster increased development, the Upper Paraná River, for example, has in operation about 130 reservoir dams (dam height >10 m), of which 14 are considered "major dams" (dam height >150 m) (Ravenga et al. 1998) that modulate its discharge, sequestrate sediments, and alter its biogeochemistry.

In this chapter on the Paraná's geochemical character, we wish to overview what we consider the major aspects that determine the geochemical nature of a river, probing into its natural functioning: the dynamics of its dissolved components, the chemical imprint of weathering, the provenance of its sediments, and its more conspicuous biogeochemical nature. In this manner, we expect to contribute to the better understanding of its natural functioning

Centro de Investigaciones Geoquímicas y de Procesos de la Superficie (CIGeS), Facultad de Ciencias Exactas, Físicas y Naturales, Universidad Nacional de Córdoba, Avenida Vélez Sarsfield 1611, X5016CGA Cordoba, Argentina, [1]e-mail: pdepetris@com.uncor.edu

M.H. Iriondo, J.C. Paggi, and M.J. Parma (Eds.)
The Middle Paraná River: Limnology of a Subtropical Wetland

as well as on the impact of anthropogenic alterations, which have been occurring in the Paraná River with increasing rate for the last 40 years.

6.2 Geological and Hydrological Framework

The Paraná River drainage basin (Fig. 6.1), with an area of 2.783×10^6 km^2 (Tossini 1959), is the most important river system in the Río de la Plata basin because it accounts for about 88% of its drainage area (3.170×10^6 km^2; Tossini 1959) and almost 80% of its total discharge to the SW Atlantic Ocean (~21,500 m^3 s^{-1}). The Paraná's drainage covers almost the entire South American continental width at 21°S. Its northernmost water sources are located in Brazil, at ~15°S, and ~45°W, whereas the western water sources are close to the Andes, at ~65°W, in the headwaters of the Pilcomayo and Bermejo Rivers. The Paraguay's upper catchments are in the Gran Pantanal (Brazilian Mato Grosso, ~15°S, and ~55° to ~60°W). The Salado (or Juramento) River is a middle-sized tributary with sources by the Eastern Cordillera that joins the Paraná in its middle course. Finally, the Carcarañá River has its headwaters in the Sierras Pampeanas of Córdoba, ~33°S, and meets the Paraná north of the city of Rosario, in the lower reach. Several authors have considered the geomorphological features of the Paraná River (e.g., Iriondo 1972, 1988; Orfeo and Stevaux 2002; Thorne 2002). Araújo et al. (1999) examined the characteristics of the Guaraní aquifer system, which is closely associated with the Paraná drainage basin. More information on other physical features of the Paraná River can be found elsewhere in this volume.

As it happened with the remaining of South America's major drainage nets, the Paraná's developed after South America's separation from Africa, 130 My ago (Early Cretaceous). In the initial syn-rift phase, extension was accompanied by strike-slip faulting and block rotation; subsequent extension went together with the extrusion of very large volumes of basaltic lava (i.e., the Serra Geral Formation), in the Late Cretaceous (Potter 1997; Potter and Hamblin 2006). Clearly, such extended flood basalts and the adjoining Cretaceous sandstones are the most conspicuous features of the Phanerozoic mantle that covers a significant portion of the Río de la Plata drainage basin (Fig. 6.1). Another outstanding attribute is the thick Quaternary layer of sediments that extends from Mato Grosso and the Gran Pantanal, in the north (~18°S), to the loess-mantled Pampa plains in the south (~35°S). The headwaters of the Bermejo and Pilcomayo Rivers exhibit thick beds of marine and continental sediments as well as outcrops of Precambrian metamorphic rocks, along with volcanic rocks of Quaternary age. Steep slopes, along with the abundance of friable sedimentary formations in humid to semiarid climates, are crucial factors determining the high sediment load that has been recorded mainly in the Bermejo drainage basin. The most important geologic characteristics are schematically presented in Fig. 6.1.

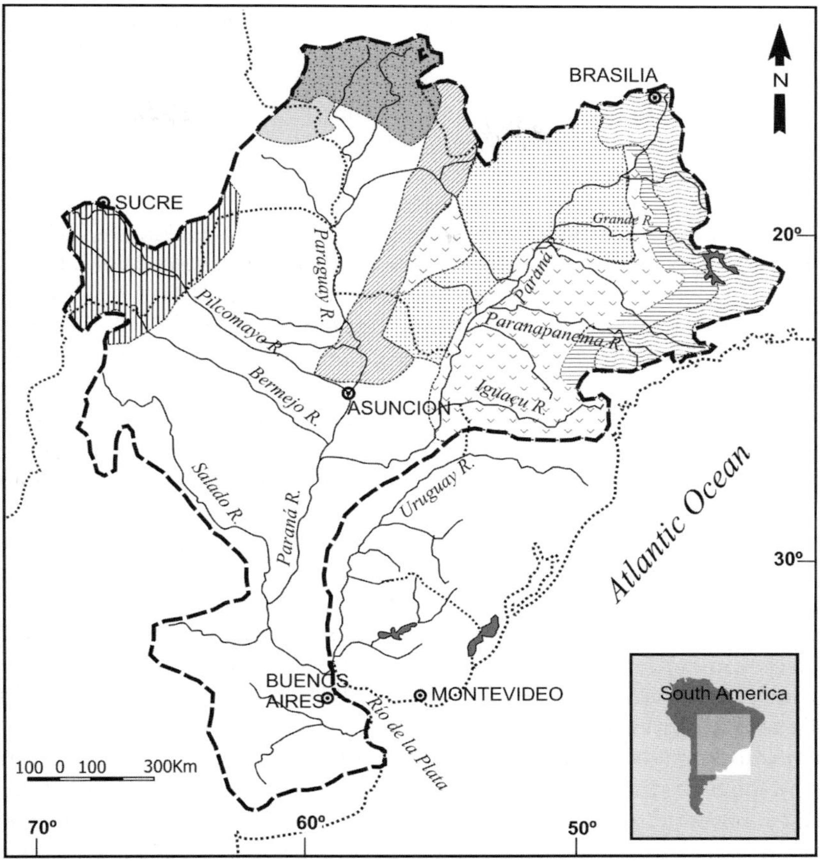

Quaternary

Tholeiitic basalts (Jurassic - Cretaceous)

Continental sands and sandstones (Cretaceous)

Continental sedimentary rocks (Carboniferous - Jurassic - Permian)

Sedimentary rocks (Devonian - Carboniferous)

Undifferenciated Precambrian - Sedimentary rocks (Cambrian - Devonian)

Marine sedimentary rocks (Cambro-Ordovician, Silurian, etc.) - Precabrian metamorphic rocks

Sedimentary rocks (Cambrian) - Metasediments (Precambrian)

Precambrian metamorphic rocks

Fig. 6.1 Schematic geological map of the Paraná River drainage basin

The Paraná River currently delivers a discharge (mean for the period 2000–2003) of ~510 km^3 (annual mean discharge of ~16,400 $m^3 s^{-1}$). Mean annual rainfall is unevenly distributed over the entire drainage basin. Maximum recorded precipitation (~2,400 mm y^{-1}) occurs along the eastern edge of the basin, near the city of Sao Paulo (Brazil) and over the upper Iguaçu drainage basin. Rainfall decreases markedly (800 to 400 mm y^{-1}) towards the western edge of the basin, along the 60° to 65°W strip. The western tributaries (Bermejo and Pilcomayo) reach maximum discharges in February–March (end of southern summer); the Paraguay River reaches maximum discharge in June–July (southern winter), and lastly, the Iguaçu River resembles the Paraná's main stem, with top discharges during February (southern summer) and low waters during the southern spring.

The mean historical (i.e., the 20th century) contribution of the Paraná's upper stretch to the total annual freshwater discharge that eventually reaches the Río de la Plata is about 73%, with the remaining fraction supplied by the Paraguay River. However, the breakdown of the Middle Paraná's mean hydrograph to unveil the relative significance of its sources shows that the Paraguay usually supplies a minimum of 19% of the Middle Paraná discharge during the Paraná's high waters (January, southern summer), and a maximum of 34% during the low waters (July, southern winter). Moreover, the complete historical time series shows that Paraguay's relative contribution to the total Paraná River discharge was as low as 1% (as in the mid-1940s) and as high as 50% (as in the early 1960s).

The summer circulation over South America is dominated by a monsoonal system whose major seasonal feature is the South Atlantic Convergence Zone (SACZ), placed along the north-eastern boundary of the Río de la Plata drainage basin (e.g., Robertson and Mechoso 2000). Another important feature in the climatic control of the region is a low-level northerly/north-easterly jet that flows east of the Andes and transports moisture along the corridor placed between the Andes and the Brazilian Altiplano (e.g., Wang and Fu 2004).

There is evidence that regional rainfall and river runoff has increased significantly in the Río de la Plata drainage basin during the 20th century. The flow increase in the Paraná and Paraguay Rivers, for example, has been pointed out by several workers (e.g., Genta et al. 1998; Robertson and Mechoso 1998; Collischonn et al. 2001; Boulanger et al. 2005; Dore 2005; García and Mechoso 2005); most authors place the breaking point around the early 1970s.

In a recent contribution (Pasquini and Depetris 2007), we employed the non-parametric Mann-Kendall trend test (Mann 1945; Kendall 1975) on discharge time series and found out that some of Paraná's tributaries (Bermejo, Paraguay, Iguaçu, and Salado Rivers) and the lowermost Paraná main stem station (at Paraná) (Fig. 6.2a) exhibit significant positive trends ($p<0.001$). Other gagging stations along the Paraná River main channel (i.e., Posadas and Corrientes) exhibited lesser levels of significance.

We also applied the seasonal Kendall test (Hirsch et al. 1982) to monthly flow data from rivers in the Río de la Plata drainage basin (Pasquini and

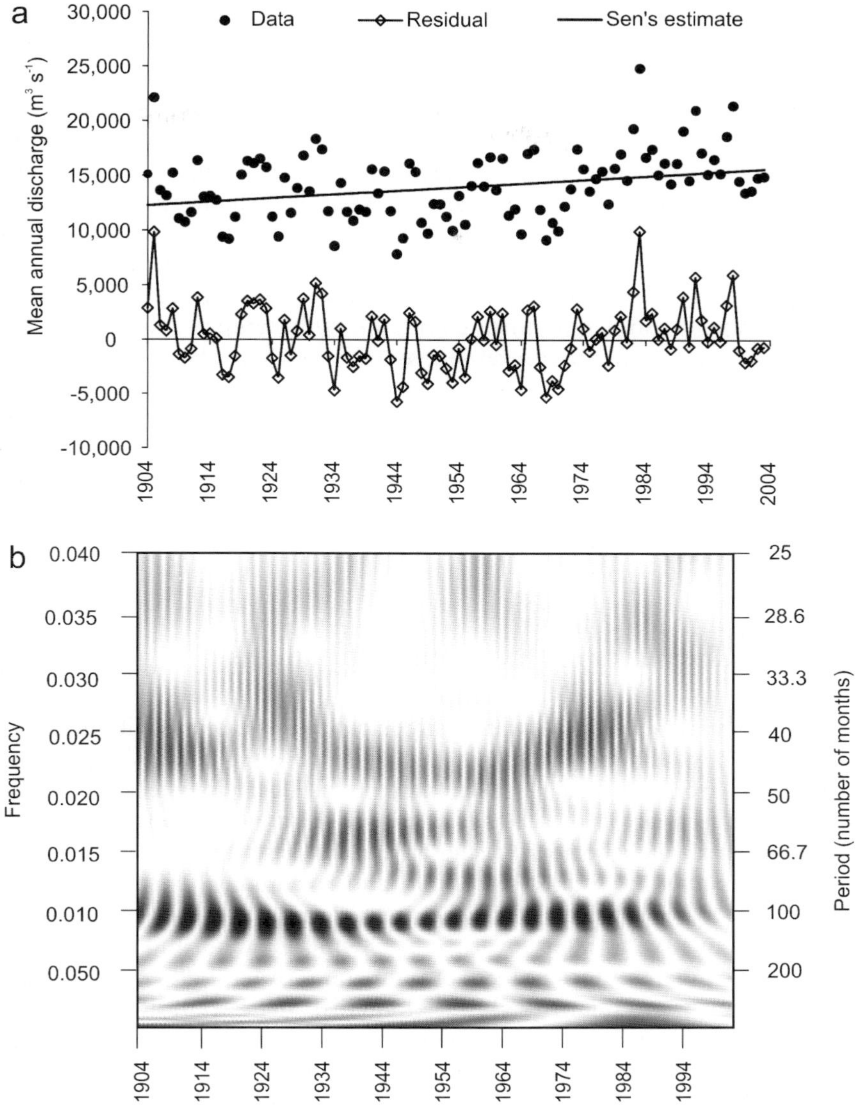

Fig. 6.2 **a** Mann-Kendall trend test for the mean annual discharge series of the Paraná River (at Paraná, ~600 km upstream the mouth), **b** Real part of the continuous Morlet wavelet spectrum of Paraná River deseasonalized monthly mean discharge. *Dark colors* represent high power

Depetris 2007). Interestingly, the months that showed a significant flow increase were those of low flow (southern winter). When high water months (southern summer) displayed an increase, they were the least statistically significant. Further, the Paraná and all its tributaries have clearly reversed a negative discharge trend, discernible until about the mid-20th century, for the currently recorded increasing trend.

Figure 6.2b also shows the application of the continuous wavelet transform (CWT) (e.g., Nakken 1999) spectral analysis to the Paraná River (at Paraná, 600 km upstream the mouth) deseasonalized discharge time series for a period of almost 100 years (1904–2003). As pointed out by other authors (e.g., Labat et al. 2005; García and Mechoso 2005) the diagram shows a strong quasi-decadal frequency that has been linked to sea-surface temperature anomalies over the northern tropical Atlantic Ocean (Robertson and Mechoso 1998). It also shows potent interannual oscillations in the 2–7 year range that are in phase with El Niño events in the equatorial Pacific. The wavelet spectra become very useful in illustrating the shifting in frequencies of interannual periodicities throughout the 20th century. The 2 to 5-year analysis shows maximum power peaks for 1928 and 1980. Power drops markedly for the range ~1935–1970 but, precisely for that time range, power increases in the 5 to 7-year band.

The occurrence of exceptional flooding events, such as those triggered by the El Niño, determines the occurrence of significant changes in the composition and concentration of dissolved and suspended phases (e.g., Depetris and Kempe 1990; Depetris and Gaiero 1998; Villar and Bonetto 2000).

6.3 The Chemistry of Dissolved Components

In the Earth's surface system, water plays a central role because it operates as both a reactant and as a transporting agent of dissolved and particulate components from land to sea (Gaillardet et al. 2005; Meybeck 2005). The processes collectively known as *chemical weathering* (see Sect. 6.4) controls the global hydrogeochemical cycle of elements transforming rocks and primary minerals to solutes and soils and eventually to sediments and sedimentary rocks. In rivers, as in the ocean, lakes, and the atmosphere, the hydrogeochemistry is largely governed by chemical equilibrium in solution or at the water–mineral interface (e.g., Stumm and Morgan 1996).

The water supply is unequally distributed in the Paraná River drainage basin, and weathering exerts an uneven control on the different upper catchments. Therefore, the dissolved chemical nature of the middle and lower Paraná River is determined not only by the dissimilar water volume contribution from the Paraná's two main tributaries (i.e., the Upper Paraná and the Paraguay, Fig. 6.1) but also by various processes that govern their dissolved species concentrations.

After Livingstone's (1963) early hydrochemical report, several authors have examined over the past 30 years the main chemical characteristics of the Paraná River dissolved load (e.g., Maglianesi 1973; Depetris 1976; Drago and Vassallo 1980; Bonetto et al. 1998; Villar et al. 1999; Villar and Bonetto 2000). In general, in Paraná River waters, the order of abundance of major anions is HCO_3^- >Cl^- >SO_4^{2-}, and of major cations is ($Na^+ + K^+$)>Ca^{2+} >Mg^{2+}. The

Middle Paraná preserves an asymmetric chemical cross section throughout its middle course because the water flowing along its western margin is saltier than the eastern one (Drago and Vassallo 1980). This characteristic is not only determined by the differing chemical signature of both the Paraguay and the Upper Paraná Rivers but also by the combined influx of surface and groundwater flow originating in the sediment-mantled Chaco-Pampa plain.

The total dissolved solids (TDS) load estimate for the Middle Paraná River is ~25 Tg y^{-1} (1 Tg = 10^9g) (Depetris and Kempe 1993), about three times lower than what appears to be its total mean sediment load. The Upper Paraná delivers more diluted waters than the Paraguay whose mean TDS concentration is usually twice that of the Paraná. If we consider their mean discharge ratio (~2.3:1), we end up with a TDS load for the Middle Paraná supplied in about equal proportions by both major tributaries. Exceptional flooding appears to trigger significant dilution and increased the TDS export rate—during the 1982–1983 ENSO-triggered flood—only up to ~26 Tg y^{-1} (Depetris and Kempe 1993).

Figure 6.3 presents a Piper diagram with chemical data from various sources that corroborates some of the above assertions by showing that the Middle Paraná River (besides departures triggered by anomalous hydrology)

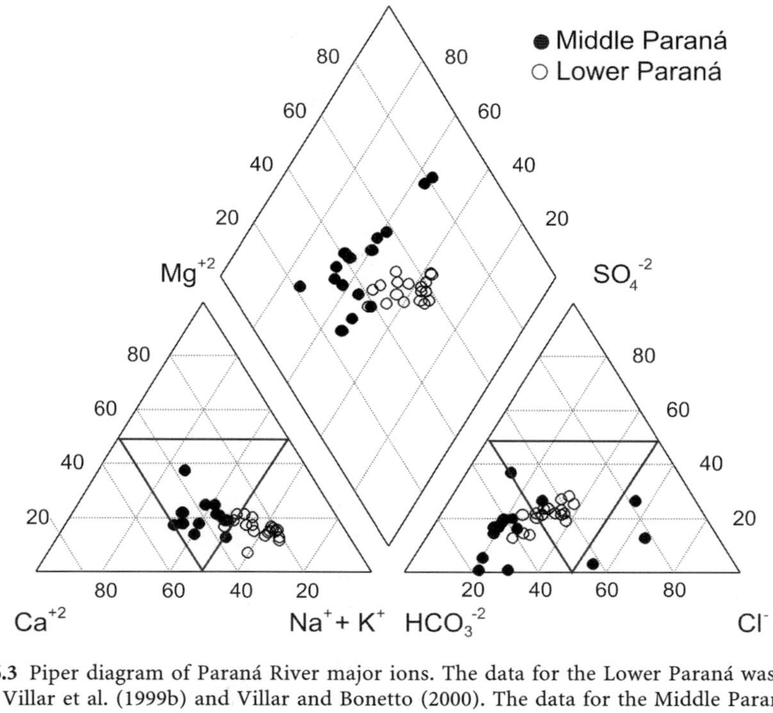

Fig. 6.3 Piper diagram of Paraná River major ions. The data for the Lower Paraná was taken from Villar et al. (1999b) and Villar and Bonetto (2000). The data for the Middle Paraná was collected during the SCOPE-UNEP International Carbon Project

is of the mixed-type in cationic terms and bicarbonate-type in anionic terms. In contrast, the river appears to switch to the mixed-anion-type and to the sodium-potassium-type in the lower reach.

Figure 6.4 shows the variation of a set of parameters measured between 1981 and 1984 at the city of Paraná (600 km upstream the mouth) within the framework of the SCOPE-UNEP International Carbon Project. The 1982–1983 El Niño took place during the sampling period (Fig. 6.4d, e) and caused significant TDS dilution, oxygen supersaturation and a discernible impact during both, the ascending and descending limbs of the hydrograph (Fig. 6.4). Oxidation and reduction reactions mediate the behavior of many chemical constituents in most aquatic compartments of the environment. The joint variation of Eh and pH throughout the flooding event showed values that are normal for rivers and streams (Fig. 6.4b). Oxygen supersaturation indicates the occurrence of photosynthesis because the equilibrium in the following reaction is displaced to the right-hand side,

$$n\mathrm{H_2O} + n\mathrm{CO_2} \Leftrightarrow (\mathrm{CH_2O})_n + n\mathrm{O_2}$$

In the Paraná, this situation was not necessarily coherent with increased water transparency (Fig. 6.4c). Likewise, low water transparency did not always convey a displacement of the equilibrium to the left-hand side of the above equation (i.e., biological respiration). Moreover, Secchi measurements (i.e., water transparency) exhibited a conflicting association with the prevailing hydrological conditions, which seems to be related to the river stage (i.e., there is a different response if a certain gage height occurs during the descending or ascending limb of the hydrograph).

Figure 6.5 shows the downriver variation of selected parameters measured in 1985 during the SCOPE-UNEP-sponsored International Carbon Project. Most variables showed a significant trend: conductivity increased, along with alkalinity, total hardness, and silica, largely due to the effect of the Paraguay River and tributaries. Dissolved oxygen failed to produce a significant trend and phosphorous was clearly consumed in the downriver direction. Drago and Vassallo (1980) found similar results during a 1975 scientific cruise.

Figure 6.6 shows a variation diagram (or spidergram) for dissolved rare earth elements (REE) in Paraná River water, main tributaries, and the Uruguay (for comparison). This group of elements is one of the least mobile in the weathering of minerals and transport by waters (e.g., Sholkovitz 1995; Gaillardet et al. 2005) and, in a qualitative way, their concentrations and fractionation is the result of pH-dependant reactions in solution or at the interface with colloids. Table 6.1 displays mean REE concentrations measured in the Paraná River basin.

Considering that the raw analytical data has been normalized, it becomes evident that concentrations fluctuate between 10^5 and 10^7 times lower that the Upper Continental Crust (Taylor and McLennan 1985). World mean dissolved REE concentrations (Gaillardet et al. 2005) are higher than the concentrations

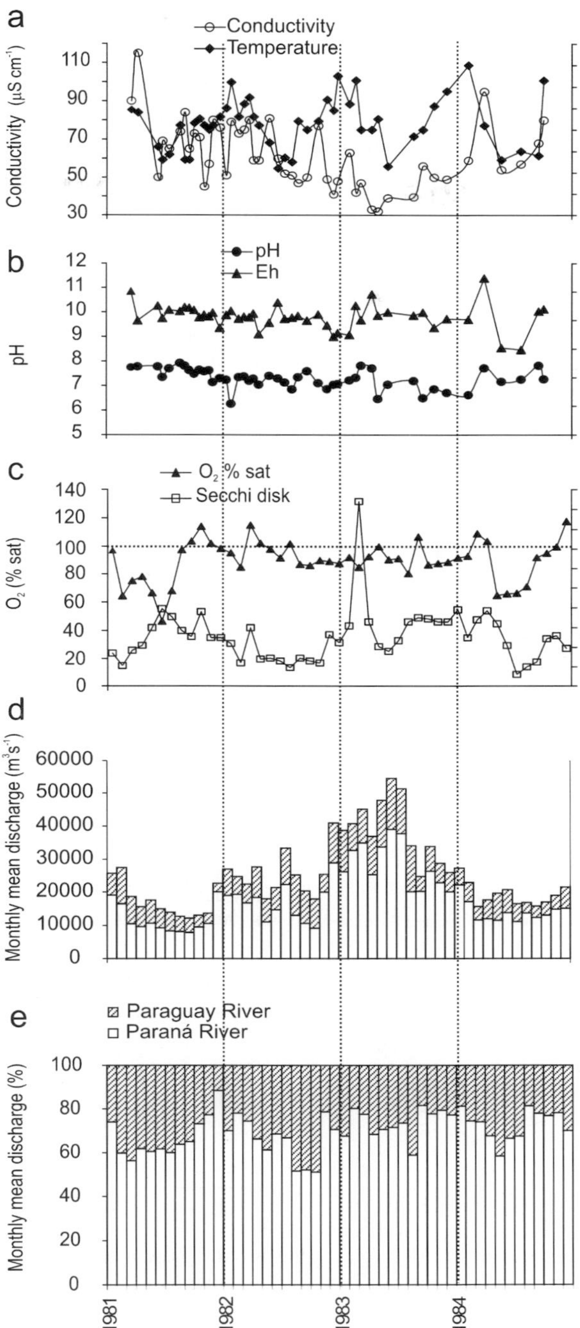

Fig. 6.4 Selected parameters determined in samples collected in the Paraná River at Paraná (~600 km upstream the mouth) during the SCOPE-UNEP International Carbon Project. Absolute and relative discharges (split into contributing tributaries) at Corrientes (~1,200 km upstream the mouth)

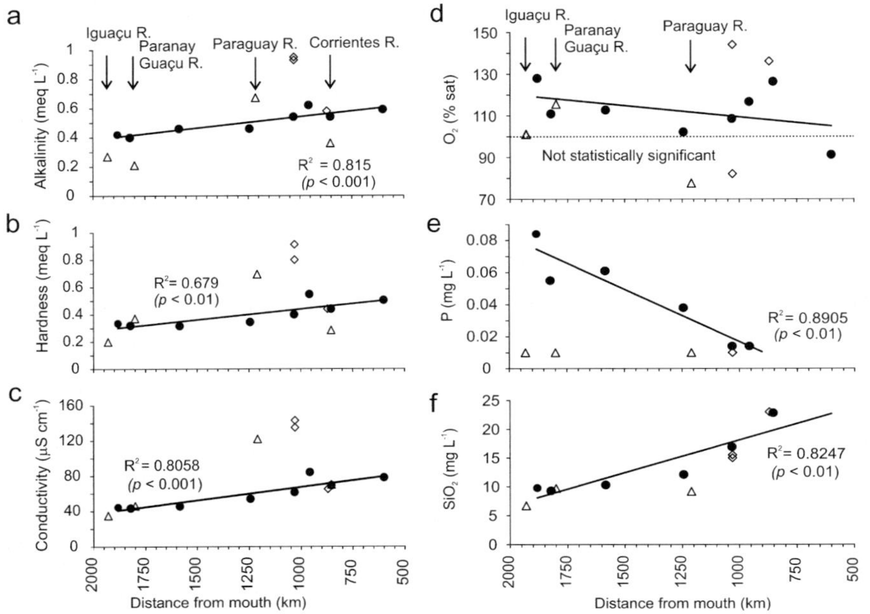

Fig. 6.5 Downstream variation of selected chemical parameters determined in the Upper and Middle Paraná River in July 1985 (SCOPE-UNEP International Carbon Project). *Filled circles* correspond to mainstream samples; *triangles* to indicated tributaries; *diamonds* to ponds and ox-bows

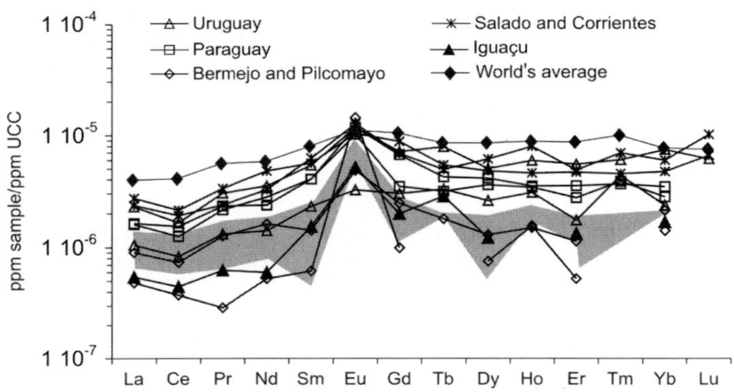

Fig. 6.6 UCC-normalized dissolved REE spidergram for the Paraná River (*shadowed area*), tributaries, the Uruguay River, and the world's average (Gaillardet et al. 2005)

measured at the Río de la Plata drainage basin. Also, with the exception of the Uruguay River, which does not show a clear europium anomaly ($Eu/Eu^* = Eu_N/(Sm_N.Gd_N)^{1/2}$, McLennan 1989), which is discernible in the suspended sediment fraction, as will be seen in Sect. 6.4), the remaining samples exhibit a sharp

Table 6.1 Dissolved REE in the Paraná River, tributaries and Uruguay River[a]

River	Middle Paraná $n = 4$	Paraguay $n = 2$	Pilcomayo $n = 1$	Bermejo $n = 1$	Iguaçu $n = 1$	Corrientes $n = 1$	Salado $n = 1$	Uruguay $n = 2$
La	0.026	0.049	0.027	0.015	0.016	0.082	0.071	0.050
Ce	0.053	0.090	0.047	0.024	0.029	0.137	0.123	0.080
Pr	0.008	0.016	0.009	0.002	0.004	0.024	0.017	0.015
Nd	0.031	0.067	0.042	0.014	0.016	0.124	0.084	0.064
Sm	0.007	0.018	0.006	0.003	0.007	0.025	0.028	0.017
Eu	0.007	0.010	0.004	0.013	0.005	0.009	0.011	0.007
Gd	0.007	0.019	0.009	0.004	0.008	0.033	0.026	0.019
Tb	0.001	0.002	0.001	<0.001	0.002	0.003	0.003	0.004
Dy	0.005	0.013	0.005	0.003	0.004	0.017	0.021	0.013
Ho	0.002	0.003	0.001	0.001	<0.001	0.004	0.006	0.004
Er	0.003	0.007	0.003	0.001	0.003	0.011	0.011	0.008
Tm	<0.001	0.001	<0.001	<0.001	<0.001	0.001	0.002	0.002
Yb	0.005	0.007	0.005	0.003	0.004	0.010	0.013	0.011
Lu	<0.001	0.002	<0.001	<0.001	<0.001	0.002	0.003	0.002

[a]Concentrations in $\mu g\ L^{-1}$

anomaly that is surely associated with the weathering of feldspars (e.g., Taylor and Mc Lennan 1985). In as much as light REE (LREE, La to Nd) are preferentially adsorbed onto colloidal particles, the Paraná and tributaries show a clear fractionation of LREE to heavy REE (HREE, Ho to Lu). Finally, dissolved REE concentrations in the Uruguay, Paraguay, Salado, and Corrientes Rivers are clearly higher than in the Paraná (Fig. 6.6).

6.4 The Signature of Weathering in the Drainage Basin

The term *weathering* is used in the earth sciences to denote the physical and chemical processes that disaggregate and leach rocks exposed on the surface of the Earth, and end up generating clays and soils that are chemically adjusted to the prevailing environmental conditions. Silicates formed at high temperatures in igneous or metamorphic environments (feldspars, olivine, pyroxenes, and micas) are metastable in such conditions, and the more soluble elements (e.g., Na^+, K^+, Ca^{2+}, Mg^{2+}) are selectively removed from the mineral debris, leaving behind soils relatively enriched in alumina and iron. It is known that in order to see the effects of weathering, the region in question must be in equilibrium with tectonics, so as to allow soil formation (i.e., *transport-limited regime*, Stallard and Edmond 1983). If, in contrast, the region is not in equilibrium with tectonics, and mineral debris is removed as soon as it is generated, the resulting sediment bears a chemical/mineralogical signature that is close to the original rocks due to insufficient weathering (i.e., *weathering-limited regime*, Stallard and Edmond 1983). Clearly, the sediments exported from continents to oceans should convey the chemical/mineralogical mark that is relevant to the dominating denudation regime. The situation, however, is complicated in systems such as the Paraná, with headwaters in tectonically active orogenic areas, often dominated by sedimentary rocks that export mineral debris that has been subjected to one or several paleoweathering cycles. There are recent studies that claim that about 75% of the Amazon's and 86% of the Paraná River's sediment load are recycled sedimentary material (Gaillardet et al. 1999). Sediments are recycled when, in the geological past, they underwent two or more passages through the exogenous cycle. In tropical climates, such repeated passing leaves a perceptible geochemical/mineralogical imprint in the sediments, but in temperate areas, such as in the southern Andes, the cannibalization of exposed sediments to form younger formations does not seem to leave a discernible geochemical signature (Pasquini et al. 2005).

Figure 6.7 includes compositional triangles (Nesbitt et al. 1996) that show the weathering evolutionary paths determined for suspended sediment samples collected in the Paraná, its tributaries, and in the Uruguay River. In Fig. 6.7a, the K-feldspar-plagioclase segment, close to the plot of the Upper Continental Crust (UCC) (Taylor and McLennan 1985), defines the starting point of arrow A, the theoretical evolutionary path of granite, which aims to the alumina-rich extreme. The Upper Bermejo's tributaries and the Bermejo and Pilcomayo Rivers plot at the beginning of the path, close to the granitic composition whereas samples from the Middle Paraná tend to concentrate on the more weathered extreme. Figure 6.7b includes a mafic term (iron + magnesium) and further shows that the evolutionary path falls halfway between the theoretical evolution of weathered granite, and the evolution of weathered basalt. Figure 6.7a includes a 0-100 CIA (chemical index of alteration) scale on the left-hand side of the figure. The CIA (Nebitt and Young 1982) was defined in molecular proportions as:

$$CIA = 100\ [Al_2O_3/(Al_2O_3 + CaO^* + Na_2O + K_2O)],$$

where CaO^* represents the Ca in the silicate fraction only. The UCC has a mean CIA of 47 (McLennan 1993), and significantly weathered materials, such as the suspended load of large river systems, have CIAs between 70 and 80. Clearly, in the CIA, soluble elements, significantly depleted in weathered sediments, are compared with Al_2O_3, which is a ubiquitous hydrolysate in deeply weathered tropical environments. The chemical index of weathering (CIW) is similar to the CIA, but it does not include K_2O in its computation (Harnois 1988).

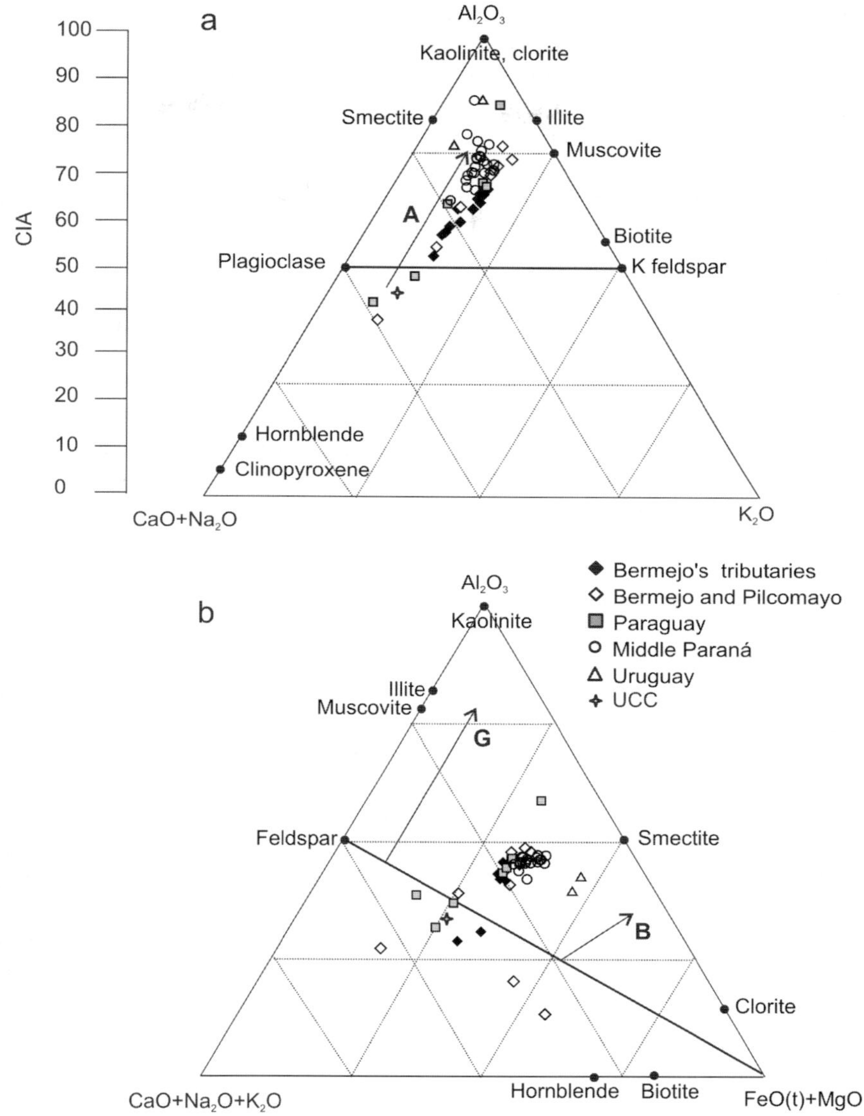

Fig. 6.7 Ternary diagrams after Nesbitt et al. (1996) showing weathering trends. **a** A-CN-K (Al_2O_3-CaO^*+Na_2O-K_2O) diagram displaying the felsic chemical composition of Paraná drainage basin sediments. The supplementary vertical axis shows the variability of CIA in the samples. **b** A-CNK-FM (Al_2O_3-CaO^*+Na_2O-K_2O-FeO+MgO) diagram, same as (a) but for mafic components. The theoretical mineral composition is from McLennan (1993) and Nesbitt et al. (1996). The UCC composition is from Taylor and McLennan (1985)

Figure 6.8a shows, by means of a box and whiskers plot, the CIA TSS variability in the Paraná system. There is a general increasing trend in the medians of each group of samples, as the sampling moves downriver. We feel that this trend is a clear effect of particle sorting in as much as the lower water velocities in the Paraná's lower stretch result in finer particle size

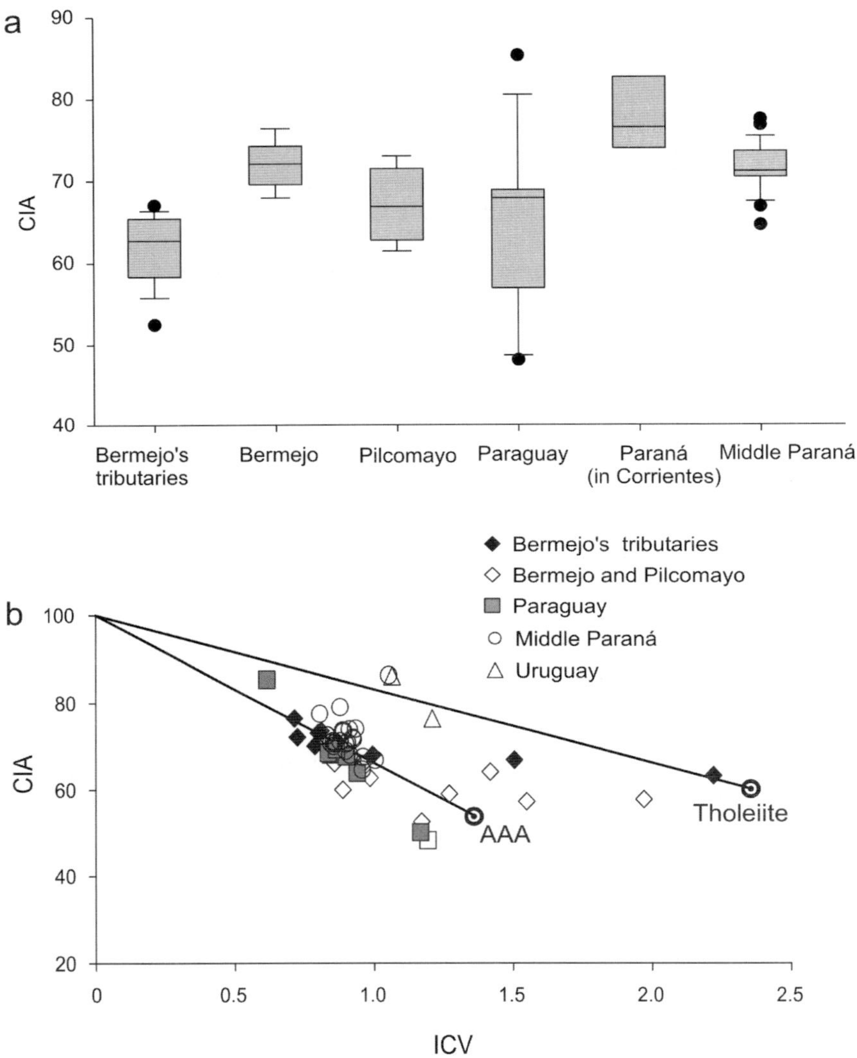

Fig. 6.8 a Box-and-whiskers diagram showing downriver CIA variability in TSS. The lower and upper boundaries in *boxes* indicate the 25th and 75th percentile, respectively. The *lines within the box* mark the median. *Whiskers* above and below the box indicate the 90th and 10th percentiles. *Dots* are outlying points. **b** TSS CIA and ICV scatter plot. *AAA* Average Andean arc (http://geokem.com)

distributions in the suspended sediment load. The median CIA of the Middle Paraná's TSS load seems to be in the vicinity of 72, which is somewhat lower than the CIAs determined for other major world rivers and corresponds to moderately weathered debris.

Another ratio, similar to the CIA, was defined as the index of compositional variability (ICV) by Cox et al. (1995), and measures the abundance of alumina relative to the other major cations in rocks or minerals:

$$ICV = (Fe_2O_3 + K_2O + Na_2O + CaO + MgO + MnO + TiO_2)/Al_2O_3$$

Silica is excluded to eliminate problems with quartz dilution.

The consideration of the above equation leads to the conclusion that nonclay silicates have a higher ICV than clay minerals do. Compositionally mature fine-grained sediments, such as mudrocks, dominated by clays and poor in nonclay silicates, will exhibit a low ICV. In contrast, immature muds or mudrocks that contain a high proportion of nonclay silicate minerals, or that are rich in clay minerals such as smectites, will have high ICVs (Cox et al. 1995).

Figure 6.8b is a bivariate plot of CIAs and ICVs determined in TSS samples collected in the Paraná River and its main tributaries. Two samples of Uruguay's TSS are also included. As a reference, the segments depict the theoretical evolution of a mean Paraná tholeiite (Iacumin et al. 2003) and a mean Andean andesite (http://www.geokem.com). In general, most Paraná samples are clustered around the theoretical andesitic line. Departures from the cluster may be explained by variations in the sediment-supplying source.

The question of measuring weathering was approached ingeniously by Gaillardet et al. (1999) by defining a separate index for each mobile element whose concentration in the sediment was compared with that of an immobile element whose magmatic compatibility is close to that of the mobile element (Hofmann 1988). For example, ratios of elements with similar magmatic compatibilities, such as Sm/Na, are expected to be less variable than ratios of elements whose magmatic compatibilities are contrasting (e.g., Th/Na). Gaillardet et al. (1999) defined indexes for all the mobile elements, but in this chapter we are going to use only the indexes for Na and for Ca:

$$\alpha_{Na} = [Sm/Na]_{sed}/[Sm/Na]_{UCC}$$
$$\alpha_{Ca} = [Ti/Ca]_{sed}/[Ti/Ca]_{UCC}$$

A value $\alpha_i = 1$ means that there is no net chemical weathering, whereas $\alpha_i > 1$ corresponds to a depletion with respect to the UCC, and $\alpha_i < 1$ to an enrichment of element *i*.

Figure 6.9 shows a bivariate plot of α_{Na} and α_{Ca}. Most values lie in the <10 interval, and several values determined for the Bermejo, Pilcomayo and their tributaries plot within the <5 framework. These values are significantly lower than the mean α_{Na} determined for tropical rivers, like the Orinoco ($\alpha_{Na} = 25$) and the Amazon ($\alpha_{Na} = 15–20$) (Gaillardet et al. 1999).

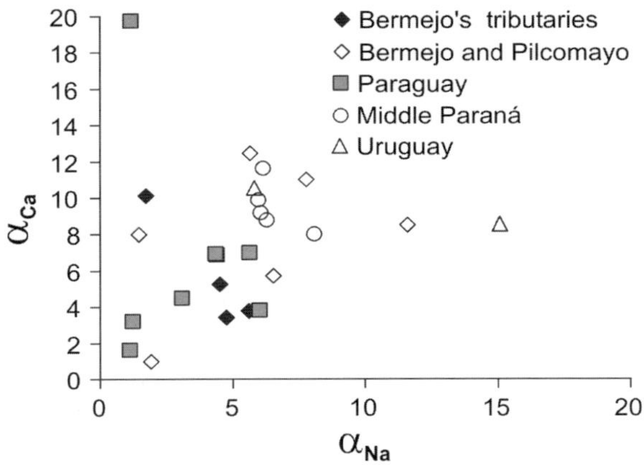

Fig. 6.9 Weathering alpha index values for Ca and Na in the Paraná drainage basin and the Uruguay River. $\alpha_i = 1$ means than river-borne particles have a composition similar to UCC

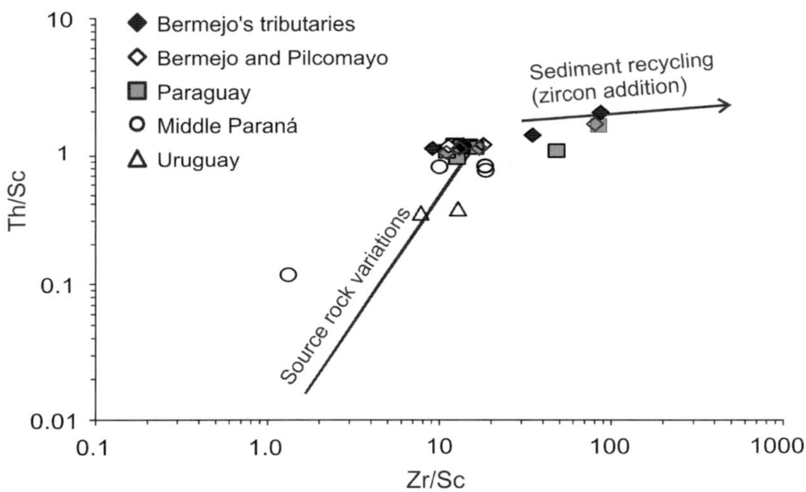

Fig. 6.10 Th/Sc and Zr/Sc ratios in Paraná River drainage basin's TSS. Anomalous zircon addition reflects sediment recycling (McLennan et al. 1993)

These rather low α_{Na} and α_{Ca} values in the Paraná River and in its headwater tributaries reinforce the notion of a dominant weathering-limited regime and suggest, further, the likelihood of significant recycling occurred in the mountainous (Andean) upper catchments.

Figure 6.10 depicts a Zr/Sc versus Th/Sc bivariate plot that has been used to show the likelihood of sediment recycling in turbidites placed along active margins (McLennan et al. 2003). In such a framework, the enrichment of Zr is often interpreted as a clear signal of recycling, as it happens with some samples collected in the Bermejo and Pilcomayo upper catchments.

6.5 The Provenance of Sediments

Although the Upper Paraná is the most significant contributor to the system's water budget, the largest proportion of the transported suspended sediment mass originates in the Bermejo River drainage basin. Various studies (e.g., Drago and Amsler 1988, 1998; Depetris et al. 2003) have shown that the Bermejo supplies most (50–70%) of the Paraná's sediment load. Moreover, the main sediment sources are located in the Bermejo's upper catchments (Upper Bermejo, Iruya-Pescado, Grande de Tarija, and San Francisco Rivers), which account for less than 2% of the Paraná's total drainage area. Total suspended sediment concentrations of 7 to 8 gL^{-1} are frequent, and even concentrations exceeding 20 gL^{-1} were recorded at the Iruya River. Consequently, sediment yields in the upper catchments reach ~12,000 t $km^{-2} y^{-1}$ (at the Iruya River), ~3,000 t $km^{-2} y^{-1}$ (at the Pescado River) and ~740 t $km^{-2} y^{-1}$ (at the San Francisco River) (http://www.corebe.org.ar).

By means of TSS's major composition determined in the Paraná River main stem and in major tributaries (Table 6.2), we inspected their geochemical classification following Herron's (1988) approach. Figure 6.11 shows that most samples collected in the Middle Paraná, and some in the Paraguay, classify as unambiguous mudstones, whereas the Upper Paraná, the Bermejo and Pilcomayo, and the Uruguay (included for comparison) plot as iron mudstones. The geochemical classification of TSS samples also allows perceiving the capacity of high-energy tributaries (i.e., Bermejo tributaries) to transport coarser mineral suites that plot as coarser-grained, more mature associations (Fig. 6.11). The Paraguay samples plotting as arenites probably convey the mature mineralogical nature of Mato Grosso's soils.

Major oxides have also been successfully used to infer the tectonic setting of sandstone-mudstone suites (Roser and Korsch 1986), interpreting that the defined categories reflect the composition of rocks in the source areas. The samples employed to produce a tectonic discrimination diagram (Fig. 6.12) are sediments transported in suspension, subjected to sorting, and—as in Patagonian rivers (Pasquini et al. 2005)—to a limited chemical weathering. Most samples from the Paraná and its tributaries plot as a tight cluster within the field ascribed to active continental margins (ACM), thus indicating that the provenance signature of the major sediment supplier (i.e. the Bermejo river) is preserved (after sorting) in the resulting fine-grained debris. Roser and Korsch (1986) made the point that this category, ACMs, includes complex active margins that may have material derived from continental margin magmatic arcs, which may have been deposited in a variety of basins (i.e., trench, fore arc, intra arc, and back arc).

Ternary diagrams that involve the use of lanthanides or trace elements allow the fine-tuning of sediment provenance in the Paraná River drainage

Table 6.2 TSS chemical composition in the Paraná River, tributaries and Uruguay River[a]

River Components	Bermejo's tributaries $n = 14$	Bermejo $n = 7$	Pilcomayo $n = 5$	Paraguay $n = 9$	Middle Paraná $n = 35$	Uruguay $n = 3$
SiO_2	54.78	62.96	52.43	61.66	59.06	54.50
Al_2O_3	12.59	15.26	14.21	13.86	16.76	17.33
Fe_3O_2	4.37	4.98	5.09	5.42	7.87	13.07
CaO	6.29	1.28	5.25	1.41	1.04	1.40
Na_2O	1.84	1.51	1.09	2.44	1.21	0.69
K_2O	2.93	3.29	2.92	3.10	2.90	0.97
MgO	4.35	1.73	6.21	1.55	1.59	1.13
TiO_2	0.56	0.69	0.69	0.74	1.11	1.72
MnO	0.08	0.08	0.07	0.08	0.12	0.21
P_2O_5	0.25	0.39	0.21	0.18	0.28	0.22
Ba	629	640	615	538	597	459
Co	14	12	14	15	17	22
Cr	54	66	98	68	159	67
Cs	8.5	9.4	19.0	9.3	9.5	4.6
Cu	28	31	45	72	109	115
Hf	7.2	6.7	6.5	5.9	5.4	4.7
Nb	15	14	20	15	16	
Ni	40	29	43	35	67	42
Pb	23	29	57	25	45	69
Rb	133	134	170	127	126	73
Sc	11	14	12	12	22	19
Sr	412	102	220	156	125	123
Ta	1.2	1.2	1.8	1.3	1.2	1.3
Th	13	14	17	13	12	6
Tl	0.4	0.6	0.6	0.4	1.1	
U	2.9	2.9	3.5	3.1	3.2	3.1
Y	26	32	28	30	32	35
Zr	197	212	222	208	234	264
La	126.8	198.6	50.6	76.7	37.2	22.4
Ce	79.2	81.5	96.6	78.7	75.7	45.9
Pr	1.2	1.2	1.8	1.2	1.2	1.3
Nd	34.1	35.0	42.3	34.7	34.3	32.0
Sm	6.4	6.7	7.9	6.5	5.9	4.4
Eu	1.2	1.3	1.5	1.3	1.4	1.2

(Continued)

Table 6.2 TSS chemical composition in the Paraná River, tributaries and Uruguay River[a] — Continued

River Components	Bermejo's tributaries $n = 14$	Bermejo $n = 7$	Pilcomayo $n = 5$	Paraguay $n = 9$	Middle Paraná $n = 35$	Uruguay $n = 3$
Gd	5.9	6.0	7.1	5.9	5.7	6.1
Tb	1.0	1.0	1.1	1.0	2.8	1.0
Dy	5.4	5.3	6.4	5.5	5.2	5.6
Ho	1.1	1.0	1.2	1.1	1.2	1.3
Er	3.2	3.1	3.9	3.2	2.9	3.2
Tm	0.5	0.5	0.6	0.5	0.5	0.4
Yb	3.0	2.9	3.5	3.0	2.6	3.2
Lu	0.5	0.4	0.5	0.4	0.4	0.5
CIA	61.6	72	67.1	65.1	72.5	81.2
ICV	1.5	0.99	1.8	0.98	0.9	1.14
α_{Na}	4.1	5.4	7.7	3.4	6.5	10.4
α_{Ca}	2.8	7.0	3.3	6.7	9.3	10.3

[a]Major oxides in %, trace and rare earth elements in $\mu g\ g^{-1}$

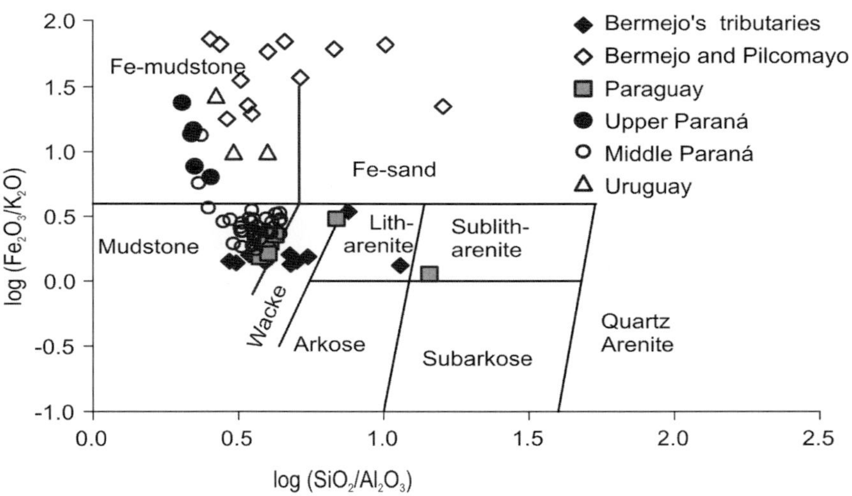

Fig. 6.11 Geochemical classification diagram for muds and mudstones (Herron 1988), applied to TSS from the Paraná River drainage basin and the Uruguay River

basin (Fig. 6.13). On the left-hand side La-Th-Sc diagram, the Paraná and tributaries form a tight cluster, along with the UCC, crossing the field borderline of continental island arcs. This sample clustering contrasts with the Uruguay River samples, dominated by the weathering products of Paraná

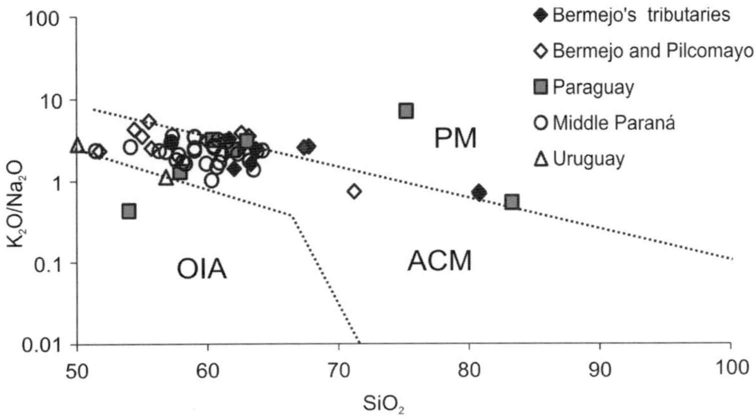

Fig. 6.12 Major oxides diagram for Paraná River drainage basin's TSS showing main fields for tectonic setting discrimination (Roser and Korsch 1986). Uruguay River samples are included form comparison. *PM* passive margin; *ACM* active continental margin; *OIA* oceanic islands arc

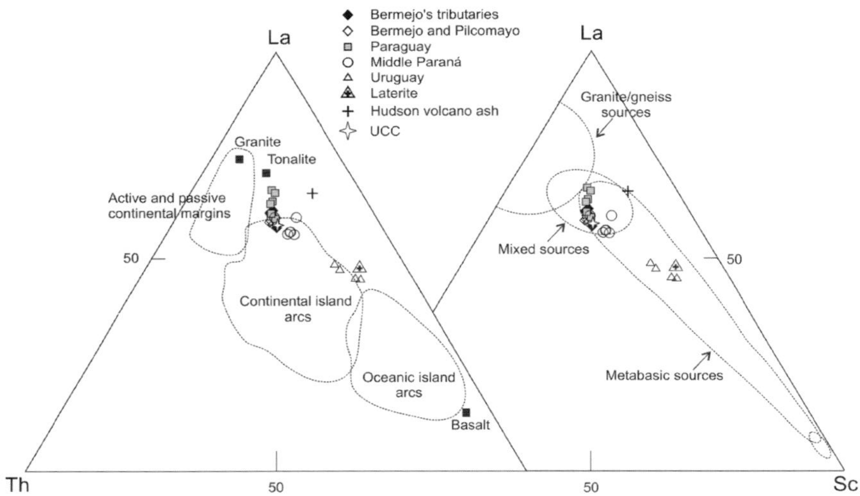

Fig. 6.13 Ternary diagrams depicting trace elemental composition of Paraná River drainage basin's TSS. Data for Uruguay River's TSS, a typical laterite, the volcanic ash of Hudson volcano (eruption of 1991), and UCC have been included for comparison. The tectonic setting discrimination fields are from Bathia and Crook (1986)

flood basalts (tholeiitic). Continental island arcs are sedimentary basins adjacent to island arcs that formed on a well-developed continental crust (e.g., Bathia 1983). The right-hand side diagram shows clearly the varied nature (i.e., from granitic to basaltic) of the ultimate igneous sources.

Rare earth elements are often used to fingerprint sediment sources (e.g., Cullers et al. 1987; Cullers 1994). Figure 6.14a shows the mean UCC-normalized

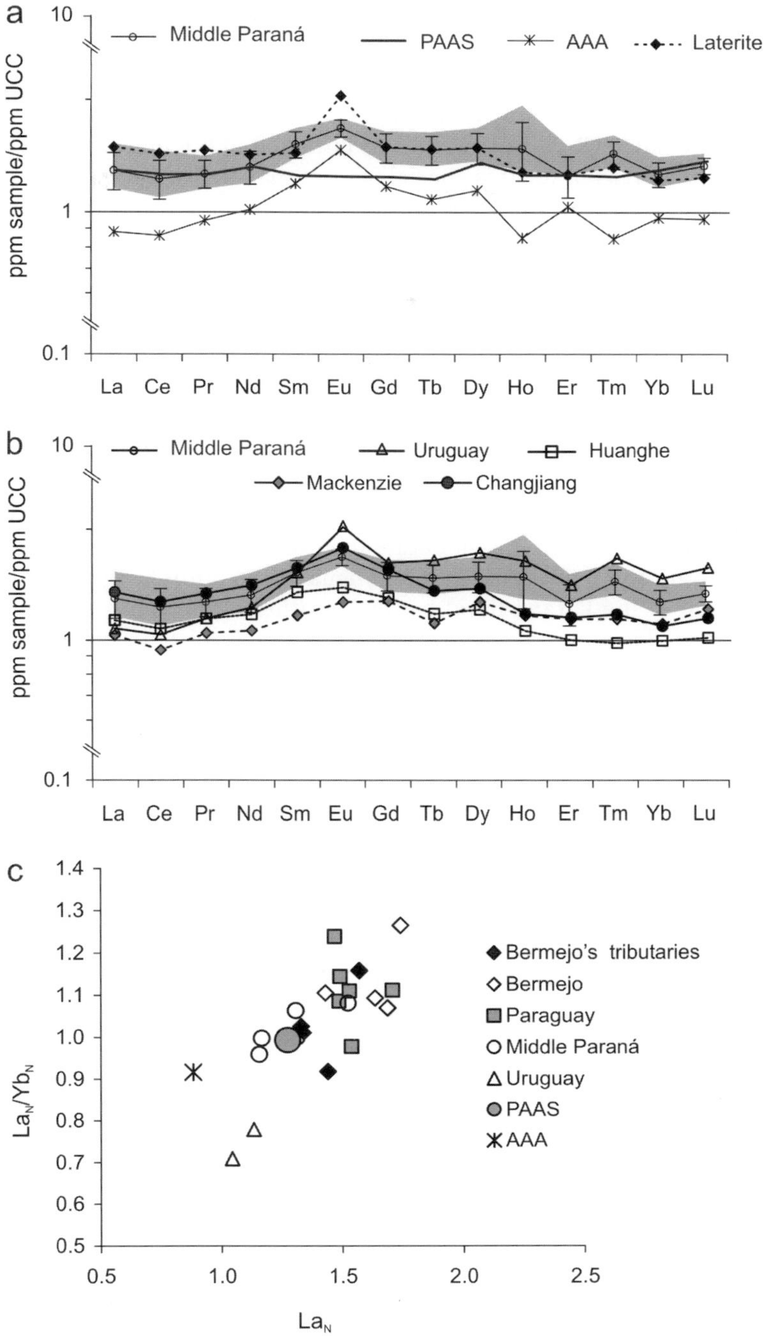

Fig. 6.14 **a** UCC-normalized REE spidergram for Paraná River sediments (*shadowed area*) including for comparison the Post Achaean Australian Shale (*PAAS*), Average Andean Arc (*AAA*), and a typical laterite. **b** UCC-normalized REE spidergram for Paraná River sediments (*shadowed area*) and for other world rivers included for comparison. **c** HREE and LREE fractionation (La_N vs. La_N/Yb_N) in the Paraná River drainage basin's TSS. Uruguay River, PAAS, and AAA are included for comparison

variation diagram of Middle Paraná River TSS samples. The pattern is similar to the PAAS (Post-Archean Australian Shale) (Taylor and McLennan 1985) for the LREE, and HREE, but Paraná's is enriched in the middle REE (MREE: Sm to Dy). The reason for this special feature may be connected to the selective weathering of phosphates (Hannigan and Sholkovitz 2001). A lateritic soil (from Misiones, Argentina) is also included in the figure for comparison, as well as an Andean andesite (http://www.geokem.com). These last two patterns show, like the Paraná's, a clear-cut Eu anomaly. The TSS's REE fingerprint of other world rivers is shown in Fig. 6.14b. Some rivers (i.e., Uruguay, Chiangjiang), like the Paraná, exhibit a significant Eu anomaly, a feature that is determined by the TSS mineralogy. Another interesting attribute is that in some rivers, due to pH-controlled differential adsorption, UCC-normalized spidergrams show depleted LREE with respect to the MREE and HREE.

Figure 6.14c shows the HREE and LREE fractionation through the ratio of UCC-normalized La and Yb (e.g., Taylor and McLennan 1985). The Paraguay and Bermejo Rivers exhibit a moderate fractionation that is not visible in the Middle Paraná and is reversed in the Uruguay River, thus suggesting that the contribution of the Paraná flood basalts' weathering products to the Paraná suspended load is unimportant. This fractionation has been also linked to the age of the rock sources: ancient rocks produce high La_N/Yb_N ratios in suspended sediments (Goldstein and Jacobsen 1988). This is also seen in Table 6.3, where mean values for several parameters linked to REE abundance are listed. All of them project a trend that suggests that the final REE signature in the Paraná is a weighted mean between the Bermejo-Paraguay signal and what appears to be the pattern delivered by the Jurassic-Cretaceous flood basalts, clearly depicted by the Uruguay River TSS. Simple calculations again reveal the importance of the Bermejo in the final Paraná geochemical TSS signature.

The isotopes of Sr and Nd have proved useful in sediment provenance studies (e.g., Mc Lennan et al. 1990).Concluding our consideration of the Paraná's sediment provenance, Fig. 6.15 shows a plot of $^{143}Nd/^{144}Nd$ and $^{87}Sr/^{86}Sr$ ratios obtained in the Paraná and Uruguay rivers' TSS samples (Henry et al. 1996). Unequivocally, the Uruguay's TSS plot next to the field defined for Paraná tholeiites, whose weathering products cover most of Uruguay's drainage basin. In contrast, Paraná's TSS clearly plot within the

Table 6.3 REE parameters in TSS from Paraná River, tributaries and Uruguay River

River	Bermejo's tributaries $n = 5$	Bermejo $n = 6$	Paraguay $n = 8$	Middle Paraná $n = 5$	Uruguay $n = 2$
La_N/Yb_N [a]	1.03±0.01	1.13±0.09	1.11±0.08	1.03±0.05	0.74±0.05
Eu/Eu^*	0.96±0.03	0.96±0.03	1.08±0.02	1.119±0.05	1.28±0.05
Σ REE	277.09±169.74	332.56±173.95	227.34±60.46	188.27±20.5	170.65±4.14
LREE/HREE	13.14±10.51	16.08±11.10	10.70±6.28	8.56±0.30	6.70±0.11

[a]Subscript $_N$ denotes UCC-normalization

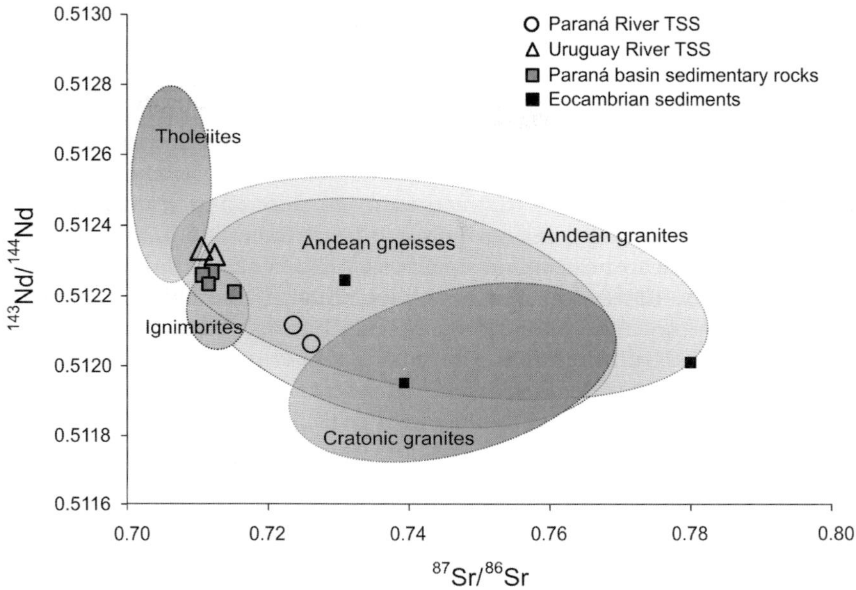

Fig. 6.15 Plot of Nd and Sr isotopes for Paraná and Uruguay TSS showing likely sediment sources. Other rocks are included for comparison: tholeiites (Iacumin et al. 2003); Paraná basin sedimentary rocks (Sallet et al. 2005); Eocambrian sediments, Andean gneisses and granites (Lucassen et al. 2001); cratonic granites (de Assis Janasi 2002; Sallet et al. 2005)

boundaries of silicic rock sources, thus suggesting that sediment beds in the Upper Bermejo drainage basin, which (as seen above) supply most of the Paraná's sediment load, appear to bear the dominant signature of an acid-intermediate source. The Upper Paraná which, like the Uruguay, also has a significant proportion of its drainage covered with soils developed on tholeiitic basalts, does not convey (as pointed out above) a tholeiitic signature to the TSS in the middle reach.

6.6 Main Biogeochemical Characteristics

The cycling of carbon involves complex processes by means of which it is transformed from organic carbon (the form found in living organisms) to inorganic carbon and back again to the organic realm. Most of the carbon in rivers originates as atmospheric CO_2 and either cycles back to the atmosphere or is finally buried, mostly as organic matter, along with marine sediments. In the majority of the world's large rivers, CO_2 exhibits a significant supersaturation, thus indicating a net flux of CO_2 towards the atmosphere (i.e., through riverine outgassing). Clearly, nutrients and carbon are essential participants in the riverine biogeochemical scenario (e.g., Degens et al. 1991).

Carbon in water bodies is usually reported as total organic carbon (TOC), which is the sum of particulate organic carbon (POC), and dissolved organic carbon (DOC). The separation of TOC into its components is accomplished by filtration through glass-fiber filters.

DOC is organic material from plants and animals broken down into such a small size that it is "dissolved" into water. Some DOC molecules have a recognizable chemical structure that can easily be defined as "labile" (such as carbohydrates, fatty acids, and proteins), often created within the river from animal feces or decomposition of organic tissue (i.e., autochthonous carbon). Most molecules, however, have no readily identifiable structure and are lumped under the term humic or fulvic substances, often yellow to dark brown in color. DOC imported from outside a river or lake is created by the decomposition of organic debris that has fallen around or in the water (i.e., allochthonous carbon). As it happens with DOC, POC can be autochthonous or allochthonous, and each fraction can be labile (i.e., available for biologic consumption) or refractory in nature. Perdue and Ritchie (2005) published a comprehensive overview on organic matter in freshwaters.

On the basis of a set of measurements performed between 1981 and 1984 within the framework of the SCOPE-UNEP International Carbon Project, it was possible to establish for the Paraná a TOC mean transport rate of 8–9 Tg y^{-1}, of which ~7 Tg y^{-1} (~85%) correspond to DOC and ~1 Tg y^{-1}(~15%) to POC. This sampling period included the extraordinary ENSO-triggered flood of 1982–1983 (Depetris and Kempe 1993). Earlier studies showed that during the exceptional flooding the river exported a higher proportion of DOC than during the normal hydrological functioning (Depetris and Kempe 1993). These findings were corroborated for the lower reach by Villar and Bonetto (2000) for the 1997–1998 ENSO flooding event.

The dynamics and sources of carbon in the Paraná River were treated in several papers published during the last decade (e.g., Depetris and Kempe 1990, 1993; Villar and Bonetto 2000; Depetris 2005). In this section, we will review what is currently known about both aspects.

6.6.1 Organic Matter Sources

Paraná's POC is, in general, dominantly autochthonous (i.e., the outcome of riverine biological production). This is confirmed by several biogeochemical variables determined in the Paraná, among which we can mention $\delta^{13}C$ in POC, which fluctuated between –20 and –25 per mil (Depetris 2005). Dissolved and particulate amino acids and carbohydrates (Depetris and Kempe 1993) showed that during the customary functioning, the relatively clear waters of the Upper Paraná mostly transported a relatively organic-rich TSS load, with POC contents of 4–10% of noticeably labile nature (20–40% of POC). Carbohydrate analyses further suggested a dominantly autochthonous nature (i.e., phytoplankton) for the organic load, with mannose:xylose ratios of

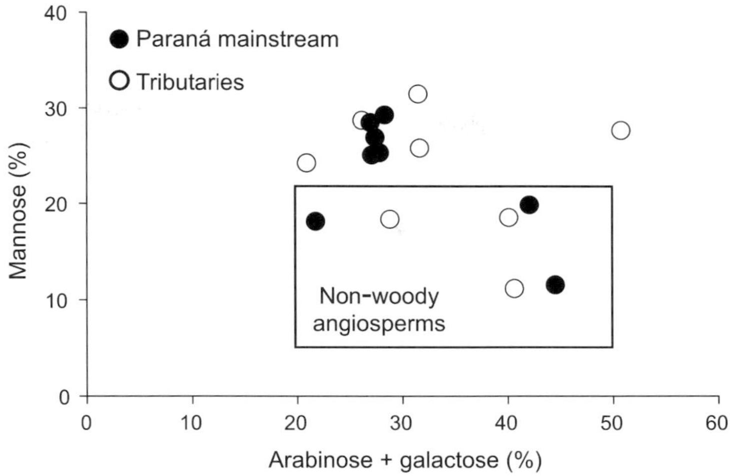

Fig. 6.16 Differentiation of non-woody angiosperms debris in Paraná's POC. Sugars are calculated on a glucose-free basis (Killops and Killops 1994)

1.8–2.6. Figure 6.16, however, also shows that POC receives a complementary contribution from non-woody angiosperms (i.e., grasses and deciduous trees), discernible in both, the mainstream and tributaries.

In contrast, the Paraguay River, which is the most significant TSS and POC supplier in the system, appears to deliver a more refractory organic load (the labile fraction is ~18% of POC) with a more significant contribution from soil-derived materials and bacteria than the Upper Paraná. The middle reach and its well-developed floodplain appear to contribute labile POC and also material that has undergone significant biological degradation, as suggested by the relative abundances of ribose (~21%) and arabinose + galactose (~19%) (Depetris and Kempe 1993).

The C:N ratio in POC can also be used to characterize the origin of organic debris. In general, ratios >20 can be taken as indicators of vascular plan material, those between 15 and 8 seem to indicate eroded soil humus, and lower values (i.e., <8) represent planktonic sources. Figure 6.17 shows the variation of POC:PN in the Paraná during 1981–1984, as a function of discharge. Data was separated into decreasing and increasing discharges. Clearly, ponds and streams were flushed out during descending water levels (i.e., discharges) and hence delivered debris derived from higher plants (POC:PN > 20) as well as plankton (POC:PN < 8). Increasingly high discharges (as during the 1982–1983 ENSO event) mostly transported soil-derived materials. Villar and Bonetto (2000) also highlighted the significance of floodplain lotic environments in the overall carbon and nutrient budgets.

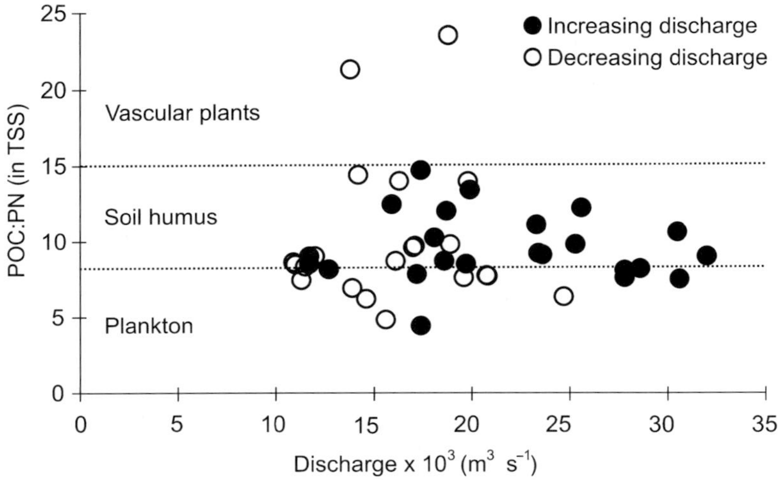

Fig. 6.17 POC:PN ratio variability as a function of discharge in the Middle Paraná River mainstream. *Dotted lines* separate likely organic sources. Data are split into increasing and decreasing discharges

6.6.2 Dynamics

Dissolved organic carbon (DOC) was the only biogeochemical variable whose mass transport rate increased significantly during the 1982–1983 ENSO flood with both, high (>20 mg L^{-1}) and low (<5 mg L^{-1}) concentrations occurring during very high discharges (>22,000 m^3 s^{-1}) (Fig. 6.18). Such oscillatory behavior was probably determined by the contrasting nature of different DOC sources, which became operative during different stages of the flood (e.g., nearby ponds, which are seasonally inundated, or far-off ponds that are only subjected to exceptional flooding).

Figure 6.19a shows the variation of POC both, in concentration and as a percent of TSS. The specialized literature shows that increasing TSS concentrations are normally accompanied by a relative decrease of POC. This seems to be the case in the Paraná at the beginning (i.e., 1981–1982) and the end of the ENSO-triggered event. During the period of high waters (i.e., 1982–1983) there is a significant covariation in both parameters.

Phosphorous and silicon are elements that mediate between the biotic and non biotic realms. P-PO_4^{3-} concentration markedly oscillated during low waters, probably as a result of variable contributions from the flood valley's lotic and lentic domains, but, as expected, maintained low concentrations throughout the flooding event due to biological consumption and dilution (Fig. 6.19b). Villar and Bonetto (2000) found a similar pattern in the Lower Paraná during the 1997–1998 ENSO flooding, although the mass transport rate of total phosphorous and soluble reactive phosphorous showed a net

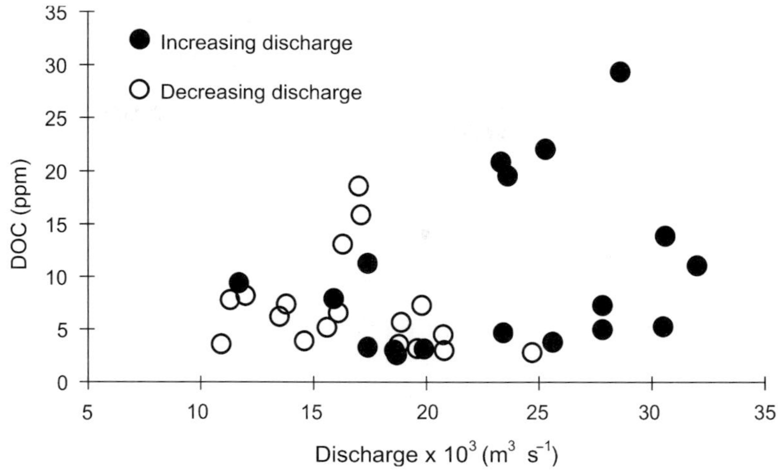

Fig. 6.18 Scatterplot for Middle Paraná River dissolved organic carbon (DOC) and river discharge. Data separated into increasing and decreasing discharges

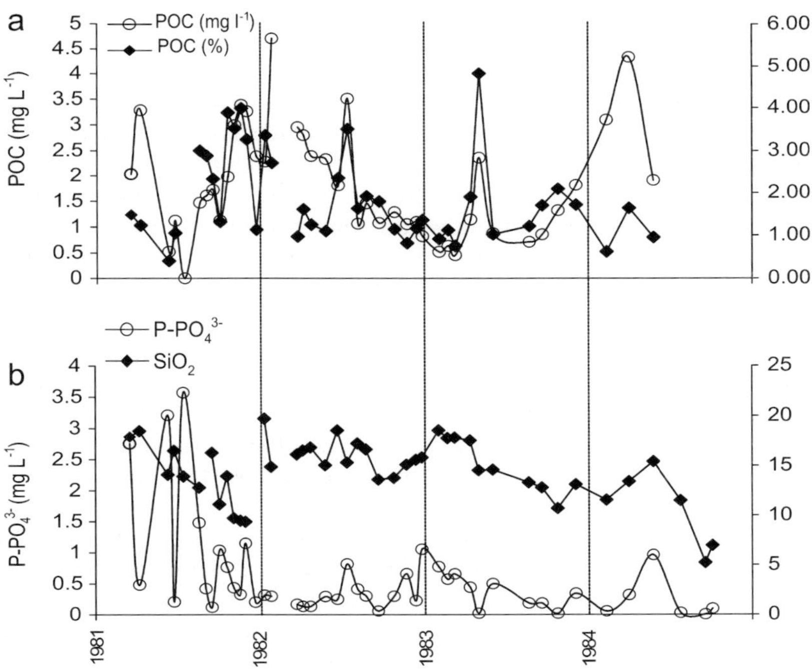

Fig. 6.19 Biogeochemical data collected in the Paraná River mainstream (at Paraná, ~600 km upstream from mouth) during the SCOPE-UNEP International Carbon Project. **a** Absolute and relative (i.e., % of TSS) POC concentrations. **b** Dissolved $P\text{-}PO_4^{3-}$ and SiO_2

increase during the high waters season. In contrast, SiO_2 is usually free from a hydrological control; in the Middle Paraná, decreasing concentrations seem to be associated with the transfer to the mainstream of silicon-depleted waters, which have been affected by the biological consumption of freshwater sponges and diatoms in ponds and ox-bow lakes (Fig. 6.19b).

6.7 Conclusions

The foremost aspect that arises in the integral investigation of the Paraná River is that the prevailing climatic system has a significant teleconnection with the Equatorial Pacific, which results in exceptional flooding that periodically occurs in coherence with the El Niño events. The hydrological feature that immediately follows in scientific importance is that the analysis of the historical discharge record shows a statistically significant increase in discharge that is discernible, according to several authors, since the early 1970s. Further, trend analysis shows that this increase in mean discharge is mostly happening during the seasonal low-water period (i.e., southern winter months). Therefore, the Paraná River is showing clear signs of the impact of change on its hydrology that, in turn, currently affects the biogeochemical dynamics of the whole fluvial system, and probably more so in the foreseeable future.

The two main upper course tributaries, the Upper Paraná and the Paraguay Rivers exhibit differing dissolved chemical signals. The latter has a TDS mean concentration that is roughly twice as high as the one measured in the former. The two rivers meet near the city of Corrientes (~1,200 km upstream from the Paraná mouth) and persist separated (i.e., the western margin has a higher TDS concentration than the eastern one) for many kilometers. The uneven chemical cross section, however, is preserved throughout the river's middle reach, not necessarily because the Middle Paraná maintains the original chemical signatures of the tributaries but, more likely, due to the significant dissolved solids supply (via surface or ground waters) that originates in the extensive sediment-mantled plain that frames the Paraná on its western side.

All the different approaches that are usually employed to evaluate de degree of weathering reached by soils and sediments (e.g., CIA, ICV, alpha indexes) show that the Paraná's suspended load is the result of moderately leached inorganic debris. Most Paraná River sediments are supplied by the Bermejo River system, which is clearly subjected to a weathering-limited denudation regime (i.e., due to tectonic disequilibrium, the rate of soil physical removal is faster than soil formation) and debris is supplied to the fluvial system by exposed friable sediment beds at a very high rate. The resulting geochemical data leads to the conclusion that Paraná River sediments, like in other drainage systems with Andean headwaters, are mostly recycled material (i.e., sediments that have passed more than once through the exogenous cycle).

The Paraná River delivers freshwaters and particulate matter that ultimately find their way to South America's Atlantic coastal zone. Following the previously mentioned findings, the geochemical signature exhibited by its sediments, however, does not correspond to a passive continental margin (such as South America's Atlantic edge) and bears the clear signature of an active continental margin. Although it may be anticipated by the extraordinary sediment yield (e.g., over 1,000 t km^{-2} y^{-1}) of the Bermejo's upper catchments, it is clear then that the weathering products of the extensive Serra Geral Formation (i.e., tholeiites) play a lesser role in the overall Paraná sediment budget. Clearly, most sediments originating at or close to the Andean backbone, preserve an intermediate geochemical signature that is discernible in the Paraná's suspended load.

Finally, closing this bio-geochemical image for the Paraná River, we can summarize that (a) although most transported organic matter is autochthonous, debris originating in grasses and higher plants is sometimes discernible in the POC load; (b) debris with a high C:N ratio (e.g., >20) is exported from the flood valley during exceptional flooding, such as ENSO-triggered floods; (c) most of the Middle Paraná's organic load is supplied by the Paraguay River, which delivers a more refractory POC than the Upper Paraná River; and (d) several biogeochemical parameters indicate that the system's behavior is different during the ascending or descending limbs of the hydrograph.

Acknowledgements. We wish to acknowledge the financial assistance received from Argentina's CONICET (PIP 5947) and FONCYT (PICT 25594) during the preparation of this manuscript.

References

Araújo LM, França AB, Potter PE (1999) Hydrogeology of the Mercolsul aquifer system in the Paraná and Chaco-Paraná Basins, South America, and comparison with the Navajo-Nugget aquifer system, USA. Hydrogeol J 7:317–336

Bathia MR (1983) Plate tectonics and geochemical composition of sandstones. J Geol 91:611–627

Bathia MR, Crook KAW (1986) Trace elements characteristics of graywackes and tectonic setting discrimination of sedimentary basins. Contrib Miner Petrol 92:181–193

Bonetto C, Villar C, de Cabo L, Vaithiyanathan P (1998) Hydrochemistry of a large floodplain river. Verh Int Ver Limnol 26:899–902

Boulanger JP, Leloup J, Penalba O, Rusticucci M, Lafon F, Vargas W (2005) Observed precipitation in the Paraná-Plata hydrological basin: long-term trends, extreme conditions and ENSO teleconnections. Climate Dynam 24:393–413

Collischonn W, Tucci CEM, Clarke RT (2001) Further evidence of changes in the hydrological regime of the River Paraguay: part of a wider phenomenon of climate change. J Hydrol 245:218–238

Cox R, Lowe DR, Cullers RL (1995) The influence of sediments recycling and basement composition on evolution of mudrocks chemistry in the southwestern United States. Geochim Cosmochim Acta 59:2919–2940

Cullers RL, Barrett T, Carlson R, Robinson B (1987) Rare-earth elements and mineralogic changes in Holocene soil and stream sediment: a case study in the Wet Mountains, Colorado, USA. Chem Geol 63:275–297

Cullers RL (1994) The chemical signature of source rocks in size fractions of Holocene stream sediments derived from metamorphic rocks in the Wet Mountains region, Colorado, USA. Chem Geol 113:327–343

de Assis Janasi V (2002) Elemental and Sr-Nd isotope geochemistry of two Neoproterozoic mangerite suites in SE Brazil: implications for the origin of the mangerite-charnockite-granite series. Precamb Res 119:301–327

Degens ET, Kempe S, Richey JE (eds) (1991) Biogeochemistry of major world rivers. SCOPE 42. Wiley, Chichester

Depetris PJ (1976) Hydrochemistry of the Paraná River. Limnol Oceanogr 21:736–739

Depetris PJ, Kempe S (1990) The impact of the El Niño event on the Paraná River, its discharge and carbon transport. Paleogeogr Palaeoclimatol Palaeoecol (Global and Planetary Change Section) 89:239–244

Depetris PJ, Kempe S (1993) Carbon dynamics and sources in the Paraná River. Limnol Oceanogr 387:382–395

Depetris PJ, Gaiero DM (1998) Water-surface slope, total suspended sediment and particulate organic carbon variability in the Paraná River during extreme flooding. Naturwissenschaften 85:26–28

Depetris PJ, Probst JL, Pasquini AI, Gaiero DM (2003) The geochemical characteristics of the Paraná River suspended sediment load: an initial assessment. Hydrolog Process 17:1267–1277

Depetris PJ (2005) Revisiting biogeochemical aspects of the Paraná River. In: Heathwaite L, Webb B, Rosenberry D, Weaver D, Hayashi M (eds) Dynamics and biogeochemistry of river corridors and wetlands. IAHS Publ. 294. International Association of Hydrological Sciences, Wallingford, pp 159–166

Dore MHI (2005) Climate change and changes in global precipitation patterns: What do we know? Environment International 31:1167–1181

Drago E, Vassallo M (1980) Campaña limnológica (Keratela I) en el río Paraná medio: características físicas y químicas del río y ambientes leníticos asociados. Ecol Argentina 4:45–54

Drago EC, Amsler ML (1988) Suspended sediment at a cross section of the Middle Paraná River: concentration, granulometry and influence of the main tributaries. In: Bordas MP, Walling DE (eds) Sediment budgets. IAHS Publ. 174. International Association of Hydrological Sciences, Wallingford, pp 381–396

Drago EC, Amsler ML (1998) Bed sediment characteristics in the Paraná and Paraguay Rivers. Water International 23:174–183

Drever JI (ed) (2005) Surface and ground water, weathering, and soils. Treatise on Geochemistry 5. Elsevier, Amsterdam

Gaillardet J, Dupré B, Allègre CJ (1999) Geochemistry of large river suspended sediments: silicate weathering or recycling tracer? Geochim Cosmochim Acta 63:4037–4051

Gaillardet J, Viers J, Dupré B (2005) Trace elements in river waters. In: Drever JI (ed) Surface and ground water, weathering, and soils, vol 5. Elsevier, Amsterdam, pp 225–272

García NO, Mechoso CR (2005) Variability in the discharge of South American rivers and in climate. Hydrolog Sci J 50:459–477

Genta JL, Perez-Irigarren G, Mechoso CR (1998) A recent increasing trend in the streamflow of rivers in Southeastern South America. J Climate 11:2858–2862

Goldstein SJ, Jacobsen SB (1988) Rare earth elements in river waters. Earth Planet Sci Lett 89:35–47

Hannigan RE, Sholkovitz ER (2001) The development of middle rare earth element enrichments in freshwaters: weathering of phosphate minerals. Chem Geol 175:495–508

Harnois L (1988) The CIW index: a new chemical index of weathering. Sediment Geol 55:319–322

Henry F, Probst JL, Thouron D, Depetris PJ, Garçon V (1996) Nd-Sr isotopic compositions of dissolved and particulate material transported by the Paraná and Uruguay rivers during high (December 1993) and low (September 1994) waters periods. Sci Géol Bull 49:89–100

Herron MM (1988) Geochemical classification of terrigenous sands and shales from core or log data. J Sediment Petrol 58:820–829

Hirsch RM, Slack JR, Smith RA (1982) Techniques of trend analysis for monthly water quality data. Water Resour 20:107–121
Hofmann AW (1988) Chemical differentiation of the earth: the relationship between mantle, continental crust and oceanic crust. Earth Planet Sci Lett 90:297–314
Iacumin M, De Min A, Piccirillo EM, Bellieni G (2003) Source mantle heterogeneity and its role in the genesis of Late Archean-Proterozoic (2.7–1.0 Ga) and Mesozoic (200 and 130 Ma) tholeiitic magmatism in the South American Platform. Earth Sci Rev 62:365–397
Iriondo MH (1972) Mapa geomorfológico de la llanura aluvial del río Paraná desde Helvecia hasta San Nicolás, República Argentina. Revista de la Asociación Geológica Argentina 27:155–160
Iriondo MH (1988) A comparison between the Amazon and Paraná River systems. In: Degens E, Kempe S, Naidu S (eds) Transport of carbon and minerals in major worlds rivers, lakes and estuaries, vol 66. Mitt Geol Palaönt Inst Univ Hamburg, Hamburg, pp 77–92
Kendall MG (1975) Rank correlation methods. Griffin, London
Killops SD, Killops VJ (1994) An introduction to organic geochemistry. Longman Scientific & Technical, Essex
Labat D, Ronchail J, Guyot JL (2005) Recent advances in wavelet analyses: part 2- Amazon, Paraná, Orinoco and Congo discharges time scale variability. J Hydrol 314:289–311
Livingstone DA (1963) Chemical composition of rivers and lakes. Data of geochemistry. US Geol Surv Prof Paper 440 G:G1–G64
Lucassen F, Becchio R, Harmon R, asemann S, Franz G, Trumbull R, Wilke HG, Romeer RL, Dulski P (2001) Composition and density model of the continental crust at an active continental margin—the Central Andes between 21° and 27°S. Tectonophysics 341:195–223
Maglianesi R (1973) Principales características químicas y físicas de las aguas del Alto Paraná y Paraguay inferior. Physis B 32 85:185–197
McLennan SM (1989) Rare earth elements in sedimentary rocks: influence of provenance and sedimentary processes. In: Liping DR, McKay GA (eds) Geochemistry and mineralogy of rare earth elements. Rev Mineral 21:169–200
McLennan SM, Taylor SR, McCulloch MT, Maynard JB (1990) Geochemical and Nd-Sr isotopic composition of deep-sea turbidites: crustal evolution and plate tectonic associations. Geochim Cosmochim Acta 54:2015–2050
McLennan SM, Bock B, Hemming SR, Hurowitz JA, Lev SM, McDaniel DK (1993) The roles of provenance and sedimentary processes in the geochemistry of sedimentary rocks. In: Lentz DR (ed) Geochemistry of sediments and sedimentary rocks: evolutionary considerations to mineral deposits-forming environments, vol. Geo Text 4. Geol Assoc Canada, Alberta, pp 7–38
McLennan SM (1993) Weathering and global denudation. J Geol 101:295–303
Mann HB (1945) Nonparametric tests against trend. Econometrica 13:245–259
Meybeck M (2005) Global occurrence of major elements in rivers. In: Drever JI (ed) Surface and ground water, weathering, and soils, vol 5. Elsevier, Amsterdam, pp 207–223
Nakken M (1999) Wavelet analysis of rainfall-runoff variability isolating climatic from anthropogenic patterns. Environ Model Softw 14:283–295
Nesbitt HW, Young GM (1982) Early Proterozoic climates and plate motions inferred from major elements chemistry of lutites. Nature 299:715–717
Nesbitt HW, Young GM, McLennan SM, Keays RR (1996) Effects of chemical weathering and sorting on the petrogenesis of siliciclastic sediments, with implications for provenance studies. J Geol 104:525–542
Orfeo O, Stevaux J (2002) Hydraulic and morphological characteristics of middle and upper reaches of the Paraná River (Argentina and Brazil). Geomorphology 44:309–322
Pasquini AI, Depetris PJ, Gaiero DM, Probst JL (2005) Material sources, chemical weathering and physical denudation in the Chubut River basin (Patagonia, Argentina): implications for Andean Rivers. J Geol 113:451–469
Pasquini AI, Depetris PJ (2007) Discharge trends and flow dynamics of South American rivers draining the southern Atlantic seaboard: An overview. J Hydrol 333:385–399

Perdue EM, Ritchie JD (2005) Dissolved organic matter in freshwaters. In: Drever JI (ed) Surface and ground water, weathering, and soils, vol 5. Elsevier, Amsterdam, pp 273–318
Potter PE (1997) The Mesozoic and Cenozoic paleodrainage of South America: a natural history. J South Am Earth Sci 10:331–344
Potter PE, Hamblin WK (2006) Big rivers worldwide. Brigham Young University Geology Studies, Provo
Ravenga C, Murray S, Abramovitz J, Hammond A (1998) Watersheds of the world: ecological value and vulnerability. World Resources Institute, Washington DC
Robertson AW, Mechoso CR (1998) Interannual and decadal cycles in river flows of Southeastern South America. J Climate 11:2570–2581
Robertson AW, Mechoso CR (2000) Interannual and interdecadal variability of the South Atlantic Convergence Zone. J Climate 11:2947–2957
Roser BP, Korsch RJ (1986) Determination of tectonic setting of sandstones-mudstone suites using SiO_2 content and K_2O/Na_2O ratio. J Geol 94:635–650
Sallet R, Moritz R, Fontignie D (2005) The use of vein fluorite as probe for paleofluid REE and Sr-Nd isotope geochemistry: the Santa Catarina fluorite district, Southern Brazil. Chem Geol 223:227–248
Sholkovitz ER (1995) The aquatic chemistry of rare earth elements in rivers and estuaries. Aquat Geochem 1:1–34
Stallard RF, Edmond JM (1983) Geochemistry of the Amazon 2. The influence of geology and weathering environment on the dissolved load. J Geophys Res 88:9617–9688
Stumm W, Morgan JJ (1996) Aquatic chemistry. Chemical equilibria and rates in natural waters. Wiley InterScience, New York
Taylor SR, McLennan SM (1985) The continental crust: its composition and evolution. Oxford, Blackwell
Thorne CR (2002) Geomorphic analysis of large alluvial rivers. Geomorphology 44:203–219
Tossini L (1959) Sistema hidrográfico y cuenca del Río de la Plata. Anales de la Sociedad Científica Argentina 167: 41–64
Villar C, de Cabo L, Vaithiyanathan P, Bonetto C (1999a) Pore water N and P concentration in a floodplain marsh of the Lower Paraná River. Hydrobiologia 392:65–71
Villar CA, Stripeikis J, D́Huicque , Tudino M, Troccoli O, Bonetto C (1999b) Cd, Cu and Zn concentrations in sediments and the invasive bivalves *Limnoperma fortunei* and *Corbicula fluminea* at the Río de la Plata basin, Argentina. Hydrobiologia 416:41–49
Villar CA, Bonetto C (2000) Chemistry and nutrient concentrations of the Lower Paraná River and its floodplain marshes during extreme flooding. Arch Hydrobiol 148:461–479
Wang H, Fu R (2004) Influence of cross-Andes flow on the South American low-level jet. J Climate 17:1247–1262

Part II
Plant Life

7 Phytoplankton

Yolanda Zalocar de Domitrovic[1], M. Devercelli[2], and M.O. García de Emiliani[2]

7.1 Introduction

Phytoplankton composition and population density are mainly controlled by water discharge, turbidity, and temperature in most large rivers, as mentioned in early studies (Welch 1952; Hynes 1970). These rivers show a similar phytoplankton community structure characterized by the dominance of centric diatoms, sub- or co-dominance of other taxonomic groups, especially coccal green algae, and lots of sporadic species (Rojo et al. 1994). Although the environmental conditions in turbid and deep river channels are not favorable for (eu-or meroplanktonic) algal growth, fluvial retentivity and fluid exchange seem to be the mechanisms that explain the potamoplankton maintenance in lowland rivers. Thus, the phytoplankton content of the main flow may be augmented by temporary recruitment of individuals from storage zones within the main river channel, side-arms or intermediate lakes (Reynolds and Descy 1996).

Concerning the Paraná River, as well as other major South American fluvial systems (Amazon, Orinoco), the numerous and diverse water bodies located in the extensive floodplain probably play a bigger role as storage zones for fluvial algae than those within the mainstream, judging by the ratio between floodplain and mainstream areas (Bonetto and Wais 1990; Lewis et al. 1995; Junk 1996). In these fluvial systems, the annual flood is the major driving force for river–floodplain interactions, including algal exchanges (Junk et al. 1989).

The studies carried out on the phytoplankton community of the Middle Paraná River system are summarized in some reviews (García de Emiliani 1990; Bonetto and Wais 1990). Other information was analyzed in unpublished papers or published in local journals in Spanish.

Consequently, this synthesis includes the whole data of the phytoplankton community, covering hydrologically normal annual periods and anomalous hydrological cycles, induced by the El Niño Southern Oscillation (ENSO). Most of the information is based on the main channel, side channels and floodplain lakes of the upstream stretch (near Corrientes city) and the downstream stretch (near Paraná city) of the Middle Paraná River.

[1]Centro de Ecología Aplicada del Litoral (CONICET), Corrientes, Argentina,
e-mail: yzalocar@arnet.com.ar
[2]Instituto Nacional de Limnología (CONICET) J. Maciá 1933, 3016 Santo Tomé, Argentina

M.H. Iriondo, J.C. Paggi, and M.J. Parma (Eds.)
The Middle Paraná River: Limnology of a Subtropical Wetland

7.2 Phytoplankton Distribution Along the Main River and at Cross Channel

Differences in biotic and abiotic variables between the left and right banks are observed in the upstream stretch of the Middle Paraná River main channel and in the downstream confluence of the High Paraná and the Paraguay Rivers. These differences are because the Paraguay River's water runs by the right bank and the High Paraná River's water runs by the left bank. The Bermejo River (principal tributary of the Lower Paraguay River) gives to the Middle Paraná River more than 60% of the suspended inorganic load (Drago and Amsler 1988). In consequence, at Corrientes city (30 km downstream of confluence), water transparency and dissolved oxygen are higher in the left bank than in the right bank, whereas the inverse situation occurs with suspended solids and conductivity.

In this sector (near Corrientes City), phytoplankton density and composition show significant differences in a transversal section. Phytoplankton density is higher on the left bank than on the opposite bank (Table 7.1). This asymmetrical distribution remains constant during the entire annual hydrological cycle and continues for about 400 km downstream of the Confluence (Corrientes City to Esquina) (Fig. 7.1). The lowest density value and the greatest transversal differences are associated with the flood period of the Bermejo River. Intermediate and variable values along the section are observed in the center of the main channel in function of the dynamics of the own community and the relative influence of the flow in both rivers.

In the lower stretch (near Paraná City), that shows homogeneous physical and chemical characteristics in a transversal section, no significant differences were found in phytoplankton density and structure between both banks (Table 7.1). Although turbulence is enough to maintain a relative homogeneity, some variations are observed at the banks in relation to the center of the channel, especially on the right bank, in periods of alluvial plain drainage.

The main channel shows typical potamoplankton, characterized by low density and concentration of chlorophyll *a*, high diversity and specific richness. Bacillariophyceae (50%, mostly represented by *Aulacoseira granulata* and its bioforms) dominate; in the upper stretch, there is a sub-dominance of Chlorophyceae, followed by Cyanobacteria (coming from the High Paraná River) (Fig. 7.2), whereas in the lower stretch, Chlorophyceae and Cryptophyceae share alternatively the sub-dominance. These changes downstream would be explained by the increasing dilution of the contributions by both rivers and the gradual influence of the alluvial plain, since the water residence time in the main channel (5–20 days), profound and turbid, is not favorable for the development of its populations. The predominance corresponds to (acclimatable) R-strategist species, adapted to variations in turbulence and light intensity (*Aulacoseira granulata*), and, in a lower number, to associations of a few quantitatively important C- or R-C- strategist species (Chlorococcales and Cryptomonadales). The constancy

Table 7.1 Mean values of some environmental water characteristics and phytoplankton attributes along the Middle Paraná River main-channel, during the sampling period (annual), and low- and high-water phases, with the number of subsurface samples (*n*) monthly obtained. *LB* left bank (*RB* right bank), *C* center, *Bac* Bacillariophyceae, *Chl* Chlorophyceae, *Cry* Cryptophyceae

Sampling sites	Secchi disk (cm)	Conduct. (µS cm^{-1})	Density (ind. ml^{-1})	Dominant-Subdominant (%)	Sampling period
1. CORRIENTES: LB (RB) - 27°27′S-58°51′W					
annual $n = 13$	52 (17)	51.3 (124)	1,140 (4319)	Bac: 48 (70)- Chl: 38 (20)	March 81–March 82
low-water $n = 10$	61 (19)	51.5 (126)	1,379 (454)	Bac: 47 (72)- Chl: 34 (22)	
high-water $n = 3$	21 (7)	50 (123)	344 (352)	Bac: 55 (75)- Chl: 27 (20)	
a. El Niño period $n = 9$	8 (6)	50 (60)	74 (66)	Bac: 33 (30)- Chl: 9 (8)	Dec 82–July 83
2. BELLA VISTA: LB (RB) - 28°30′S-59°04′W					
annual $n = 13$	35 (18)	63 (99)	621 (354)	Bac: 49 (68)- Chl: 27 (20)	March 81–March 82
low-water $n = 10$	41 (229)	65 (100)	758 (403)	Bac: 53 (65)- Chl: 27 (21)	
high-water $n = 3$	12 (6)	56 (93)	169 (193)	Bac: 62 (77)- Chl: 22 (15)	
3. GOYA: LB (RB) - 29°04′S-59°14′W					
annual $n = 13$	27 (21)	71 (96)	575 (375)	Bac: 55 (64)- Chl: 33 (21)	March 81–March 82
low-water $n = 10$	32 (24)	74 (96)	703 (444)	Bac: 53 (2)- Chl: 29 (22)	
high-water $n = 3$	11 (9)	60 (97)	229 (145)	Bac: 74 (79)- Chl: 7 (5)	
4. ESQUINA: LB (RB) - 29°59′S-59°35′W					
annual $n = 12$	23 (17)	75 (97)	395 (283)	Bac: 66 (65)- Chl: 21 (21)	March 81–March 82
low-water $n = 9$	26 (19)	76 (99)	466 (323)	Bac: 66 (62)- Chl: 20 (23)	
high-water $n = 3$	16 (11)	70 (90)	183 (164)	Bac: 64 (79)- Chl: 24 (9)	
5. PARANÁ: C - 31°31″S-60°16″W					
annual $n = 15$	23	94	810	Bac: 72- Chl: 12	June 84–March 86

(*Continued*)

Table 7.1 Mean values of some environmental water characteristics and phytoplankton attributes along the Middle Paraná River main-channel, during the sampling period (annual), and low-and high-water phases, with the number of subsurface samples (*n*) monthly obtained. *LB* left bank (*RB* right bank), *C* center, *Bac* Bacillariophyceae, *Chl* Chlorophyceae, *Cry* Cryptophyceae—Continued

Sampling sites	Secchi disk (cm)	Conduct. (μS cm^{-1})	Density (ind. ml^{-1})	Dominant-Subdominant (%)	Sampling period
low-water $n=5$	20	103	1,095	Bac: 79- Chl: 7	
high-water $n=10$	26	88	667	Bac: 68- Chl: 15	
6. PARANÁ : C - 31°42′S-60°29′W					
annual $n=14$	27	88	516	Bac: 72- Chl: 11	Nov 77–Dec 78
low-water $n=12$	28	90	565	Bac: 72- Chl: 12	
high-water $n=2$	17	75	224	Bac: 67- Chl: 15	
a. El Niño period $n=14$	36	66	194	Cry: 57- Bac: 30	Nov 97–Dec 00
b. La Niña period $n=12$	22	99	655	Cry: 38- Bac: 26	Nov 99–Dec 00
7. PARANÁ: C - 31°44′S-60°40′W					
annual $n=11$	21	91	457	Bac: 60- Cry: 18	March 81–June 82
low-water $n=5$	22		688	Bac: 67- Cry/Chl: 12	
high-water $n=6$	20	91	265	Bac: 44- Cry: 30	

References: *1–4* Bonetto et al. (1982), Zalocar de Domitrovic (unpublished); *1a* Zalocar de Domitrovic and Maidana (1997); *5*: Devercelli (2000); *5*, *7*: García de Emiliani (unpublished); *6*: Perotti de Jorda (1984), García de Emiliani (1990); *6a* García de Emiliani and Devercelli (2003, unplublished); *6b* Devercelli (2006)

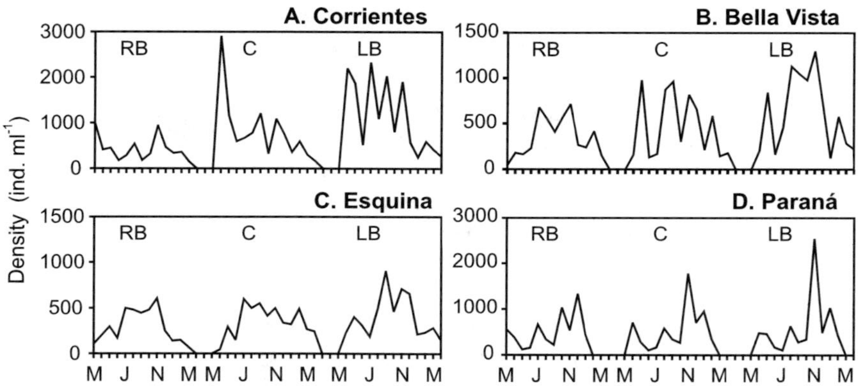

Fig. 7.1 Phytoplankton density in the Middle Paraná River main channel at downstream cross sections (*RB* right bank, *C* center, *LB* left bank, at Corrientes (**a**), Bella Vista (**b**), Esquina (**c**) (March 1981–March 1982), and Paraná (**d**) (March 1978–January 1979)

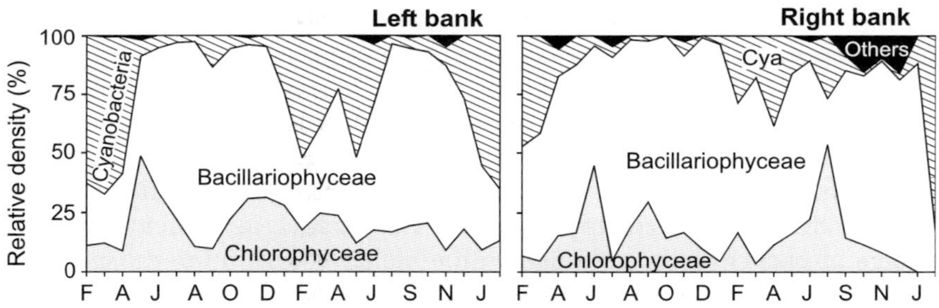

Fig. 7.2 Relative density of the main taxonomic phytoplankton groups in the left and right banks of the Middle Paraná River main channel near Corrientes (February 1978–January 1980)

of a few abundant species with the same life strategy is a common feature in large rivers, where there are strong selective influences for algal growth.

Mean annual biomass (between 0.1 and 1.2 $mm^3\ l^{-1}$) is similar throughout the stretch, mainly due to the high biovolume of *Aulacoseira*. Rojo et al. (1994) pointed out high maximum biomass values (between 0.06 and 25 $mm^3\ l^{-1}$) for other rivers of the world, in relation to that registered in the Paraná River until now. Near Corrientes City, the diversity and specific richness also show differences between the two banks. The diversity index oscillates between 2.9 and 4.1 bits $ind.^{-1}$ at the left bank, and between 1.1 and 3.7 bits $ind.^{-1}$ at the right bank. The mean value for the lower stretch (near Paraná City) is 3.2 bits $ind.^{-1}$.

7.3 Temporary Variations during Normal Hydrological Cycles and Extraordinary Periods in the Main Channel

Temporary variations in phytoplankton structure are more or less constant due to the persistence of potamoplanktonic species adapted to develop under restrictive conditions in large rivers. However, fluctuations in the hydrological cycle produce changes in the structure and abundance of the community. The dominance of centric Bacillariophyceae, and the sub-dominance of Chlorococcales or Cryptophyceae, coincides with what was observed for other large rivers (Rojo et al. 1994; Reynolds and Descy 1996), except for the sub-dominance of Cryptophyceae, characteristic of the Middle Paraná River last stretch.

Density is frequently inversely related to flux by effects of dilution and physical changes produced by this factor (in the velocity, turbulence, water residence time and degree of connectivity with the alluvial plain). Moreover, the coupling of hydrosedimentological pulse and thermal cycle with flow are important combinations that act on phytoplankton development.

In low waters of normal hydrological cycles there is a higher density of Bacillariophyceae, with a high proportion of *Aulacoseira granulata* (and its bioforms) in relation to *A. distans* and *Cyclotella*. During high waters, density of Chlorococcales increases (Fig. 7.3). The participation of these (invasive) C-strategists would be associated to their inoculation from other habitats, since they are small species with a high reproductive rate but that require a high radiation and nutrients to develop. Cryptophyceae present a low percentage of participation, although with a higher contribution during higher flows. Their variations are due to several species of *Cryptomonas*, in high waters, and to *Rhodomonas minuta,* in low waters.

During extraordinary hydrological fluctuations, the flow does not always predict the changes in phytoplankton density through a direct relationship. This is due to the fact that high flow episodes, where plankton is transported at a high velocity, and low flow episodes, where restrictions occur (lack of inoculum, loss of organisms by sedimentation), act by limiting phytoplankton development (Reynolds 2000). The physical changes in the system and the effects of temperature and sediment transport on the community increase or decrease as a result of the extreme flows.

During El Niño, phytoplankton is homogeneous throughout the transversal section (Table 7.1). During La Niña, in the lower stretch, differences in density are observed between the center of the river and the banks. This would be due to a higher proliferation of small flagellates and Cyanobacteria at riverbank sites where the hydraulic retention time increases.

During the ENSO phases, phytoplankton presents a similar specific composition than in normal periods, but shows a higher richness due to the increase in the number of sporadic species. Density (Table 7.1) and algal

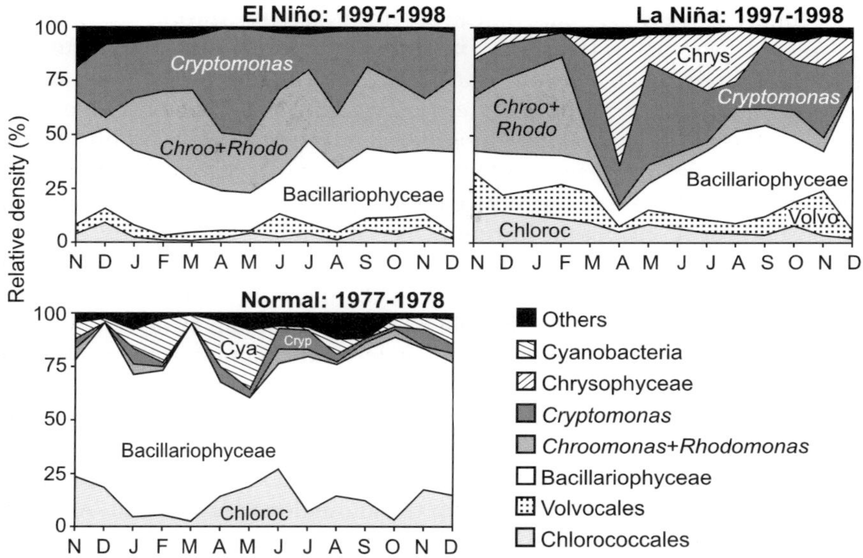

Fig. 7.3 Relative density of the main taxonomic phytoplankton groups in the Middle Paraná River main channel (central site), near Paraná City, during "normal" (**a**) and anomalous hydrological periods (**b**, **c**)

mean transport are lower during El Niño and higher during La Niña, compared with the normal period (3,890, 9,030, and 6,340 $\times 10^9$ ind. s^{-1}, respectively). However, when analyzing temporal variations, the highest density is observed during mid waters, when the exchange between the main channel and the alluvial plain is permanent.

A distinctive characteristic of the two periods is the higher development of unicellular Volvocales and Cryptophyceae (Fig. 7.3). Their high density during the flood, at the beginning of the pulse, could be due to their incorporation from alluvial plain environments where they develop efficiently. After a prolonged high intensity flooding or an isolation period, the plain does not act as recruitment site of organisms any more, and its maintenance depends on the life strategies they have to survive. The high reproduction rate of these flagellated forms allows them to counteract the dilution effect, and their high area/volume relationship contributes to decrease losses by sedimentation. During La Niña, unicellular Volvocales (C-strategists) are favored by the better conditions of transparency and hydraulic stability. The dominance of Cryptophyceae shows the capacity of these species to exploit and adapt to varied environmental conditions (C-R-S-strategists). Moreover, the alternative nutritional strategy they have could constitute a competitive advantage in relation to other species (Jones 2000), although not much is known on heterotrophic feeding mechanisms in rivers.

Bacillariophyceae decrease markedly their density during periods with hydrological anomalies (Fig. 7.3). They are sub-dominant during El Niño and numerically not much relevant during La Niña. In the latter, small centric diatoms (smaller than 7 μm) and *Skeletonema* spp. predominate. The low density of *Aulacoseira* is explained by sedimentation processes in both periods and by retention of organisms by topographic obstacles and/or vegetation when water displacement is produced in the plain during the flood. However, the absence of *A. granulata* during peaks of floods and droughts, its lower density after prolonged hydrological anomalies, and its constant presence in low flow side channels indicate that it needs the input from other environments to develop in the main channel (García de Emiliani 1990).

Considering that El Niño and La Niña provoke changes in the proportion of species characteristic of normal hydrological cycles and not their substitution by others, it is possible to think that the community can be re-structured rapidly when the event ends, due to its structural resilience (Reynolds 1997). However, variations in the structure and, particularly, the scarce participation of *Aulacoseira* persist, even when the anomalous periods end. An explanation could be that the magnitude and duration of extreme flows govern the intensity of physical changes and, consequently, the duration of changes in the community structure. The selection would operate favoring C-strategists and species adapted to live in heterogeneous conditions (C-R-S-strategists), and not favoring the typical R-strategists that predominate during normal hydrological conditions. However, the information is not enough to evaluate if anomalous hydrological phases affect temporarily the density and structure of phytoplankton or if they act as a disturbance, altering the community during a more prolonged time lapse (Sparks et al. 1990).

7.4 Temporary Changes in Side Channels of Different Size and Connectivity Degree

Side channels in the lower section of the Middle Paraná River cover a wide range of sizes, but because of their differences in flow they can be divided into small- and medium-size. The physical and chemical characteristics show wider ranges of variation than in the main channel (Table 7.2). Nutrients (as in the main channel) are not limiting for phytoplankton development.

As was previously observed, the turbid plain rivers do not offer appropriate environmental conditions for growth of potamophytoplankton, and the origin and surviving mechanisms of eu or meroplanktonic algal populations remain in doubt. The algal storage riverbank zones in the own channel and the incorporation of populations to the main flux when the water level varies explain the maintenance of phytoplankton in European rivers with a strong anthropogenic impact (Reynolds and Descy 1996).

Table 7.2 Mean values of some environmental water characteristics and phytoplankton attributes in medium- and small-side channels of the Middle Paraná River, during the sampling period (annual), and low- and high-water phases, with the number of subsurface samples (*n*) monthly obtained. *Bac* Bacillariophyceae, *Cry* Cryptophyceae, *Cya* Cyanobacteria, *wd* without data

Side-channels	Q ($m^3 s^{-1}$)	Conduct. ($\mu S\ cm^{-1}$)	Density (ind. ml^{-1})	Dominant-subdominant (%)	Sampling period
Small-size					
1. CORRENTOSO - 31°40′58″S-60°42′36″W					
a. normal period					
annual *n*=17	77	wd	574	Bac: 57- Cry: 17	April 74–Nov 75
low-water *n*=15	66	wd	611	Bac: 59- Cry: 13	
high-water *n*=2	164	wd	295	Bac: 46- Cry: 44	
b. normal period					
annual *n*=17	102	166	874	Bac: 72- Cry: 15	Feb 87–July 88
low-water *n*=15	84	191	1,155	Bac: 77- Cry: 11	
high-water *n*=2	127	143	528	Bac: 64- Cry: 19	
c. El Niño period	279	89	225	Cry: 54- Bac: 29	Nov 97–Dec 98
2. CONCHILLAS - 31°29′25″S-60°20′1″W					
annual *n*=16	80	208	809	Bac: 59- Cry: 22	June 84–March 86
low-water *n*=6		265	1,423	Bac: 83- Cry: 8	
high-water *n*=10		168	441	Bac: 46- Cry: 31	
3. CHIPÁ - 31°27′33″S-60°20′58″W					
annual *n*=17	53	212	918	Bac: 62- Cry: 20	June 84–March 86
low-water *n*=7		271	1,612	Bac: 82- Cry: 8	
high-water *n*=10		172	433	Bac: 49- Cry: 28	

(*Continued*)

Table 7.2 Mean values of some environmental water characteristics and phytoplankton attributes in medium- and small-side channels of the Middle Paraná River, during the sampling period (annual), and low- and high-water phases, with the number of subsurface samples (*n*) monthly obtained. *Bac* Bacillariophyceae, *Cry* Cryptophyceae, *Cya* Cyanobacteria, *wd* without data—Continued

Side-channels	Q ($m^3 s^{-1}$)	Conduct. ($\mu S\ cm^{-1}$)	Density (ind. ml^{-1})	Dominant-subdominant (%)	Sampling period
4. CORDOBÉS - 31°42′51″S-60°45′49″W					
annual n=11	31	Wd	425	Bac: 61- Cry: 17	April 74–April 75
low-water n=9	27	Wd	470	Bac: 64- Cry: 12	
high-water n=2	51	Wd	227	Bac: 46- Bac: 39	
5. COLASTINECITO - 31°31′20″S-60°17′39″W					
annual n=17	30	109	818	Bac: 63- Cry: 17	June 84–March 86
low-water n=7		124	1,107	Bac: 60- Cry: 16	
high-water n=10		99	616	Bac: 65- Cry: 17	
6. MENDIETA - 31°29′39″S-60°19′59″W					
annual n=17	20	150	871	Bac: 61- Cry: 21	June 84–March 86
low-water n=7		166	1,490	Bac: 77- Cry: 11	
high-water n=10		140	438	Bac: 49- Cry: 28	
Medium-size					
7. SANTA FE - 31°40′26″S-60°43′11″W					
annual n=11	1378	wd	512	Bac: 62- Cry: 13	April 74–April 75
low-water n=9	1227	wd	582	Bac: 64- Cya: 12	
high-water n=2	2060	wd	193	Bac: 53- Cry: 23	
8. CORONDA - 31°41′38″S-60°42′36″W					
annual n=10	104	wd	373	Bac: 58- Cry: 24	June 74–April 75

low-water $n=9$		wd	403	Bac: 60- Cry: 20	
high-water $n=1$		wd	103	Cry: 54- Bac: 31	
9. COLASTINÉ - 31°30′3″S-60°20′53″W					
annual $n=10$	1840	142	1,217	Bac: 69- Cry: 16	April 85–March 86
low-water $n=4$		190	2,091	Bac: 70- Cya: 12	
Side-channels	Q (m^3s^{-1})	Conduct. (µS cm^{-1})	Density (ind. ml^{-1})	Dominant-subdominant (%)	Sampling period
high-water $n=6$		109	635	Bac: 68- Cry: 22	
10. SAN JAVIER - 31°28′52″S-60°21′00″W					
annual $n=16$	595	215	922	Bac: 59- Cry: 23	June 84–March 86
low-water $n=6$		277	1,756	Bac: 80- Cry: 9	
high-water $n=10$		171	421	Bac: 46- Cry: 30	

References: 1a, 4, 7, 8: Schiaffino (1977); 1b: Anselmi de Manavella and García de Emiliani (1995), García de Emiliani (1997); 1c: García de Emiliani (unpublished); 2, 3, 5, 6, 9, 10: Devercelli (2000), García de Emiliani (unpublished).

The water bodies of the Paraná's alluvial plain could act as wide zones of fluvial algal storage. The side channels, with a lower dilution power and more distant from the main channel, could receive the higher impact of algal recruitment. The higher exchange of organisms between channels and plain is produced during the main flood pulse, but small fluctuations in the hydrogram also produce exchanges with adjacent zones (Junk et al. 1989; Lewis et al. 1995). The repetition of these mechanisms of temporary loss and reincorporation of larger size populations would explain the maintenance of phytoplankton in large rivers.

Phytoplankton of main and side channels shows a similar structure (Fig. 7.4), with an increasing mean density in side channels as the degree of connection with the main channel decreases. The increasing z_{eu}/z_{max} relationship as depth decreases could be the most favorable environmental change for algal growth, particularly in low waters, when the water residence time is higher.

The specific richness is high, although only a low proportion of species is relatively constant (35%). It increases from the main channel to the small side channels, as well as from low waters to high waters, probably due to the gradual recruitment of species from close storage zones (in the connected channels and lakes) until the most distant and diverse zones. Specific diversity remains high (~3 bits ind.$^{-1}$), with increments in periods of plain drainage (~4), and with less pronounced decreases in side channels (~2.5) than in the main channel (~2), in high waters.

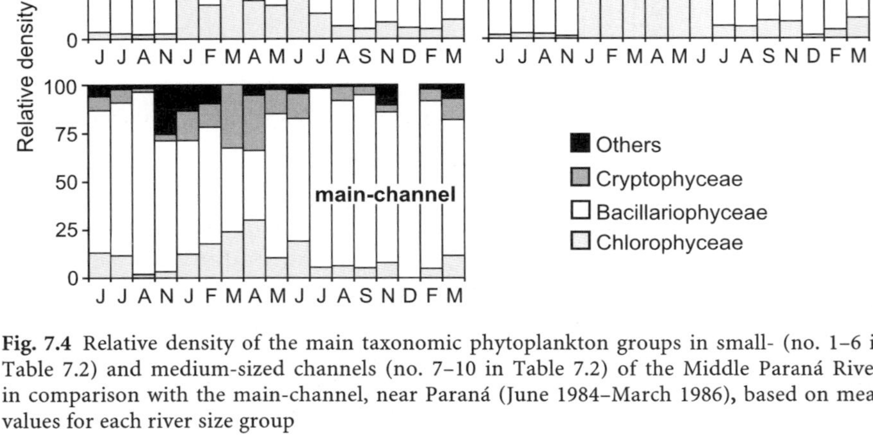

Fig. 7.4 Relative density of the main taxonomic phytoplankton groups in small- (no. 1–6 in Table 7.2) and medium-sized channels (no. 7–10 in Table 7.2) of the Middle Paraná River, in comparison with the main-channel, near Paraná (June 1984–March 1986), based on mean values for each river size group

Annual mean density in side channels does not show significant differences in relation to the main channel. Changes in density between low and high waters are more marked in side channels (where dilution is of approximately 3 times) than in the main channel (dilution lower than 1) (Table 7.2). These differences in the side channels are due to the increase of *Aulacoseira granulata* in low waters and its reduction in high waters (probably by sedimentation and/or retention among alluvial plain macrophytes). Small algae increase their density from low to high waters. This would probably be due to their capacity to counteract the dilution effects with a higher growth and reproduction velocity, as have the Chlorococcales (C-strategists) and Crytophyceae (R-C-strategists).

Algal mean transport in medium channels (Santa Fe = 860×10^9 ind. s^{-1}; Coronda = 610×10^9 ind. s^{-1}) and small channels (Cordobés = 20×10^9 ind. s^{-1}; Correntoso = 40×10^9 ind. s^{-1}) is significantly lower than in the main channel (6,340 ind. s^{-1}, November 1977–December 1978). In the Correntoso River, mean density (Table 7.2) and transport values were higher (82×10^9 ind. s^{-1}) for another normal hydrological period (February 1987–July 1988). When comparing this period to El Niño, changes are more marked in the Correntoso River than in the main channel, decreasing mean phytoplankton density by four times (225 versus 904 ind. ml^{-1}). The flood favors the dominance of Cryptophyceae in both channels, although with the same mean density than in the normal period. Bacillariophyceae (mainly *A. granulata*) decrease in almost one order of magnitude in relation to the normal period (82 versus 695 ind. ml^{-1}).

These results show that side channels offer better conditions for algal growth than the main channel ($>z_{eu}/z_{mix}$), and that turbulence allows to maintain great diatoms in suspension, even in low waters of the smallest rivers. Since phytoplankton density in alluvial plain environments increases with increasing distance from the main channel, side channels would have a greater capacity of species recruitment. Most abundant and typically fluvial species are from the family Bacillariophyceae and orders Cryptomonadales and Chlorococcales.

7.5 Temporary Variations in Floodplain Lakes with Different Water Residence Time and Vegetation

Several studies emphasize the influence of the river hydrological cycle on limnology and aquatic communities of floodplain water bodies. Although floods produce the most important disturbances in these systems, the impact on phytoplankton may be extremely variable as a result of the interaction between river waters and local environmental conditions, such as the lake distance to the main channel, topographic position, type of connection with the river (direct or indirect), lake morphometry, water residence time, and development of macrophyte cover and biomass (García de Emiliani 1997; Zalocar de Domitrovic 2003).

In normal hydrological cycles, the lake limnetic zone shows a low depth (z_{max} = 2 or 3 m) and varied physical and chemical water conditions (Table 7.3). Most limnological differences among these water bodies are explained by variations in the hydrometric level which determine the intensity and duration of isolation and connection phases.

Density, biomass, diversity and specific richness of phytoplankton are higher in isolation periods (higher water residence time) than in connection periods (in relation to the disturbance and dilution produced by the river).

During isolation, there is a higher number of dominant groups in relation to the connection periods (5 versus 2), since there is a wide environmental variability. In general, there is a predominance of Chlorophyceae (lakes no. 3, 4, and 8, Table 7.3) or Bacillariophyceae (lakes no. 6 and 7, Table 7.3), with sub-dominance of Cryptophyceae, except for some environments where Cryptophyceae, Euglenophyceae or Cyanobacteria predominate (lakes no. 1, 2 and 5, Table 7.3, respectively). During connection periods, in contrast, only two groups are dominant: Cryptophyceae (lakes no. 2, 4–7) or Chlorophyceae (lakes no. 1, 3 and 8, Table 7.3). The sub-dominance of Bacillariophyceae, with *Aulacoseira*, is more related to environments with a higher degree of connection with the river (lakes no. 1, 2, 6 and 7, Table 7.3).

In the El Puente and San Nicolás lakes (no. 3 and 4, Table 7.3), located upstream in the Middle Paraná River (near Corrientes City), Chlorophyceae show high density in both hydrological phases (Fig. 7.5), represented by unicellular Volvocales and small Chlorococcales. The former show predominance of *Chlamydomonas* (*Ch. leptobasis* and *Ch. microsphaera*), and the latter show predominance of *Monoraphidium minutum*, *M. contortum*, *Schroederia antillarum*, *Choricystis cylindracea* and *Chlorella vulgaris*. Cryptophyceae, represented by *Cryptomonas* (*C. marssonii* and *C. ovata*), *Chroomonas* and *Rhodomonas*, are dominant and sub-dominant in density. These species, due to their small size, are not dominant in biomass. Only exceptionally both groups can reach an important percentage of biomass when, in addition to their high density, they are associated to other large species, such as *Pyrobotris squarrosa* (e.g., in the San Nicolás Lake), during the connection period, or *Cryptomonas rostratiformis*, during the isolation period (Fig. 7.5).

Euglenophyceae (*Euglena* spp. and *Trachelomonas* spp.), represented in low density, reach important percentages in biomass, particularly during the gradual decrease of waters in the connection period. This group, characteristic of rich organic matter environments, reaches higher biomass in environments farther from the main channel and with a higher macrophyte cover (e.g., El Puente Norte and San Nicolás lakes, Fig. 7.5).

A low density of Cyanobacteria is found in the upper section, in contrast with the lower section, where blooms are more frequent. Only exceptionally they increase in density and biomass with the river input (Fig. 7.5), probably

Table 7.3 Mean values of some environmental water characteristics and phytoplankton attributes in the Middle Paraná River floodplain lakes, during the sampling period (annual) and the lotic isolation and connection phases, with the number of subsurface samples (*n*) monthly (*M*), biweekly (*BW*) or weekly (*W*) obtained in the central or littoral areas (*L*) of the lakes. *Ec Eichhornia crassipes*. *wd* without data. *Chl* Chlorophyceae, *Cry* Cryptophyceae, *Bac* Bacillariophyceae, *Eug* Euglenophyceae, *Cya* Cyanobacteria

Name - Coordinates	Localization, lake form and area	Secchi disk (cm)	Conduct. (µS cm^{-1})	Density (ind. ml^{-1})	Richness (No. of taxa)	Dominant-Subdominant (%)	Sampling period (frequency) (% vegetation-main sp.)
1. LOS PÁJAROS - 27°25′S-58°45′W					113		
annual *n*=17	Nueva Cerrito island	46.5	66	432	87	Chl: 37-Cry: 30	Dec 78–June 80 (M) (10–35% *E. crassipes*)
isolation *n*=4	Round oval	48	70	1,095	50	Cry: 38-Chl: 37	
connection *n*=13	4.6 hm^2	41	64	228	23	Chl: 38-Bac: 24	
2. LLANTA - 27°30′S-58°55′W					152		
annual *n*=7 (7)	Barranqueras island	41 (53)	95 (108)	485 (606)	83	Eug: 29 (Bac:45)-Cry: 28 (Eug:15)	Dec 81–Feb 83 (M) (45% *Ec*+*E. azurea*, 25% *Salvinia* sp.)
isolation *n*=5 (5)	Round oval	42 (57)	97 (120)	609 (864)	29	Eug: 31 (Bac:58)-Chl: 27 (16)	
connection *n*=2 (2)	2.3 hm^2	33 (41)	86 (84)	58 (260)	9	Cry: 48 (Bac:44)-Bac: 22 (Cry:29)	
3. EL PUENTE NORTE - 27°27′45″S-58°51′50″W					201		
annual *n*=17	Chaco province	42.24	135	2,235	43	Chl: 56-Cry: 22	Sept 95–June 96 (M, W) (45–95% *E. crassipes*)
isolation *n*=8	Wide elongated	59.38	151	3,343	44	Chl: 57-Cry: 23	
Connection *n*=9	21 hm^2	27	250	1,251	44	Chl: 54-Cry: 18	
a. El Niño period *n*=17		78.5	134	591	83	Cry: 40-Chl: 33	Dec 82–Jul 83

(*Continued*)

Table 7.3 Mean values of some environmental water characteristics and phytoplankton attributes in the Middle Paraná River floodplain lakes, during the sampling period (annual) and the lotic isolation and connection phases, with the number of subsurface samples (*n*) monthly (*M*), biweekly (*BW*) or weekly (*W*) obtained in the central or littoral areas (*L*) of the lakes. *Ec Eichhornia crassipes. wd* without data. *Chl* Chlorophyceae, *Cry* Cryptophyceae, *Bac* Bacillariophyceae, *Eug* Euglenophyceae, *Cya* Cyanobacteria—Continued

Name - Coordinates	Localization, lake form and area	Secchi disk (cm)	Conduct. (µS cm^{-1})	Density (ind. ml^{-1})	Richness (No. of taxa)	Dominant-Subdominant (%)	Sampling period (frequency) (% vegetation-main sp.)
4. SAN NICOLÁS - 27° 27′40″S-58°51′55″W					185		
annual *n*=17	Chaco province	29.88	247	3,615	50	Chl: 43-Cry: 41	Sept 95–June 96 (M, W) (50–95% *E. crassipes*)
isolation *n*=8	Wide elongated	35.63	195	4,571	59	Chl: 52-Cry: 34	
connection *n*=9	5 hm^2	24.78	294	2,755	42	Cry: 54-Chl: 27	
5. LAS CHILCAS - 27° 27′47″S-58°51′55″W					107		
annual *n*=12 (7)	Sirgadero island	54 (50)	88 (96)	5,190 (7,372)	75 (61)	Chl: 30 (Cya: 43)-Cya: 29 (Chl: 31)	May 73–March 74 (M) (20–30% *E. crassipes*)
isolation *n*=7 (5)	Round oval	36 (15)	93 (99)	8,634 (10,032)	61 (50)	Cya: 44 (60)-Chl: 28 (28)	
connection *n*=5 (2)	18 hm^2	80 (120)	78 (88)	439 (724)	51 (34)	Cry: 36 (Chl: 40)-Chl: 33 (Cry: 39)	
6. LOS MATADORES - 31°41′S-60°42′W					115		
annual *n*=22 (18)	Clucellas island	30	wd	1,434 (1,148)	98 (80)	Cry: 36 (36)-Bac: 34 (25)	April 74–July 75 (BW) (10–20% *E. crassipes*)
isolation *n*=8 (4)	Wide elongated 3.3 hm^2	20	wd	2,889 (2,012)	73 (50)	Bac: 48 (38)-Cry: 16 (Chl: 30)	
connection *n*=14 (14)		29	wd	602 (90)	80 (74)	Cry: 47 (42)-Bac: 26 (22)	

a. El Niño period $n=14$		54	88	429	118	Cry: 53- Bac: 23	Nov 97–Dec 98 (M) (50–0% *Ec*)
b. La Niña period $n=11$		76	206	4,199	101	Chl: 44- Cry: 22	Nov 99–Dec 00 (M) (30–50% *Ec*)
7. CUARENTENA - 31°42′15″S-60°37′30″W							
Main body (side-arms)					135		
annual $n=48$ (48)	Carabajal Island Irregular	35 (56)	100 (109)	643 (874)	76 (83)	Cry: 38 (50)- Bac: 30 (Chl: 18)	March 81–August 82 (M) (10–20% *E. crassipes*)
isolation $n=23$ (23)	274 hm^2	19 (23)	108 (123)	906 (1,322)	63 (64)	Bac: 37 (Cry: 44)- Cry: 28 (Chl: 20)	
connection $n=25$ (25)		49 (84)	92 (98)	397 (514)	60 (75)	Cry: 48 (54)- Bac: 23 (Chl: 17)	
8. EL TIGRE - 31°41′S-60°40′W					218		
annual $n=34$	Los Mellados Island	69	192	6,321	119	Chl: 45- Cry: 18	Nov 86–July 88 (BW) (20–30% *E. crassipes*)
isolation $n=24$	Round oval	53	201	8,032	117	Chl: 48- Cya: 18	
connection $n=10$	3.6 hm^2	108	171	2,213	78	Chl: 37- Cry: 29	
a. El Niño period $n=13$		74	271	548	103	Cry: 48- Chl: 21	Nov 97–Dec 98 (M) (20–90% *Ec*)
b. La Niña period $n=1$		32	271	1,294	42	Chl: 64- Cry: 16	Nov 99 (M) (90–100% *Ec*)

References: *1* Zalocar de Domitrovic (1990); *2* Zalocar de Domitrovic (1992); *3,4* Zalocar de Domitrovic (1993, 2003); *5* Huszar and García de Emiliani (unpublished), García de Emiliani (1990); *6* Perotti de Jorda (1977), García de Emiliani (1980, 1990); *6a*, *6b*, *8* Devercelli and García de Emiliani (unpublished); *7* García de Emiliani (unpublished); *8* García de Emiliani (1993, 1997)

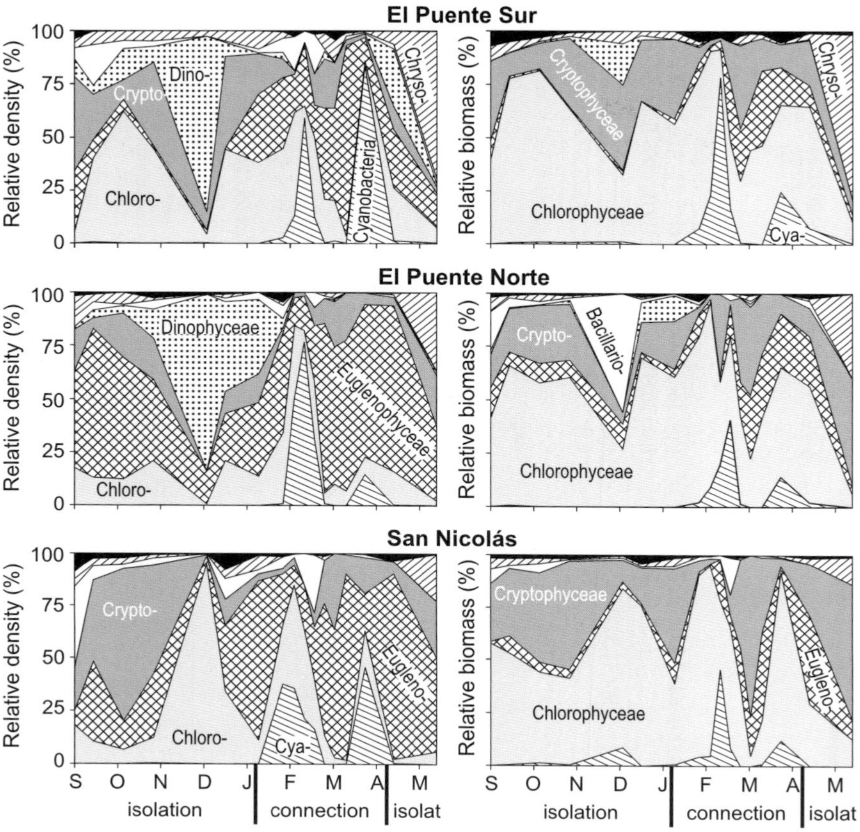

Fig. 7.5 Relative density and biomass of the main taxonomic phytoplankton groups in flood-plain lakes (near Corrientes City) during isolation and connection periods (September 1995–June 1996)

incorporated by horizontal dragging from other environments (represented by *Merismopedia tenuissima*, *Aphanocapsa elachista*, *Microcystis pulverea* and *M. aeruginosa*).

In general, there is a predominance of nanoplanktonic organisms, with sizes between 1 and 20 μm (C-strategists), mainly Chlorophyceae and Cryptophyceae. These organisms fluctuate through time and alternate with larger ones (between 20 and 100 μm), such as Euglenophyceae, Dinophyceae and Chrysophyceae (R- and S-strategists). These groups show a gradual increase during isolation, due to nutrient enrichment, produced during the flood. Growth is produced rapidly and "explosively" (Chrysophyceae), or slowly and gradually, as waters decrease (e.g., dinoflagellates, Euglenophyceae or Cyanobacteria, in the lower section).

Phytoplankton attributes are reduced during river connection periods, showing a relative homogeneity. This is not always due to a dilution effect or to a reduction in transparency, but to profound changes in the physical and chemical conditions during the plain filling, generally adverse for life of organisms, as, for example, for Cryptophyceae, Chrysophyceae and Dinophyceae during the first two weeks of flooding (Zalocar de Domitrovic 1999). In such a situation, only organisms morphologically and/or physiologically adapted to particular environmental conditions can survive: e.g., Cyanobacteria and small flagellated and coccoid species with a high competitive ability (C-strategists).

In addition to nutrient enrichment during flooding, typically fluvial algae that get into river waters (R- or R-C-strategists) reach high density when water residence time and temperature increase. The river transports vegetative, reproductive and/or dispersal cells, acting as inocula in plain environments, where they contribute to biodiversity conservation when they find appropriate environmental conditions to flourish. The concept of "ecological memory" (sensu Padisák 1992), applied to fluvial phytoplankton (Rojo et al. 1994), indicates that rivers interacting with the plain are receptors and distributors of such species memory.

Variations in the hydrogram allow the enrichment of river waters with more abundant algal populations than those introduced previously, so these environments act as storage zones. The gradual increase in algal concentration as the hydrometric level decreases, and the alternation of dominant taxonomic groups are compatible with what was proposed by Huszar and Reynolds (1997) as a "gradual hydroclimatic change", conducted by internal processes, as the plankton analogous of the "gradual climatic change", proposed by Wilson (1994).

Anomalous hydrological states affect phytoplankton structure and density. For example, these phases are compared to normal periods in two lakes, one of them (Los Matadores Lake, N° 6, Table 7.3) directly connected to the Correntoso River (small size, no. 1, Table 7.2), and the other one (El Tigre Lake, no. 8, Table 7.3) indirectly connected to it. These environments, located in the lower section of the Middle Paraná River, illustrate the extreme variations in the degree and type of connection to the river.

In the Los Matadores Lake, phytoplankton composition is similar to that of the river during a normal period, but density is ~3 times higher. In general, Cryptophyceae dominate in high waters and Bacillariophyceae dominate in low waters (Fig. 7.6a). During El Niño, density is notably reduced in relation to the normal period (Table 7.3), and is similar to those of the Correntoso River and the main channel. In both types of environments, the dominant groups are replaced by Cryptophyceae (Fig. 7.6b). The highest density is registered during La Niña, with dominance of Chlorophyceae, and subdominance of Cryptophyceae (Fig 7.6c).

In the El Tigre Lake, phytoplankton density shows marked variations between the connection and isolation periods during a normal period (Fig. 7.6d). During El Niño, mean density decreases in one order of magnitude in relation to the normal period (Table 7.3). Chlorophyceae, the

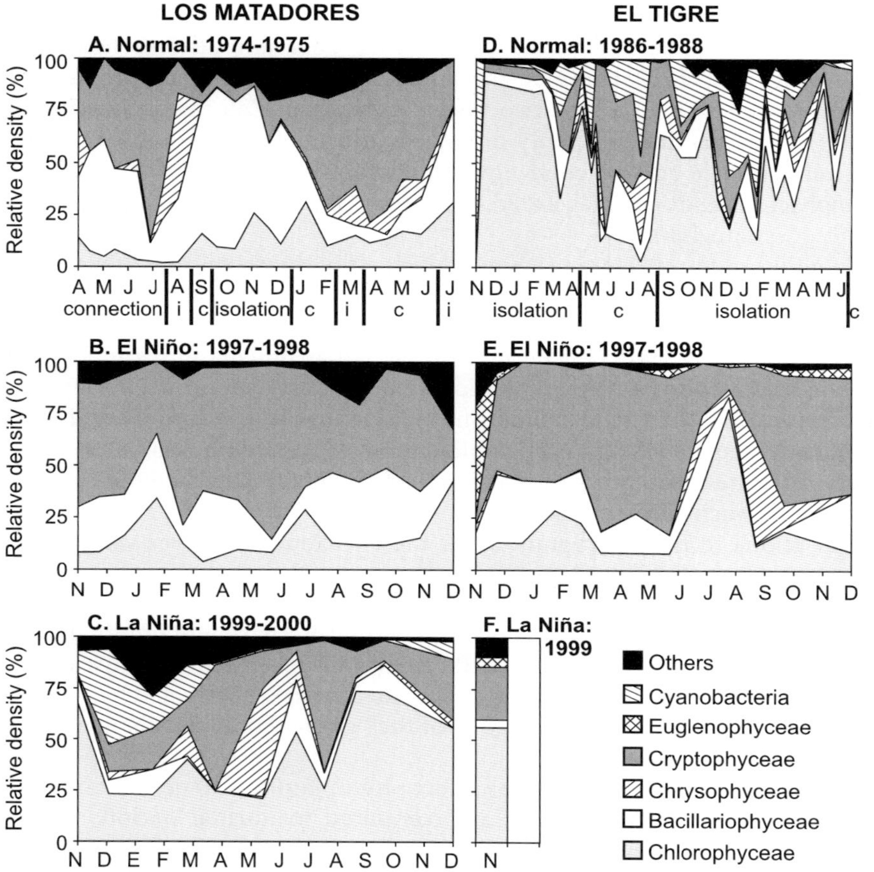

Fig. 7.6 Relative density of the main taxonomic phytoplankton groups in floodplain lakes near Santa Fe and Paraná (Los Matadores and El Tigre lakes) during normal hydrological cycles (**a**, **d**) and anomalous hydrological periods (**b**, **c**, **e**, **f**)

dominant group during the normal period, are replaced by Cryptophyceae (*Cryptomonas* and *Chroomonas*) (Fig. 7.6e). In this situation, phytoplankton density and composition are similar to those of the river. During La Niña, in contrast, density values are intermediate, with dominance of Chlorophyceae, and sub-dominance of Cryptophyceae (Fig. 7.6f). In this period, the most drastic changes in the community are produced due to the scarce water level of this environment, silted by vegetation.

In normal periods, and during the scarce lotic influence, the dominance of Bacillariophyceae in the Los Matadores Lake is due to the high concentrations of centric forms of *Aulacoseira granulata*, *A. distans* and *Cyclotella meneghiniana* (connected floodplain: CFL(6), Table 7.4). The large populations of these species coincide with high temperatures and a

Table 7.4 Main species or algal groups during an annual period (June 84–March 86) in sampling sites of the main-channel (*MCH*), medium (*MSC*: 5–10 in Table 2) and small side-channels (*SSC*: 1–4 in Table 2) and connected and isolated floodplain (*CFL*, *IFL*), selected by their occurrence: + ≥ 30; □ ≥ 50; ■ ≥ 80% of the samples, and mean annual density (mad): × ≥ 50; ○ ≥ 100; ● ≥ 500 org.ml^{-1}. Presence: _

	MCH % - mad	MSC % - mad	SSC % - mad	CFL(6) % - mad	IFL(8) % - mad
Chroococcales	-	-	+ -	+ -	□ - ●
Aphanocapsa delicatissima West & West	-	-	-	+ -	+ - ○
Merismopedia tenuissima Lemm.	-	-	+ -	-	-
Nostocales	□ -	+ -	+ -	+ -	- ●
Anabaena planctonica Brunth.	-	-	-	+ -	- ●
Raphidiopsis spp.	□ -	-	+ -	-	-
Chlorococcales	■-×	■-×	■-×	■ - ○	■ - ●
Chlorella spp.	□ -	□-	□ -	-	-
Crucigenia+Crucigeniella	+ -	+ -	□-	+ -	□- ○
Crucigenia quadrata Morr.	-	-	-	_	_
Crucigenia tetrapedia (Kirchn.) West & West	+ -	+ -	+ -		_
Crucigeniella rectangularis (Näg.) Kom.	-	-	-	-	_
Dictyosphaerium spp.	-	-	-	+ -	+ - ○
Dictyosphaerium ehrenbergianum Näg.	-	-	-	_	_
Dictyosphaerium pulchellum Wood.	-	-	-	_	_
Monoraphidium spp.	■ -	■ -	□ -	□ - ×	■ - ●
Monoraphidium circinale (Nyg.) Nyg.	-	-	-	-	_
Monoraphidium contortum (Thur.) Kom.-Legn.	■ -	■ -	□ -	_	_
Monoraphidium griffithii (Berk.) Kom.-Legn.	-	-	-	_	_
Monoraphidium minutum (Näg.) Kom.-Legn.	-	-	-	_	_
Scenedemus spp.	+ -	+ -	+ -	■ - ×	■ - ●
Scenedesmus ecornis (Ehr.) Chod.	-	-	-	_	_
Scenedesmus intermedius Chod.	-	-	-	_	_
Scenedesmus nanus Chod.	-	-	-	_	_
Scenedesmus quadricauda (Turp.) Bréb.	-	-	-	_	_
Volvocales	+ -	+ -	+ -	□ - ×	□ - ○
Chlamydomonas spp.	+ -	+ -	+ -	_	_
Spermatozopsis exsultans Kors.	-	-	-	-	+ - ×

(Continued)

Table 7.4 Main species or algal groups during an annual period (June 84–March 86) in sampling sites of the main-channel (*MCH*), medium (*MSC*: 5–10 in Table 2) and small side-channels (*SSC*: 1–4 in Table 2) and connected and isolated floodplain (*CFL*, *IFL*), selected by their occurrence: + ≥ 30; □ ≥ 50; ■ ≥ 80% of the samples, and mean annual density (mad): × ≥ 50; ○ ≥ 100; ● ≥ 500 org.ml^{-1}. Presence: _—Continued

	MCH % - mad	MSC % - mad	SSC % - mad	CFL(6) % - mad	IFL(8) % - mad
Ulothricophyceae	-	+ -	-	-	-
Planctonema lauterbornii Schmid.	-	+ -	-	-	-
Chrysophyceae	-	-	-	□ - ×	□ - ○
Synura echinulata Kors.	-	-	-	+ -	_
Mallomonas akrokomos Ruttner + 2 species n.i.	-	-	-	□ -	+ - ×
Bacillariophyceae Centrales	■ -●	■ -●	■ -●	■ - ●	□ - ○
Aulacoseira distans (Ehr.) Sim.	■ -×	□ -○	□ -○	■ - ○	□ - ○
Aulacoseira granulata (Ehr.) Sim. + var.	■ -●	■ -●	■ -●	□ - ○	-
Cyclotella meneghiniana Kütz.	-	□ -×	+ -	□ - ○	+ - ×
Bacillariophyceae Pennales	□ -	□ -	□ -	□ - ○	■ - ○
Navicula spp.	-	+ -	+ -	+ -	+ -
Nitzschia spp.	-	-	-	□ - ○	□ - ○
Nitzschia acicularis (Kütz.) W. Smith	-	-	-	_	_
Nitzschia apiculata (Greg.) Grun.	-	-	-	_	_
Nitzschia reversa W. Smith	-	-	-	_	_
Synedra acus Kütz.	-	+ -	-	-	-
Cryptophyceae	■ -×	■ -○	■ -○	■ - ○	■ - ●
Chroomonas + *Rhodomonas*	□ -	□ -	□ -	■ - ○	□ - ○
Chroomonas acuta Uterm.	-	+ -	+ -	_	_
Rhodomonas minuta Skuja	□ -	□ -	□ -	_	_
Cryptomonas (small size)	-	-	-	■ - ○	■ - ○
Cryptomonas tenuis Pascher	-	-	-	-	_
Cryptomonas pusilla Bachm.	-	-	-	_	_
Cryptomonas (large size)	■ -×	■ -×	-	■ -×	■ -○
Cryptomonas ovata Ehr.	+ -	+ -	-	_	_
Cryptomonas erosa Ehr.	■ -	■ -×	-	_	_
Dinophyceae	+ -	+ -	+ -	-	-
Gymnodinium sp.	-	+ -	+ -	-	-
Euglenophyceae	+ -	+ -	+ -	■ - ×	■ - ○
Euglena spp.	-	-	-	+ -	+ - ×
Euglena acus Ehr.	-	-	-	-	_
Euglena pusilla Playf.	-	-	-	_	_
Strombomonas verrucosa (Daday) Defl.	-	+ -	+ -	-	-

(*Continued*)

Table 7.4 Main species or algal groups during an annual period (June 84–March 86) in sampling sites of the main-channel (*MCH*), medium (*MSC*: 5–10 in Table 2) and small side-channels (*SSC*: 1–4 in Table 2) and connected and isolated floodplain (*CFL*, *IFL*), selected by their occurrence: + ≥ 30; □ ≥ 50; ■ ≥ 80% of the samples, and mean annual density (mad): × ≥ 50; ○ ≥ 100; ● ≥ 500 org.ml^{-1}. Presence: _—Continued

	MCH % - mad	MSC % - mad	SSC % - mad	CFL(6) % - mad	IFL(8) % - mad
Trachelomonas spp.	-	-	-	□ -	□ - ×
Trachelomonas oblonga Lemm.	-	-	-	_	_
Trachelomonas volvocina Ehr.	-	-	-	_	_
Trachelomonas volvocinopsis Swir.	-	-	-	_	_

decrease in nutrient concentrations when water level decreases. In contrast, high biovolume (S-strategists) and nitrogen fixing Cyanobacteria (*Anabaena planctonica*) predominate in the El Tigre Lake (isolated floodplain: IFL(8), Table 7.4).

Constant and/or dominating Chlorophyceae, according to their annual mean in both lakes (CFL (6) and IFL (8), Table 7.4), are represented by small Chlorococcales (C-strategists) of genera *Monoraphidium* and *Scenedesmus*. The sub-dominance corresponds to Cryptophyceae (*Cryptomonas* and *Chroomonas*), during the periods of connection, and to Cyanobacteria (S-strategists, e.g., *Anabaena planctonica*), during isolation. Sporadic high densities of Chrysophyceae, Euglenophyceae and Dinophyceae, related to variations in the hydrometric level and nutrient availability, alternated. The former are associated to an activation of statospores, probably distributed by the river (i.e., *Synura equinulata*, *Mallomonas* spp. and *M. akrokomos*) (Table 7.4).

Specific diversity also increases significantly in the period of relative isolation and decreases to similar values than those from the river when the water input increases, so the annual mean is slightly higher in lakes (3 bits ind.$^{-1}$) than in the river (2.8 bits ind.$^{-1}$).

During the isolation phases, "microsuccessions" are registered, with their own characteristics for each water body, depending on the environmental conditions. Every successional process is interrupted by the flooding, perceived as an intense disturbance by the community, and allowing the input of R-strategist species, especially Cryptophyceae and Bacillariophyceae. During a new period of isolation the community begins a new successional process, which starts with the dominance of C-strategists and pulses of S-strategists, with species belonging to different taxonomic groups, controlled by their particular nutritive requirements. Moreover, lower magnitude hydrological pulses produce the reversion of the successional process to a former state during this period (García de Emiliani 1993).

Extreme high waters states (El Niño) affect the community density and structure. The permanence of species representative of normal hydrological cycles and the rapid recovery of the own characteristics of these environments when the water level decreases induce to think that it is a temporary alteration of the community and cannot be interpreted as a severe disturbance (Sparks et al. 1990). In contrast, in environments that are generally isolated from the main channel, as the case of the El Tigre Lake, the flooding is perceived as an intense disturbance due to the profound changes produced by El Niño, being more severe during La Niña.

The daily rate of change in alluvial plain environments is higher during connection periods (with a strong increase at the beginning of the pulse: maximum 0.33 day^{-1}) than during isolation (lower than 0.1 day^{-1}) (García de Emiliani 1997; Zalocar de Domitrovic 2003), which suggests a great environmental variability (Reynolds 1992) and a higher velocity of biogeochemical processes (Estéves 1988).

Approximately 600 algal taxa were observed in the Middle Paraná alluvial river-plain system during more than 30 years of study records (Zalocar de Domitrovic 2005). In the Middle Paraná River lower section, there is a higher number of species of Chlorophyceae and Chlorococcales. Upstream of the Middle Paraná River, however, where there is a mixed hydrological regime, there is a higher number of species of Euglenophyceae and Chlorococcales, according to the influence of the Paraguay or Paraná rivers, respectively. Rojo et al. (1994) pointed out that Desmidiaceae are typical of tropical environments. However, in the Paraná River (where Chlorococcales predominate qualitatively), they never reach a qualitative dominance as that described, for example, for the Amazon River (Uherkovich 1984) or for the alluvial valley of the Magela Creek, a tributary of the Alligator River in Australia (Filayson et al. 1990).

7.6 Conclusions

In normal hydrological periods, the asymmetry of the phytoplanktonic community between the two banks of the main channel, upstream of the Middle Paraná River, is related to the different physical and chemical water quality of the Paraguay and High Paraná rivers. In the lower section, where mixing of both rivers is completed, there is a relative homogeneity of the community throughout the transversal section.

In the main channel and the alluvial plain, the spatiotemporal changes of the phytoplanktonic community are associated to the hydrosedimentological regime. Regular variations in the hydrological phases, and shorter permanent fluctuations are essential for phytoplankton maintenance, since they allow the lateral exchange with zones of higher algal accumulation in the own channel and with alluvial plain environments.

Between the main channel and the external limit of the alluvial plain there is a complex combination of lenitic and lotic environments with different conditions of flux and water quality. During isolation, these water bodies show clear differences with the main channel due to the higher number of species, diversity, density and biovolume. These environments, temporarily linked to the river flux, act as refuge areas for species storage.

Waters that drain from the upper sections of the river and that run towards plain environments by overflow, find appropriate environmental conditions (scarce horizontal flux, high thermal amplitude, high organic matter content, etc.) that favor a rapid re-organization in the pattern of integration of the phytoplanktonic community. The alluvial plain plays an essential role in maintaining algal populations and incorporating them towards or from the main flux, depending on the hydrological fluctuations and on the occurrence of appropriate environmental conditions for their development.

During normal hydrological periods, the main channel shows typical fluvial phytoplankton, dominated by central Bacillariophyceae (*Aulacoseira granulata*) and, alternatively, with sub-dominance of Chlorococcales, Cryptophyceae or Cyanobacteria.

In anomalous hydrological cycles, there are differences in the abundance and structure of the community in relation to normal hydrological cycles, as a consequence of the ENSO phenomenon, attributed to the loss of normal exchange of organisms between the main flux and the areas with higher water residence time, that function as storage zones of fluvial species.

During El Niño, the main channel maintains a relative homogeneity of the community, although mean density values are significantly lower than those registered in normal hydrological cycles. The dominant and sub-dominant classes change to Cryptophyceae and Bacillariophyceae, respectively. During La Niña, the scarce contribution of large Bacillariophyceae species (*Aulacoseira*) to density demonstrates a profound change in the community structure, probably due to the inaccessibility of inocula from other sites.

In high waters of normal periods and in extraordinary flood and drought periods, the predominance of Cryptophyceae and small unicellular green algae would be related to the capacity of these species to adapt to conditions of variable intensity of mixing and light, maintaining a high reproductive rate, so they would have intermediate features between the (acclimatable-invasive) R-C primary strategists.

During La Niña, the differences between the center of the main channel and the banks in the lower section can be attributed to a higher proliferation of species in the banks and a higher water retention time. In the side channels and environments of the alluvial plain, La Niña provokes a more drastic change in the community, due to the relative hydraulic stability, mainly in the lakes, where the variation in the proportion of species is caused by their adaptations to a great environmental variability, related to a reduction in the water level and the advance of vegetation.

References

Anselmi de Manavella MI, García de Emiliani MO (1995) Composición y dinámica del fitoplancton en una sección transversal del río Correntoso (llanura aluvial del río Paraná). Rev Asoc Cienc Nat Litoral 26:39–54

Bonetto AA, Wais IR (1990) The Paraná River in the framework of modern paradigms of fluvial systems. Acta Limmol Brasil 3:139–172

Bonetto AA, Zalocar de Domitrovic Y, Vallejos ER (1982) Contribución al conocimiento del fitoplancton del Paraná Medio. I. Ecosur 9:189–212

Devercelli M (2000) Efectos de las fases de aguas altas y bajas sobre el fitoplancton de ríos de distinto tamaño (Paraná Medio, Argentina). Tesis de Licenciatura. Universidad Nacional del Litoral, Santa Fe, Argentina, pp 37

Devercelli M (2006) Phytoplankton of the Middle Paraná River during an anomalous hydrological period: a morphological and functional approach. Hydrobiologia 563:465–478

Drago EC, Amsler ML (1988) Suspended sediment at a cross section of the Middle Paraná River: concentration, granulometry and influence of the main tributaries. Int Assoc Sci Hydrol 174:381–396

Estéves FA (1988) Fundamentos da Limnología. Interciência/FINEP, Río de Janeiro, Brazil

Filayson CM, Cowie YD, Bailey BJ (1990) Characteristics of a seasonally flooded freshwater system in monsoonal Australia. In: Whigham DE, Good RE, Kuet J (eds) Wetland ecology and management: case studies. Kluwer Academic Publishers, Dordrecht, pp 141–162

García de Emiliani MO (1980) Fitoplancton de una laguna del valle aluvial del Paraná Medio ("Los Matadores", Santa Fe, Argentina). I. Estructura y distribución en relación a factores ambientales. Ecología 4:127–140

García de Emiliani MO (1990) Phytoplankton ecology of the Middle Paraná river. Acta Limnol Bras 3:391–417

García de Emiliani MO (1993) Seasonal succession of phytoplankton in a lake of the Paraná River floodplain, Argentina. Hydrobiologia 264:101–104

García de Emiliani MO (1997) Effects of water level fluctuations on phytoplankton in a river–floodplain lake system (Paraná River, Argentina). Hydrobiologia 357:1–15

García de Emiliani MO, Devercelli M (2003) Influencia del fenómeno "El Niño" sobre el transporte y la estructura del fitoplancton en el cauce principal del río Paraná medio, Argentina. Bol Soc Argent Bot 38:29–38

Huszar VLM, Reynolds CS (1997) Phytoplankton periodicity and sequences of dominance in an Amazonian floodplain lake (Lago Batata, Pará, Brazil): responses to gradual environmental change. Hydrobiologia 346:169–181

Hynes HBN (1970) The ecology of running waters. Liverpool University Press, Liverpool

Jones RI (2000) Mixotrophy in planktonic protists: an overview. Freshw Biol 45:219–226

Junk WJ (1996) Ecology of floodplains: a challenge for tropical limnology. In: Schiemer F, Boland KT (eds) Perspectives in tropical limnology. SPB Academic Publishing, Amsterdam, pp 255–265

Junk WJ, Bayley PB, Sparks RE (1989) The flood pulse concept in river–floodplain systems. Can Spec Publ Fish Aquat Sci 106:110–127

Lewis WM Jr, Hamilton SK, Saunders JF (1995) Rivers of northern South America. In: Cushing CE, Cummins KW, Minshall GW (eds) River and stream ecosystems. Elsevier, Amsterdam, pp 219–256

Padisák J (1992) Seasonal succession of phytoplankton in a large shallow lake (Balaton, Hungary) a dynamic approach to ecological memory, its possible role and mechanisms. J Ecol 80:217–230

Perotti de Jorda NM (1977) Pigmentos y producción primaria en el Paraná Medio: laguna "Los Matadores" (Santa Fe, Argentina). Physis 36:89–113

Perotti de Jorda NM (1984) Estudios limnológicos de una sección transversal del tramo medio del río Paraná. IX: Biomasa y productividad del fitoplancton. Rev Asoc Cienc Nat Litoral 15:117–133

Reynolds CS (1992) Dynamics, selection and composition of phytoplankton in relation to vertical structure in lakes. Arch Hydrobiol 35:13–31
Reynolds CS (1997) Vegetation processes in the pelagic: a model for ecosystem theory. Excellence in Ecology, 9. Ecology Institute, Oldendorf
Reynolds CS (2000) Hydroecology of river plankton: the role of variability in channel flow. Hydrolog Process 14:3119–3132
Reynolds CS, Descy JP (1996) The production, biomass and structure of phytoplankton in large rivers. Archiv für Hydrobiologie/Supplement 113, Large Rivers 10:161–187
Rojo C, Alvarez Cobelas M, Arauzo M (1994) An elementary, structural analysis of river phytoplankton. Hydrobiologia 289:43–55
Schiaffino M (1977) Fitoplancton del río Paraná. I. Sus variaciones en relación al ciclo hidrológico en cauces secundarios en la llanura aluvial. Physis 36:115–125
Sparks RE, Bayley PB, Kohler SL, Osborne LL (1990) Disturbance and recovery of large floodplain rivers. Environ Manage 14:699–709
Uherkovich G (1984) Phytoplankton. In: Sioli H (ed) The Amazon: limnology and landscape ecology of a mighty tropical river and its basin. Junk Publishers, Dordrecht, pp 295–310
Welch PS (1952) Limnology, 2nd edn. McGraw-Hill, New York
Wilson JB (1994) The intermediate disturbance hypothesis is based on patch dynamics. N Z J Ecol 18:176–181
Zalocar de Domitrovic Y (1990) Efecto de las fluctuaciones del nivel hidrométrico sobre el fitoplancton en tres lagunas isleñas en el área de confluencia de los ríos Paraná y Paraguay. Ecosur 16:13–29
Zalocar de Domitrovic Y (1992) Fitoplancton de ambientes inundables del río Paraná (Argentina). Estudio comparativo entre áreas libres y vegetadas. Rev Hydrobiol trop 25:177–188
Zalocar de Domitrovic Y (1993) Fitoplancton de una laguna vegetada por Eichhornia crassipes en el valle de inundación del río Paraná (Argentina). Amb Subtrop 3:39–67
Zalocar de Domitrovic Y (1999) Estructura y dinámica del fitoplancton en la cuenca del eje potámico Paraguay-Paraná (Argentina). Tesis doctoral, Universidad Nacional de Córdoba, pp 375
Zalocar de Domitrovic Y (2003) Effect of fluctuations in water level on phytoplankton development in three lakes of the Paraná river floodplain (Argentina). Hydrobiologia 510:175–193
Zalocar de Domitrovic Y (2005) Biodiversidad del fitoplancton en el eje fluvial Paraguay-Paraná. In: Aceñolaza FG (ed) Temas de la Biodiversidad del Litoral Fluvial Argentino II. INSUGEO, Miscelánea 14, pp 229–242
Zalocar de Domitrovic Y, Maidana NI (1997) Taxonomic and ecological studies of The Paraná River diatom flora (Argentina). Bibliotheca Diatomologica 34, J Cramer, Berlin

8 Aquatic Macrophytes

Rafael A. Sabattini[1] and Victor H. Lallana

8.1 Introduction

The vegetation of Middle Paraná River flooding environments has been studied from different points of view. In particular, there are numerous studies related to composition, structure and successional dynamics (Neiff 1975, 1979, 1986a; Lewis and Franceschi 1979; Franceschi and Lewis 1979; Sabattini et al. 1983; Aceñolaza et al. 2004). These studies show a great variety of methods, approaches, and working scales, which make the interpretation and delimitation of communities difficult. Few studies approach methodological aspects for these environments (Sabattini and Lallana 1998).

Since 1980, studies on floristic surveys and cover of aquatic, marshy and flooding-land plant communities (levees, sand banks), carried out using aerial pictures and fieldwork, became important in this area due to the perspective of a dam construction in the Middle Paraná River (Bayo et al. 1981; Lallana and Sabattini 1989; Brizuela et al. 1992). The new working tools, i.e., satellite images, multispectral scanning, and color slides, allowed new appreciations of the distribution of plant communities at other scales, from generation and uses of similarity patterns, approaching the landscape description based on homogeneous plant units (Lallana and Sabattini 1989; Brizuela et al. 1992; Sabattini and Lallana 1993; Kandus 1997; Aceñolaza et al. 2005; Marchetti and Aceñolaza 2005; Ginsburg et al. 2005).

A wide revision on the use of these technologies for the delimitation of communities and identification of species was carried out by Lallana (1997), where it was established reflectance and visual patterns for 42 species, being the most studied ones the water hyacinth *Eichhornia crassipes* ("camalote"), *Hydrilla verticillata*, *Potamogeton* spp., the water milfoil *Myriophyllum spicatum* ("cola de zorro"), *Ceratophyllum demersum,* the tall grass *Panicum prionitis* ("paja brava"), the Humboldt's willow *Salix humboldtiana*, the river alder *Tessaria integrifolia*, and the "canutillos": the water paspalum *Paspalum repens*, the elephant panicgrass *Panicum elephantipes,* and the aleman grass *Echinochloa polystachya.*

Facultad de Ciencias Agropecuarias, UNER, C.C. N° 24, E3100WAA Paraná, Entre Ríos, Argentina, [1]e-mail: rsabatti@fca.uner.edu.ar

M.H. Iriondo, J.C. Paggi, and M.J. Parma (Eds.)
The Middle Paraná River: Limnology of a Subtropical Wetland

One of the basic studies of community mapping at a large scale was carried out in 1979 by Masera (unpublished), covering an extensive area of 300 km in length by approximately 31 km in width, including a great part of the Middle Paraná River. In this study, 45% of the total area (9,070 km^2) was found to be totally flooded or with some water, water channels, lagoons and floating aquatic vegetation, and the remaining 55% maintained conditions of variable hydromorphy that in some seasons, were subjected to floods. In these areas, gallery forests grow in the levees and extensive pastures (savannahs or flooding scrublands) cover the low areas, where the dominant species is *P. prionitis*, always maintaining some water on the surface or groundwater very close to it. These pastures occupy 30.7% of the cover, and forests, scrubs (bushes), and forests close to channel islands cover 23.9% of the area.

At another detail level, plant communities from two large channel islands have been identified using Landsat 7 images (Brizuela et al. 1992; Marchetti and Aceñolaza 2005). In the study of Carabajal Island, covering 3,868 ha, two hydrometric situations were compared: high waters (Brizuela et al. 1992) and low waters (Bayo et al. 1981), with 40.8 and 72% of emergent lands for both situations. These hydrometric changes lead to great modifications in the island landscape in relation to environments and plant covers (Table 8.1), significantly modifying the mainland areas, lakes and flooding planes (scrublands). The Chapetón Island, located 30 km north of the Carabajal Island, showed a greater forest area and a lower lake area (Table 8.1).

Since 1960, river hydrometric fluctuations constitute the most important macrofactor to explain richness, distribution, and abundance of organisms (Junk 1970; Bonetto 1975; Junk and Howard-Williams 1984; Lewis et al. 1987), although procedures to relate quantitatively changes in biodiversity with the hydrological variation regime have been tested only recently (Neiff 1999; Casco et al. 2005). On the other hand, as pointed out by Neiff (1996) and Casco (2003), floristic surveys in different hydrological moments change the number and type of species, and only a few can bear both phases (flood and drought).

From a floristic point of view, there are records for the Paraná River and related basins (Iberá marshes and Los Saladillos system). Species richness has

Table 8.1 Percentage of cover for environment and plant communities in the Carabajal and Chapetón Islands according to hydrometric situations

Carabajal Island	Lake	Flood plane	Trees	Floating aquatic plants	Ingrained vegetation	Others
High water	33	15	25.81	14.5	10.2	1.7
Low water	21	43	26.2	16.7	nd	24
Chapetón Island	6	24	45	19		6

nd not determined

been determined by different authors, and even today it remains in revision. There are 277 species for the Upper Paraná River (Neiff 1986b), 75 species for the Middle Paraná River, in a Santa Rosa (province of Santa Fe) - Pueblo Brugo (province of Entre Ríos) transect (Manzi and Fritschy, unpublished), and 94 species in an island in front of the city of Paraná (Peltzer 1998), whereas 259 species were surveyed in the Lower Paraná River (Franceschi et al. 1985). Aceñolaza et al. (2004) pointed out that more than 372 species have been registered in the Pre-Delta National Park (Diamante, province of Entre Ríos), the lower limit of the Paraná River middle section. In the Iberá marshes (province of Corrientes), 126 species were cited (Neiff 1981). A more recent study (Arbo and Tressens 2002) registered 1,516 species of vascular plants, from which hydrophytic and marshy plants were similar to those of the Middle Paraná River. On the other hand, 133 species were cited for the system of Saladillos rivers (province of Santa Fe), showing low similarity with the flora of the Paraná River.

From a phytogeographical point of view, the area of the Middle Paraná River belongs to the Amazon dominion, "Paranense" (i.e., Paraná River) province, and Mixed Rainforests district. In Argentina, it covers the whole province of Misiones, the northeast of the province of Corrientes, and it extends to the south, constituting narrow galleries along the Paraná and Uruguay Rivers and their tributaries (Cabrera 1976). The climax rainforest is the maximum expression of vegetation, characterized by the presence of lianas, epiphytes and ferns. As it advances to the south, the rainforest becomes poor in species and only constitutes fringes along the riverbanks. The rainforest southern limit is found in Punta Lara, in the province of Buenos Aires (Cabrera 1976). Among the tree species of the Middle Paraná River sector, the most representative ones are *S. humboltiana* (the native willow), *Erytrina crista-galli* (the cockspur coral tree or "seibo", national flower), *Myrsine laetevirens* ("canelón"), *Sapium haematospermun* ("curupí"); *Cathornium polyantum* (white "timbó"), *Inga vena* (red "ingá"), *Croton urucurana* ("dragon's blood"), *Nectandra falsifolia* ("laurel"), *T. integrifolia* (river alder) and *Acacia caven* ("espinillo").

8.2 Plant Communities and Succession

Numerous descriptions and longitudinal profiles of the Middle Paraná River have been carried out (Bonetto 1975; Manzi unpublished), but a description or general model of the environments and their main associated communities is proposed with functional purposes and following the criterion by Lugo et al. 1986 (Fig. 8.1). Lotic and lentic environments of the Middle Paraná River show a great taxonomic and ecological diversity of hydrophytes (Burkart 1957; Schulz 1961; Franceschi et al. 1985; Neiff 1986a), constituting an important portion of the system biomass (Bayo et al. 1980; Sabattini et al.

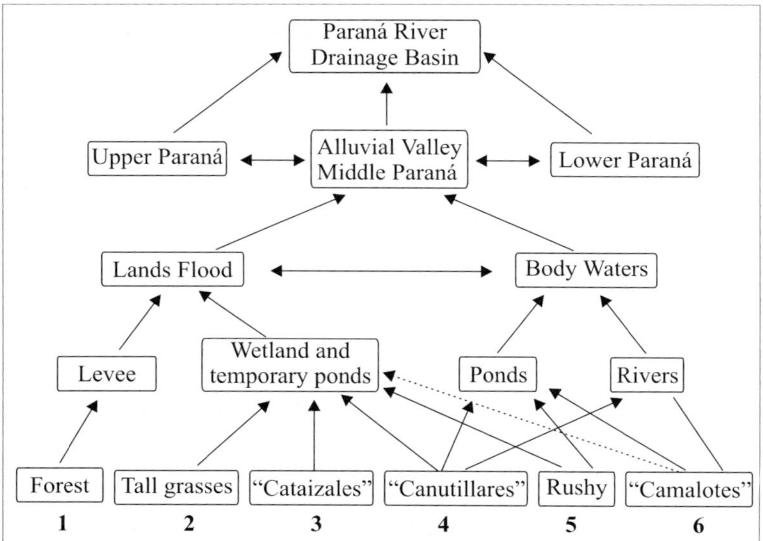

1	*Sapium haematospermun, Cathormion polyantum, Erythrina crista-galli, Nectandra falcifolia, Salix humboltiana, Myrsine lactenvirens, Acacia caven, Tessaria integrifolia, Inga vena*
2	*Panicum prionitis*
3	*Polygonum hydropiperoides, P. ferruginum, P. punctatum*
4	*Panicum elephantipes, Paspalum repens, Echinochloa polystachya*
5	*Schoenoplectum californicus*
6	*Eichhornia crassipes, E. azurea, Pistia stratiotes, Ludwigia peploides, Salvinia herzogii, Enhidra anagallis, Azolla caroliniana, Hydrocotile* spp.

Fig. 8.1 Hierarchical characterization of ecosystems of the Middle Paraná River basin at the main plant community level (1–6) and their relative location in the island environments

1983). According to Burkart (1957), in the Paraná's Delta floodplains, two large geomorphological units are differentiated: a high one (levee), and a low one, characterized physiognomically by the white forest and the scrubland, respectively. This situation, in turn, is found in the Middle Paraná River, but the second area has received other denominations, as flooding plane and accumulation spring (Tur and de Orellana 1970).

Lallana and Sabattini (1986) characterized the environments of the Middle Paraná River, stressing the scrublands (*P. prionitis*), "cataizales" (*Polygonum, ferrugineum, P. acuminatum*), "verdolagales" (*Ludwigia peloides*), "canutillares" (*P. elephantipes, P. repens* and *E. polystachya*) and water hyacinth beds or "camalotales" (*E. crassipes*) as the most conspicuous communities, constituting different areas along a topographic gradient from the levee to the lagoons or old river channels.

From a functional point of view, the hydrosedimentological dynamics of the Paraná River, with its annual periods of high and low waters, is the determining factor of visible and/or quantifying changes in the biotic communities and, particularly, in the aquatic macrophytes. Other authors call the high water period potamophase and the low water period limnophase, and indicate the flood periods as river "pulses"; this theory is widely documented by Neiff (1990, 1996, 1999), and its influence on plant communities has also been modeled (Casco et al. 2005).

These phenomena of periodic or extraordinary recurrence, with periods of decades or greater, provoke the displacement of a great quantity of plant biomass that drifts by the island environments until reaching the main river current and continue their trajectory even until the Río de la Plata.

The dispersing units of aquatic plants play an important role in the island ecosystems of the Middle Paraná River, since they assure life continuity, allow the establishment of new plant populations in the pioneer stages of the riparian succession, and give highly energetic food to numerous fish (Lallana 1990). Hydrocores and phytohydrocores predominate as the dispersing mechanisms in these environments (Lallana 1990). The quantity of seeds (of their own and from other species of the alluvial plain) that can be transported by the aquatic plants among their roots varies according to the environments, type of substrate and season of the year: the range goes from a few seeds per m^2 (100–200) to several thousands (2,000–5,000). The highest seed diversity is found in the roots of *E. crassipes*, followed by *Pistia stratiotes* and *Salvinia* spp. (Lallana 1990).

The recurrent phenomenon of the flood pulses produces a profound renovation in the lentic and lotic water bodies of the floodplain, which implies the beginning, from an initial stage, of new successional processes of riparian vegetation. These processes are very dynamic and shorter than terrestrial ones.

Sabattini and Marta (1988 unpublished), through a phytosociological study carried out in a 5.75-ha lake (El Tigre, province of Santa Fe), between March 1987 and April 1988, evaluated the successional changes of aquatic vegetation in three areas (aquatic, marginal and levee), according to the topographic gradient.

At the beginning of the samplings, the lagoon remained isolated from the river (maximum depth, $Z=0.94$ m), and vegetated in all its sides (650 m), with the exception of a sector (50 m). Forty-five percent of vegetation was found in the water and the rest in the marginal area between the aquatic area and the levee. In the aquatic area, *P. repens*, *Ludwigia peploides* and *E. crassipes* were mostly observed, representing 30% of the total lagoon area.

The dominant vegetation of the marginal area was *Polygonum acuminatum*, *E. polystachya*, *Echinochloa helodes*, *Hymenachne amplexicaulis* and *Panicum rivulare*. The situation in May 1987 ($Z=2.21$ m) is represented in Fig. 8.2.

Afterwards, the water level increased until June ($Z = 3.10$ m), exceeding in 0.50 m the height of the island levee; before this situation, there was only a little area emerging from the levee ($Z = 2.21$ m) and a lower number of species was observed, since some of them could not adapt to the new hydrometric situation. Others, that were also flooded, began a period of decomposition, for example, *P. repens*, *L. peploides* and *Enhydra anagalllis*, which are substrate-rooted species.

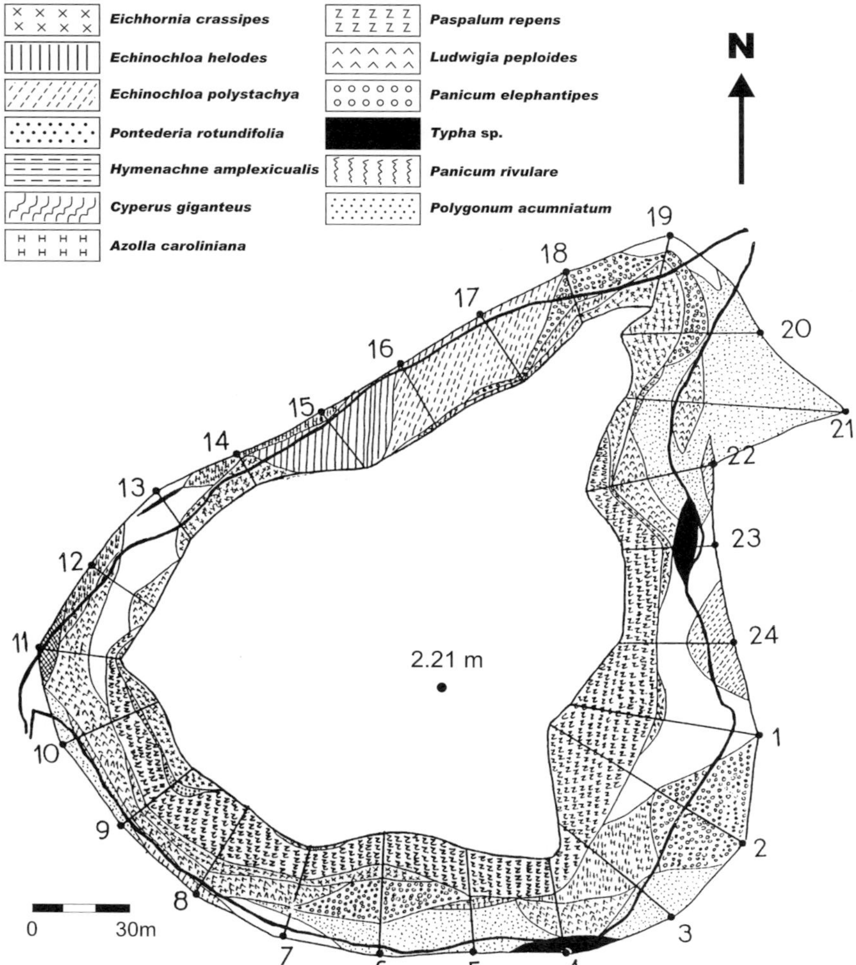

Fig. 8.2 Plant zonation of the El Tigre Lake (province of Santa Fe) in an hydrometric situation of low waters

Since July and until the end of the studied cycle, the water level decreased. Table 8.2 shows the dominant species, according to season and hydrometric level, observed by areas (topographic gradient) in autumn-winter. A differentiation by areas was observed with increasing hydrometric levels; *P. repens*, *E. crassipes* and *L. peploides* predominated from the water mirror to the bank, respectively, with different variants. In turn, there were other less important species (covers <5%), indicated in parentheses in Table 8.2. On the contrary, when the low water began, *L. peploides* dominated in the three areas, together with other free-floating hydrophytes.

Table 8.2 Physiognomic characteristic of aquatic and marsh vegetation of a pond in relation to hydroperiod and topographic gradient, (*z* depth)

	Towards the mirror	Intermediate	Towards the margen
May Z=2.21 m	*Paspalum repens*	*Eichhornia crassipes*	*Ludwigia peploides + Polygonum acuminatum (Salvinia herzogii)*
Mean z (m)	1.55	1.33	0.92
June Z=3.10 m	*P. repens + (S. herzogii)*	*E. crassipes + (L. peploides, P. repens)*	*L. peploides + Azolla caroliniana*
Mean z (m)	2.46	1.90	1.82
July Z=2.92	*P. repens + A. caroliniana*	*E. crassipes +* Lemnaceae	*L. peploides + (Enhidra anagallis, Panicum elephantipes)*
	… … … … … … … … … *(S. herzogii)* … … … … … … … … … …		
		… … … … *(A. caroliniana)* … … … … …. ….	
Mean z (m)	2.05	1.91	1.62
August Z=2.34 m	… … … … … … … … … … .. *(L. peploides)*		
	(A. caroliniana, S. herzogii, Pistia stratiotes, Lemnaceae)		
	P. repens	*L. peploides*	*P. acuminatum + E. crassipes*
Mean z (m)	1.35	1.24	1.15
September	.... *L. peploides + A. caroliniana*		
Z=1.69 m	*P. repens + S. herzogii*	*S. herzogii*	*P. acuminatum + E. crassipes*
Mean z (m)	0.78	0.70	0.42

In May, *P. repens, E. crassipes, L. peploides* and *P. acuminatum* dominated in each area; the last two species dominated in the area towards the bank, with *Salvinia herzogii* as accompanying species. Afterwards, small free-floating plants were retained in the lagoon due to the flood, most of all in the intermediate to marginal area. The area towards the water mirror was dominated by *P. repens*, and the marginal area by *L. peploides* and *Azolla caroliniana*; the intermediate one, in addition to *E. crassipes*, showed physiognomic features of other areas, since *L. peploides* and *P. repens* appeared as accompanying species, whereas *P. acuminatum* remained totally flooded until August.

In July, *P. repens, E. crassipes* and *L. peploides* still dominated, but *S. herzoggi* and *A. caroliniana* were observed as accompanying species in the three areas, the latter in the intermediate to marginal area. In both areas, other macrophytes, as Lemnaceae, *E. anagallis* and *P. elephantipes,* were found as accompanying species.

Since August, *L. peploides* (with low cover) dominated the three areas and small free-floating plants were also observed. The differentiation of each area was observed by the presence of *P. repens, L. peploides* and *P. acuminatum*, together with *E. crassipes*. This species, because of the low water effect, was occupying the marginal position.

In September, the dominance of *L. peploides* and *A. caroliniana* was more conspicuous in the three areas and the differentiation per areas was similar to the previous month, except for the intermediate area, where the high cover of *S. herzogii* delimited this sector, extending towards the water mirror, together with *P. repens*.

We concluded that rooted-plant species showed changes in cover, depending on the topographic position where they were rooted, and on the effect of their growth cycle and the variations in the hydrometric level. On the other hand, the free-floating plants occupied intermediate positions or showed displacements towards the bank or the water mirror, according to the hydrometric level and wind effect.

8.3 Habitat Characteristics and Ecological Factors

The main plant communities of the Middle Paraná River were already pointed out previously; thus, in this section, we include a data analysis of the main environmental physical factors for a particular situation of mid and high waters. Sabattini and Lallana (1998) surveyed more than 40 island environments of the Middle Paraná River, including 20 lagoons, 12 brooks and 11 rivers, with a total of 152 floristic censuses to establish a minimum sampling area for this type of environments. The greatest number of censuses per sampling station occurred in lagoons, together with the greatest specific richness. This might be partly related to the water retention time, the gradual slope in those environments and the shallow depths. In rivers and brooks, richness is low, and rooted plants predominate (e.g., *P. elephantipes* and *Polygonum ferrugineum*) since they can bear their high stream velocities. The development of a sampling method based on transect length and on the presence of new species, might allow the quantification, in a systematic and reliable way, of the aquatic vegetation in flooding areas, establishing a representative sampling size, resulting in three censuses per sampling station for rivers and brooks, and in 11 censuses per sampling station for lagoons.

The ten most frequent species were the following: *S. herzogii* (58.5%); *P. elephantipes* (53%); *A. caroliniana* (49%); *P. repens* (46.5%); *Polygonum punctatum* (46.5%); *P. ferrugineum* (42%); *E. crassipes* (37%); *Mikania* sp. (35%); *Eichhornia azurea* (28%) and *L. peploides* (28%) (Table 8.3). The relative frequency of environments (rivers, brook and lagoons) allows to see the trends of species distribution (Table 8.3). For example, *P. elephantipes* and *P. ferrugineum* are more frequent in lotic environments, *E. crassipes* and

Table 8.3 Total and relative frequency of species per kind of environment

		Relative frequency per environment			
Species	Frequency (%)	Rivers (n=11)	Streams (n=12)	Lake (n = 20)	*Growth form
Salvinia herzogii	58	18.2	58.3	80	FF
Panicum elephantipes	53.5	81.8	58.3	35	E
Azolla caroliniana	49	18.2	41.6	70	FF
Paspalum repens	46.5	18.2	41.6	65	E
Polygonum puntactum	46.5	18.2	41.6	65	E
Polygonum ferrugineum	42	54.5	41.6	35	E
Eichhornia crassipes	37	36.6	50	30	FF
Mikania sp.	35	36.6	25	40	E
Eichhornia azurea	28	0	8.3	55	RF
Ludwigia peploides	28	9	16.6	45	RF
Hydrocotyle sp.	20.9	9	16.6	30	RF
Myriophyllum sp.	18.6	0	0	40	S/E

* *S* submerged; *FF* free floating; *RF* rooted; floating leaves; *E* emergent

Mikania spp. have the same distribution in all environments, and the rest of species increase their presence in lagoons (Table 8.3).

The sampled brooks were generally located along the left bank of the Paraná River. They were characterized by reduced width (<40 m), slow to zero flow velocity, a water depth of no more than 1.5 m at the center, and no dominant vegetation. The alluvial valley rivers were 2–5 m deep, 30–300 m wide, with a mean flow velocity of 0.50 m s^{-1} (Table 8.4).

A great variation in vegetation was noticed in the lagoons according to the topographic gradient, usually showing two co-dominating species; the "cataizal" community (*P. ferrugineum* and *P. punctatum*), the "canutillar" community (*P. elephantipes* and *P. repens*), the "verdolagal" community (*L. peploides*), and other free-floating species (*E. crassipes, A. caroliniana* and *S. herzogii*) were specially important. The physical characteristics of the lagoons (Table 8.4) were the following: variable width, up to 500 m; depth at the center: between 1 and 2 m, and stream velocity: less than 0.40 m s^{-1}.

Eight significant correlations for lagoons and four significant correlations for rivers and brooks (Table 8.5) were observed through the correlation matrix analysis among the physical variables of the environment and vegetation (Table 8.4). In the case of rivers and brooks (Table 8.5), number of censuses by station was positively correlated with species richness, and negatively correlated with vegetation height. On the other hand, height was positively correlated with depth. In the lagoons, the number of censuses by station was positively correlated with species richness and vegetation zone length. Species

Table 8.4 Physical characteristics of environment and vegetation

	RIVERS										
1	Number of sampling station	2	3	4	8	9	10	11	12	19	22
2	Number of censuses	2	2	2	1	1	7	2	5	1	1
3	Specific richness	2	5	5	1	1	14	3	12	2	3
4	Height (m)	1	0.5	0.8	1	0.95	0.37	1	0.41	0.3	0.5
5	Depth (m)	2.55	1	1.49	0.8	1.95	1.32	1.85	1.29	0.68	1.16
6	Vegetated zone length (m)	8	8	4.5	11	11	38	12	30	10	8
7	Width of streams (m)	30	35	30	300	300	78	50	50	1000	1000
8	Z (m) center of streams	2.8	2.18	2.8	3	5	5.06	5.72	4.2	8	8
	STREAMS										
<- - - References - - ->	Number of sampling station	7	18	20	21	26	32	33	36	38	42
	Number of censuses	3	3	1	1	1	2	5	3	2	4
	Specific richness	6	8	7	4	6	4	10	7	3	4
	Height (m)	0.77	0.2	0.25	0.25	0.5	0.9	0.3	0.36	1.03	0.64
	Depth (m)	1.19	0.84	0.8	0.97	0.5	1.42	0.7	1.63	1.54	0.76
	Vegetated zone length (m)	4.5	5	5	5	15	40	32	5	8	13
	Width of streams (m)	150	5	5	5	15	167	56	10	35	58
	Z (m) center of streams	13.9	0.43	0.8	0.97	0.5	3	3.6	2.9	3.95	5.46

	LAKES																		
1	5	6	13	14	15	16	17	23	24	25	28	29	30	31	34	39	40	41	43
2	3	3	9	19	11	9	23	1	9	4	10	2	4	4	8	12	6	6	9
3	5	6	11	15	6	11	17	6	7	9	8	5	6	7	10	14	8	8	11
4	0.75	0.85	0.37	0.36	0.92	0.26	0.35	0.45	0.18	0.57	0.33	0.8	0.3	0.3	0.31	0.17	0.18	0.38	0.2
5	0.61	1.43	1.7	1.57	1.98	1.56	1.16	0.72	1.33	1.51	1.23	1.54	1.34	0.52	0.42	0.77	0.85	0.62	0.74
6	5.5	6.5	43	148	103	65.5	185	20	66.2	74.6	68.1	29	29.6	94	210	120	68	64	57
7	80	60	150	162	478	150	269	40	216	374	168	100	89	114	317	320	100	189	69
8	2.32	1.22	2.3	2.35	3.42	2.25	1.8	0.87	0.83	1.64	2	1.64	1.5	1.1	0.66	0.72	1.18	1.03	1.1

Table 8.5 Significant correlations between pairs of variables (other correlations NS: $P>0.05$). *NCE* number of censuses; *SRI* species richness; *HEV* height of vegetation; *DEP* depth; *VZL* vegetated zone length; WID, width of river, stream or lake

	Variable 1	Variable 2	R	Significance
Rivers and streams	SRI	NCE	.8616	***
	SRI	HEV	–.4968	*
	SRI	VZL	.6958	***
	NCE	VZL	.7236	***
	VZL	WID	.6119	**
	HEV	ZCE	.5031	*
	DEP	ZCE	.6903	**
Lake	NCE	SRI	.8296	***
	HEV	SRI	–.5953	**
	HEV	DEP	.5915	**
	WID	ZCE	.4980	*

Key: *($P<0.05$); **($P<0.01$) and ***($P<0.001$)

richness was negatively correlated with height and positively correlated with depth of the vegetated zone. Height of vegetation was positively correlated with depth of vegetation zone and Z (centre of the water body). Furthermore, there was a positive correlation between depth and Z (centre of the lagoon) and between the vegetated zone length and the lagoon width (Table 8.5).

Therefore, when the number of censuses increases, the species richness for rivers, brooks and lagoons also increases (Table 8.4). Species richness for the analyzed environments showed a negative correlation with vegetation height. This may be explained by the predominance of tall species that would suppress the specific diversity, especially in the case of rivers and brooks (Table 8.5).

One of the visible consequences of all floods and/or low waters of the Paraná River is the drift of quite compact aquatic vegetation masses, constituted by one or more species, known as camalotes, "camalotales" (i.e., floating islands on river) or "embalsados" (i.e., masses of floating water plants) (Fig. 8.3). Their shape and area are very variable, as well as their specific composition. The following species are the most commonly observed constituting "camalotales": *E. crassipes, E azurea, Pontederia rotundifolia,* and these species were observed associated or alone: *Salvinia* spp., *Polygonum* spp., *P. stratiotes* and *Azolla* sp., in a lower amount. Associations of aquatic gramineae ("canutillos") of genera *Echinochloa, Paspalum* and *Panicum* were also observed, usually covering large extensions in lentic environments and secondary rivers, and drifting together with "camalotales" or independently. The dragging of these "camalotales" occurs periodically and is related to periods of river floods, the morphology of water bodies and islands, and the action of climatic agents, mainly winds and precipitations (Boneto 1975; Fernández et al. 1990). The daily fluxes are not constant and fluctuate in a few hours, so it is difficult to establish mean times, since they do not reflect what occurs during the day. Anyway, and to illustrate the situation, minimum and maximum values for 1979/80/81, oscillating between 9 and 2,000 $m^2 h^{-1}$, are presented, with an estimation of the daily mean values of 0.15 ha d^{-1}, and minimum and maximum values of 0.01 and 4.51 ha d^{-1} (Fernández et al. 1990). The periodical displacements of the "camalotales"

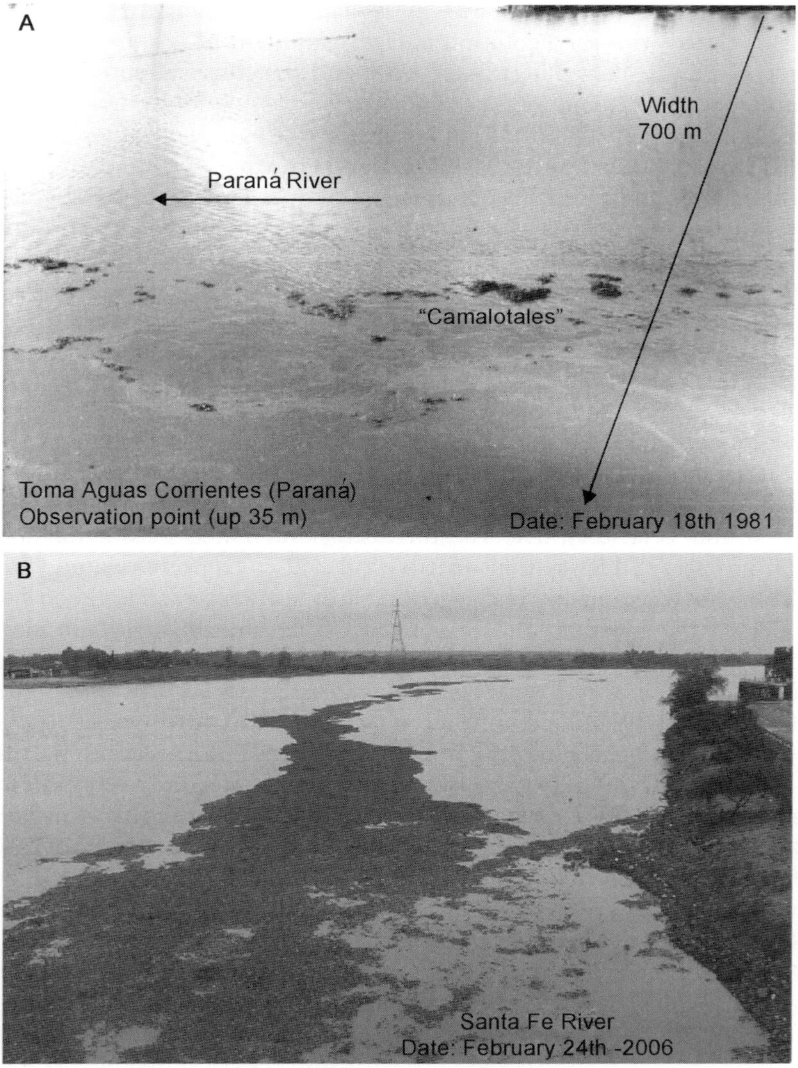

Fig. 8.3 **a** Water hyacinth "camalotales" of *E. crassipes* in the Paraná River (18/02/81) and **b** "camalotales" of *Salvinia* sp. in the Santa Fe River, after a morning wind storm (24/02/06)

and "embalsados" produce navigation problems in the Paraná River and its tributaries, sometimes blocking the passage in low flow brooks and rivers, and even putting small bridges or civil works in danger (Fernández et al. 1990).

8.4 Plant Dynamics and Growth

The Middle Paraná River shows an aquatic vegetation dynamics that is closely related with the evolution and development of islands, where flood pulses and low waters determine shapes and structures adopted by the different areas of an island (levee and flooding areas). Studies carried out in an island near the main channel (San Javier River, province of Santa Fe) are presented to visualize vegetation dynamics and growth.

The island environments present abundant aquatic and marshy vegetation, constituting the basic sustenance of the food chain; thus, it is important to know the cover and productivity of aquatic plant species in a lagoon in relation to hydrometric fluctuations, because this is the main factor conditioning the ecosystem. In a 1.5-ha lagoon near Santa Rosa (province of Santa Fe), biomass and plant cover were evaluated during an annual period (September/1986–August/1987). The depth to the center of the lagoon, the area covered by vegetation and its biomass were measured monthly on three transects perpendicular to the lagoon-side to estimate productivity.

During the first months, the water level decreased from 2.05 to 1.48 m, and then it increased from the beginning of the summer until reaching 4.75 m in winter (Fig. 8.4). The vegetated area fluctuated between 1,154 and 7,693 m^2 ha^{-1}, and its spatial distribution and cover were related with the hydrometric level: the highest vegetated area corresponded to the low water period; on the contrary, the vegetation began to loose structure and was fractioned during the flood, so those fractioned parts drifted as "embalsados" (Fig. 8.4).

The biomass fluctuated between 4.18 and 59.04 t fresh weight ha^{-1}, and between 0.30 and 8.67 t dry weight ha^{-1}. The highest biomass was observed in the following species (60 to 80%): *P. repens, P. stratiotes* and *L. peploides,* and, to a lower extent (10% or lower), in *A. caroliniana, E. anagalis, Limnobium laevigatum, P. elephantipes, H. amplexicaulis, E. azurea, E. crassipes, P. rotundifolia, S. herzogii, Scirpus cubensis* and *Utricularia* spp. Species richness varied between 6 and 21.

The minimum and maximum productivity values were 0.0308 and 0.0817 t dry weight ha^{-1} day^{-1}, and the primary productivity was 0.023 t dry weight ha^{-1} day^{-1}, representing 8.39 t dry weight ha^{-1} $year^{-1}$. There was no correlation ($P<0.95$) between depth (Z) and fresh weight ($r = 0.20$), depth and dry weight ($r = -0.43$), and between Z and the vegetated area ($r = 0.10$). Growth was independent from the hydrometric level (Fig. 8.5), depending on the plant species cycle (spring-summer-autumn), mostly constituted by *P. repens* and *L. peploides,* both rooted to the substrate.

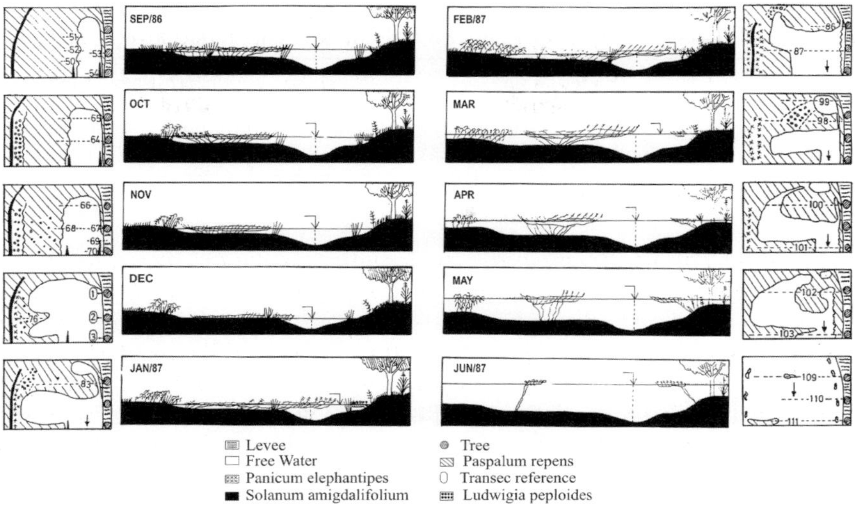

Fig. 8.4 Relative plant cover and plant section in a lake of the Middle Paraná River floodplain (Santa Rosa, province of Santa Fe)

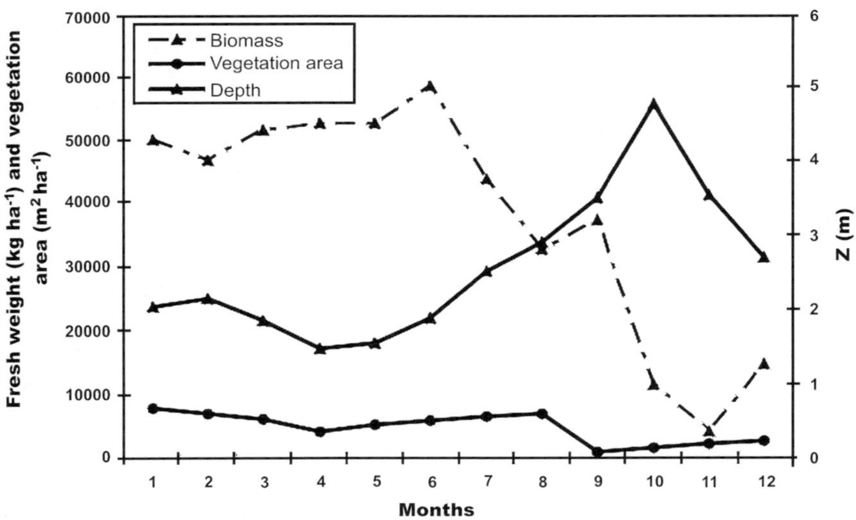

Fig. 8.5 Annual fluctuation of the area (m^2), plant biomass (*WW* wet weight) and depth (*Z*) of a lake-(Santa Rosa, province of Santa Fe)

8.5 Biomass and Productivity

E. crassipes is one of the most representative species within the hygrophilous plants in island environments of the Middle Paraná River, according to Bayo et al. (1981). These authors have done a detailed study of its population

Table 8.6 Number and surface of water bodies and biomass (t ha^{-1}) of water hyacinth "camalotales" (*Eichhornia crassipes*) in three islands of the Middle Paraná River

	Water bodies			*E. crassipes* biomass	
Island	Number	Total surface (ha)	Average surface (ha)	Minimum (t ha^{-1})	Maximum (t ha^{-1})
Carabajal	35	741	21.17	0.31	1.10
Sirgadero	20	249	12.45	0.42	1.44
Clucellas	72	187	2.60	2.31	6.15

growth in three islands of the Middle Paraná River, evaluating bimonthly hydrophyte cover in 127 water bodies. The area of the surveyed water bodies reached 1,177 ha, with mean lagoon areas varying between 2.6 and 21.17 ha, according to the islands (Table 8.6). Maximum biomass values for *E. crassipes* (90% of surveyed hydrophytes) occurred in January and February (summer), and minimum values in May and July (winter), showing an inverse relationship between the mean area of the islands water bodies and the biomass values (Table 8.6).

The net production of aquatic macrophytes in the Middle Paraná River has been considered in numerous studies, and the minimum and maximum values are shown in Table 8.7. Two to three species usually dominate the biomass of aquatic plant communities in Paraná River lentic environments, which is observed in Table 8.8, referring to ten samples of different plant communities.

The ash contents of the main macrophyte species are very variable according to the sampling sites, the condition of the environment and the state of plant development. The minimum and maximum values (% ashes on dry matter) have been determined in nine macrophytes of the Middle Paraná River: *L. peploides* (9.99–18.05%), *Schoenoplectus californicus* (5.57–18.05%), *E. crassipes* (17.29–27.40%), *S. herzogii* (14.06–29.12%), *Hydrocotyle ranunculoides* (18.44–27.00%), *Nymphoides humboltianum* (13.13–20.07%), *Limnobium laeivigatum* (17.59–32.77%), *Polygonum hidropiperoides* (8.98–24.53%) and *P. stratiotes* (22.76–28.72%).

In the particular case of aquatic gramineae, mean ash values were 19.48, 10.09, and 10.74 for *P. repens*, *P. elephantipes* and *E. polystachya*, respectively. Determinations of productivity in natural conditions have been done in a few species, in particular in the largest and most representative ones, as the "canutillares", with mean values of 0.036 t ha^{-1} d^{-1} in *P. repens* and 0.067 t ha^{-1} d^{-1} in *P. elephantipes* (Sabattini 1985), and in water hyacinth "camalotales" of *E. crassipes*, with values of 0.050 t ha^{-1} d^{-1} (Lallana 1980, 1981).

Table 8.7 Net production (minimum and maximum values) of aquatic macrophytes in lentic and lotic environments of the Middle Paraná River, according to data of various authors

	Net production dry weight (g m^{-2})			
Species	Minimum	Maximum	Location	Author
Paspalum repens	215	867	Carabajal Island	Sabattini 1985; Sabattini y Lorenzatti 1987
Panicum elephantipes	350	1,520		
Echinochloa polystachya	420	1,600		
Panicum rivulare	600	1,800		
Eichhornia crassipes				
Plants (<30 cm tall)	315	1,076	Island lake	Lallana 1980
Plants (>30 cm tall)	1,020	2,213		
Eichhornia azurea	275	832	Aquatic vegetation floating Middle Paraná River	Lallana unpublished
Salvinia herzogii	70	202		
Hydrocotyle ranunculoides	42	192		
Ludwigia peploides	78	712		
Polygonum ferrugineum	546	1,267		
Eichhornia azurea	180	1,000	Middle Paraná River alluvial valley	Sabattini et al. 1983
Paspalum repens	700	1,020		
Polygonum acuminatum	600	2,040		
Panicum elephantipes	340	2,640		
Polygonum ferrugineum	744	2,000	La Cuarentena Carabajal Island	Lallana unpublished
Salvinia herzogii	102	229		
Ludwigia peploides	70	533		
Paspalum repens	271	481		
Miriophyllum sp.	141	561		
Scirpus californicus	271	601		
Pistia stratiotes	192	230		
Nymphoides indica	74	227		
Limnobium laevigatum	126	457		
Hydrocotyle ranunculoides	243	427		
Polygonum acumninatum	466	1,264		
Panicum prionitis	(n=45) 5,690		Middle Paraná River islands	Manzi unpublished
Solunum glaucophyllum	(n=10) 6,140			
Eichhornia crassipes	(n=30) 2,000			
Paspalum repens	(n=20) 2,200			
Polygonum ferrugineum	4,500	5,500		

Table 8.8 Biomass as dry weight (g m^{-2}) of ten samples (*M*) of aquatic plant communities in a Middle Paraná River lake (Carabajal Island, Santa Fe, Argentina). Lallana, unpublished data, 1981–1983

Species	M1	M2	M3	M4	M5	M6	M7	M8	M9	M10
Azolla carolianana		99								
Eicchornia azurea		337							517	336
Eichhornia crassipes			821	387		513			10	421
Hydrocotyle ranunculoides	40		5	27		44	16		126	119
Limnobium laevigatum		55		51					87	
Ludwigia peploides	37		6	51		360	173			
Mikania periplocifolia			24							
Miriophyllum brasiliense	141									
Panicum elephantipes					147			70		
Paspalum repens										
Pistia stratiotes				12	2					
Polygonum ferrugineum					1264			1472		
Polygonum punctatum			504			65				
Salvinia herzogii	97	9	216	81	137	105	57		6	585
Scirpus cubensis		12	25			8			146	
TOTAL	**315**	**512**	**1,601**	**609**	**1,550**	**1,095**	**246**	**1,542**	**892**	**1,461**

8.6 Use of Resources

The islands and flooded lands of the Middle Paraná River represent an important natural resource for cattle and forest production. Management of these environments mainly depends on the location of the island in relation to the distance of the Paraná River main channel. The channel islands present a high topography and are very dynamic. Marginal accumulation of sediments and erosion are permanently observed. Islands far from the main channel usually show low topography but are linked to cattle management of mainland fields. Thus, "channel-bank fields" are adjacent to them, subjected to the hydrological regime, and showing plant communities of island environments.

In the Middle Paraná River area, there are approximately 1,500,000 ha of islands, flooded lands and alluvial lands, where the preponderant activity is the extensive and seasonal cattle raising (winter or fattening up cattle). The particular environmental conditions of the Gallery Rainforest zones (Cabrera 1976) favor the implantation of an hygrophilous flora and, specially, of an abundant specific richness of herbaceous plants, constituting grasslands, "canutillares" and "camalotales", very valuable natural resources from the foraging point of view.

Due to the conditions of flooding environments, we worked on studies and projects to improve productivity, based on the rational use of flooding pastures. One of the objectives is land recovery, with greater fertility and cover of foraging species, providing the environment with a higher probability and security of use and management. Thus, experiences for grass harvest and forage conditioning on "canutillares" have begun in the province of Entre Ríos, obtaining rolls with high yield and quality.

Moreover, scrublands of *P. prionitis* and *Elionurus* spp. appear as marshy communities of greater extension in islands and flooded lands; the latter have an important role in island fires (Ragonese 1967). The burning-off as a cleaning practice, carried out at the end of winter and coinciding with low waters, is a common activity used by islanders, that usually favors the rejuvenation of "enriedales" (*Solanum amigdalifolium*) and scrubland zones, allowing to decrease the plant cover and favoring the entrance of light and, thus, the growth of other plant species in those areas.

The scrubland bushes of *P. prionitis*, which are thermal insulators and waterproof, are used for housing construction in the island environments, usually in roofs. A great part of the vegetation, in turn, has ornamental, craft, and medicinal uses. Thus, the plant species as the native willow, the "timbó", the laurel, the wicker and reeds are relevant in the production of furniture, canoes, drawers, baskets and curtains.

8.7 Conclusions

Aquatic macrophytes constitute habitat and refuge for a great part of aquatic communities and contribute with biomass and nutrients to the trophic chains of the Middle Paraná River ecosystem. Emerging rooted plant species and free-floating plants stand out for their biomass, productivity, and plant cover. Although species richness is important and varies according to the topography, the season of the year and the hydrological regime, there are only a few dominant species that determine the zones of perilacustrine vegetation and the successional processes in lentic environments. The Middle Paraná River, due to the latitude where it is located, shows vegetal species common to the High Paraná River and Delta (Lower Paraná River), establishing a continuum or biological corridor, but biotic components show a higher number and diversity of habitats than in the High Paraná River. The drifting aquatic vegetation ("camalotales" and "embalsados") represents an important biomass, mobilized by flood pulses and climatic factors, and its drift moves organic matter, disseminules, insects or other organisms from one place to another in the ecological system. In this sense, the displacements of aquatic vegetation contributes to the dispersion of species, a primary stage for the beginning of riparian plant successions and that, on the other hand, can cause problems for navigation and for civil works in the Middle Paraná River.

References

Aceñolaza PH, Provedano A, Manzano J, Muñoz JD, Areta y Ronchi Virgolini AI (2004) Biodiversidad del Parque Nacional Pre-Delta. In: Temas de la biodiversidad del Litoral fluvial Argentino. Coorinador F.G. Aceñolaza (ed) INSUGEO, Misceláneas 12:169–184

Aceñolaza PH, Siones WF, Kalesnik F, Serafíni MC (2005) Determinación de unidades homogéneas de vegetación en el Parque Nacional Pre-Delta (Argentina). In: Temas de la biodiversidad del Litoral fluvial Argentino II. Coorinador FG Aceñolaza (ed) INSUGEO, Misceláneas 14:81–90

Arbo MM, Tressens SG (eds) (2002) Flora del Iberá. EUDENE. Corrientes, Argentina, 613 pp

Bayo V, Lallana VH, Lorenzatti E, Marta MC (1981) Evaluación cuantitativa de la vegetación acuática en islas del valle aluvial del río Paraná medio. Parte 1. Ecología 6:67–72

Bonetto AA (1975) Hydrologic regime of the Paraná River and its influence on ecosystems. In: Hasler AD (ed) Coupling of land and water systems. Springer, Berlin Heidelberg New York, pp 175–197

Burkart A (1957) Ojeada sinóptica sobre la vegetación del Delta del río Paraná. Darwiniana 11(3):457–561

Brizuela AB, Sabattini RA, Lallana VH, Kimsa J (1992) Identificación de las comunidades vegetales en un ambiente insular en el río Paraná mediante el análisis digital de una imagen satelitaria. SELPER 8(1):31–35

Cabrera AL (1976) Regiones Fitogeográficas Argentinas. In: Parodi M, Dimitri M (eds) Enciclopedia Argentina de Agricultura y Jardinería. 2da. Ed. Fasc. 1. Editorial A.C.M.E. Buenos Aires, Argentina, pp 1–85

Casco SL (2003) Poblaciones vegetales y su variabilidad espacio-temporal en una sección del Bajo Paraná influenciada por el régimen de pulsos. Tesis Doctoral. Universidad Nacional del Nordeste. pp 1–189 y anexos

Casco SL, Neiff M, Neiff JJ (2005) Biodiversidad en ríos del litoral fluvial. Utilidad del software PULSO. In: Temas de la biodiversidad del Litoral fluvial Argentino II. Coorinador F.G. Aceñolaza (ed) INSUGEO, Misceláneas 14:105–120

Fernandez OA, Sutton DL, Lallana VH, Sabbatini MR, Irigoyen JH (1990) Aquatic weeds problems and management in South and Central America. In: Pieterse AH, Murphy KJ (eds) Aquatic weeds. The ecology and management of nuisance aquatic vegetation. Oxford Science Publications, New York, pp 406–425

Franceschi E, Lewis JP (1979) Notas sobre la vegetación del valle santafecino del río Paraná (República Argentina). Ecosur 6(11):55–82

Franceschi EA, Prado DE, Lewis JP (1985) Comunidades vegetales y mapa de vegetación de la reserva "El Rico" e islas aledañas (Provincia de Santa Fe, República Argentina). U.N.R., Rosario, 40 p., 2 tablas, 1 mapa

Ginsburg R, Adamoli J, Herrera P, Torella S (2005) Los Humedales del Chaco: clasificación, inventario y mapeo a escala regional. In: Temas de la biodiversidad del Litoral fluvial Argentino II. Coorinador F.G. Aceñolaza. INSUGEO, Misceláneas 14:121–138

Junk WJ (1970) Investigations on the ecology and production biology of the "floating meadow" (*Paspalo-Echinochloetum*) on the middle Amazon. I. The floating vegetation and its ecology. Amazoniana 2(4):449–495

Junk WJ, Howard-Williams C (1984) Ecology of aquatic macrophytes in Amazonia. In: Sioli H (ed) The Amazon, limnology and landscape ecology of a mighty tropical river and its basin. Junk Publishers, Dordrecht, pp 269–293

Kandus P (1997) Análisis de los patrones de vegetación a escala regional en el Bajo Delta Bonaerense del Río Paraná (Argentina). Tesis doctoral, Universidad Nacional de Buenos Aires, Bs. As., 241 pp
Lallana VH (1980) Productividad de Eichhornia crassipes (Mart.) Solms. en una laguna isleña de la cuenca del río Paraná Medio. II. Biomasa y dinámica de población. Ecología (Argentina) 5:1–16
Lallana VH (1981) Productividad de *Eichhornia crassipes* (Pontederiaceae) en una laguna isleña de la cuenca del río Paraná medio. I. Análisis del crecimiento. Bol Soc Arg Bot 20:99–107
Lallana VH (1990) Dispersal units in aquatic environments of the middle Paraná River and its tributary, the Saladillo River. Proceedings EWRS Symposium on Aquatic Weeds. Uppsala, Sweden, pp 151–159
Lallana VH (1997) Relevamiento de vegetación acuática y palustre con fotografías aéreas y tecnologías digitales. Una revisión. Actas IV Simposio Argentino de Teledetección. San Juan, Argentina, pp 278–287
Lallana VH, Sabattini RA (1986) Consideraciones sobre la vegetación de las áreas inundables del Paraná medio. Bol Asoc Arg Limnol 4:9–12
Lallana VH, Sabattini RA (1989) Análisis preliminar de datos del barredor multiespectral y su aplicación en recursos naturales renovables. Actas del IV Simposio Latinoamericano en percepción remota y IX Reunión plenaria de SELPER. San Carlos de Bariloche, Argentina. Tomo I, pp 279–289
Lewis JP, Franceschi EA (1979) Notas sobre la dinámica de la vegetación del valle del río Paraná. Ecosur 6(12):145–163
Lewis JP, Franceschi EA, Prado DE (1987) Effects of extraordinary floods on the dynamics of tall grasslands of de river Paraná valley. Phytocoenología 15(2):235–251
Lugo AE, Brinson MM, Brown S (1986) Development of an integrated ecological research program for the Paraná River in Argentina. Rev Asoc Cienc Nat Litoral 17(2):137–155
Marchetti Z, Aceñolaza PH (2005) Detección satelital y descripción de patrones de vegetación en islas del Paraná medio. In: Temas de la biodiversidad del Litoral fluvial Argentino II. Coordinador F.G. Aceñolaza (ed) INSUGEO, Misceláneas, 14:151–158
Neiff JJ (1975) Fluctuaciones anuales en la composición fitocenótica y biomasa de la hidrofitia en lagunas isleñas del Paraná medio. Ecosur 2(4):153–183
Neiff JJ (1979) Fluctuaciones de la vegetación acuática en ambientes del valle de inundación del Paraná medio. Physis (Buenos Aires) B 38:41–53
Neiff JJ (1981) Vegetación acuática y anfibia del Iberá. In: Estudio del Microsistema Iberá. Tomo V. Ecología Vol. 2, 2da. Parte. Gobierno Pcia. Corrientes - INCYTH, pp 86–120
Neiff JJ (1986a) Aquatic plants of the Paraná system. In: Walker KF, Davis B (eds) The ecology of river systems. Junk Publishers, pp 557–571
Neiff JJ (1986b) Las grandes unidades de vegetación y ambiente insular del río Paraná en el tramo Candelaria–itá Ibaté. Rev Asoc Cienc Nat Litoral 17(1):7–30
Neiff JJ (1996) Large rivers of South America: toward the news approach. Verb. Intenat, Verein. Limnol 26(1):167–181
Neiff JJ (1999) El régimen de pulsos en ríos y grandes humedales de Sudamérica. In: Malvárez AI (ed) Tópicos sobre humedales subtropicales y templados de Sudamérica. Universidad de Buenos Aires. Oficina Regional de Ciencia y Tecnología de la UNESCO para América Latina y el Caribe, ORCYT, Montevideo, Uruguay, pp 97–146
Neiff JJ (1990) Ideas para la interpretación ecológica del Paraná. Interciencia 15(6):424–441
Peltzer PM (1998) Inventario de la flora y fauna del área natural protegida: islote municipal del valle aluvial del Paraná. Natura Neotropicalis 29(1):74–78
Ragonese AE (1967) Vegetación y Ganadería de la República Argentina. Colección Científica del INTA. 218 pp
Sabattini RA (1985) Dinámica y productividad de *Paspalum repens* ("canutillo") en un ambiente lenítico del valle aluvial del río Paraná. Rev Hydrobiol Trop 18(1):3–11
Sabattini RA, Lorenzatti EA (1987) Estudio fitoquímico en Gramíneas de ambientes acuáticos. I) *Paspalum repens* Berg. "canutillo". Rev. Iheringia serie Botánica 36:65–73

Sabattini RA, Lallana VH (1993) Identificación de patrones vegetales según el gradiente topográfico en islas y anegadizos, con la utilización de fotografías aéreas (La Azotea, Diamante). 5 pp. Actas del Congreso Forestal Argentino y Latinoamericano. Paraná, Argentina. Relatorios y Trabajos Voluntarios. Comisión III, Tomo II

Sabattini RA, Lallana VH (1998) Optimum sampling size of the aquatic vegetation in the alluvial valley of the Middle Paraná River (Argentina). Proceedings of the 10th EWRS Symposium on Aquatic Weeds, Lisbon, pp 111–114

Sabattini RA, Lallana VH, Marta MC (1983) Inventario y biomasa de plantas acuáticas en un tramo del valle aluvial del río Paraná Medio. Rev Asoc Cienc Nat Litoral 14(2):179–191

Schulz AG (1961) Nota sobre la vegetación acuática chaqueña "esteros y embalsados". Bol Soc Arg Bot 9:141–150

Tur NM, de Orellana J (1970) Relación suelo vegetación en la zona isleña del Paraná Medio. Actas 5ª. Reunión Argentina de la Ciencia del Suelo, Santa Fe, pp 482–491

Part III
Invertebrate Animal Life

9 Zooplankton

Susana José de Paggi[1] and Juan César Paggi

9.1 Introduction

The information on zooplankton of the Middle Paraná River comes from studies based on samplings of different intensity carried out in its 707-km reach. Studies were carried out in sections of the main channel, secondary channels, tributaries and lentic environments of the floodplain valley. The longitudinal distribution of zooplankton along the reach and some shorter extensions was investigated considering different hydrometric levels.

The taxonomic composition has been analyzed in diverse studies on rotifers (José de Paggi 1978, 1979, 1982; Martínez and José de Paggi 1988; Paggi and José de Paggi 1973; Paggi 1973a, 1973b, 1973c; Paggi 1978a, among others), cladocerans and copepods (Martínez de Ferrato 1966, 1967; Paggi 1972, 1973b, 1973c, 1978b, among others). A total of 230 species of rotifers, 125 species of cladocerans and 87 species of copepods were estimated in the main channel and aquatic floodplain environments (José de Paggi 1990, 2004a; Paggi 2004).

Among rotifers, the highest diversity was in genera *Brachionus, Lecane* and *Trichocerca*. Most species are cosmopolitan in distribution, but there are some species considered endemic to the Neotropical region, such as *Brachionus dolabratus* Harring, *B. mirus* (Daday), *B. zahniseri* Ahlstrom, *Lecane proiecta* Hauer, *Hexarthra intermedia braziliensis* (Hauer). *Testudinella ohlei* Koste and *Lepadella donneri* Koste, among others. Most species that were initially considered as endemic Amazonian species, as *Lecane amazonica* (Murray) and *L. braziliensis* Segers, were gradually appearing in the Paraná River, as more studies were being carried out.

Among cladocerans, the highest diversity corresponded to *Diaphanosoma*, followed by *Ceriodaphnia, Moina, Bosmina* and *Daphnia*. In contrast with rotifers, most cladoceran species are endemic to the Neotropical region, including the Caribbean and the southeast of North America; for example, *Diaphanosoma brevirreme* Sars, *D. fluviatile* Hansen, *Moina minuta* Hansen, *M. reticulata* (Daday) *Daphnia gessneri* Herbst, *Bosmina hagmani* Stingelin and *B. huaronensis* Delachaux.

Instituto Nacional de Limnología (CONICET-UNL), José Maciá 1933, 3016 Santo Tomé, Santa Fe, Argentina, [1]e-mail: sjdepaggi@gmail.com

M.H. Iriondo, J.C. Paggi, and M.J. Parma (Eds.)
The Middle Paraná River: Limnology of a Subtropical Wetland

Among calanoid copepods, the highest diversity corresponded to *Notodiaptomus*, an endemic genus of the Neotropical region, while among cyclopoid copepods, the most diverse genera were *Mesocyclops, Thermocyclops*, and *Microcyclops*.

The three large South American basins, Orinoco, Amazon, and del Plata, from which the Paraná River is part, are found in the same biogeographical Guyano-Brazilian sub-region and share many zooplankton species. It is probable that the communication among the three fluvial systems does not allow the existence of many endemisms restricted to one basin. Their floodplains are hydrologically connected, since the water sheds do not constitute permanent barriers and can communicate themselves, at least during some hours (Morello 1984). Connections among basins and connections intra-basins, activated by the flood pulse, would be an excellent dispersal mechanism for rotifers (José de Paggi 2004a), as well as for other groups. Among cladocerans, there are also many species in common, except for a few taxa which seem to be restricted to the Orinoco and Amazon basins, as *Bosminopsis negrensis* Brandorff, *Moina rostrata* MacNair and *Holopedium amazonicum* Stingelin, and are not present in the Paraná River. The three basins share most copepod genera, with the exception of *Rhacodiaptomus* and *Oithona*, which are absent in the Paraná River and are present in the Amazon and Orinoco Rivers.

The qualification of a taxon as endemic depends, to a great extent, on the relative knowledge of the fauna in diverse areas, e.g., *Onchobunops tuberculatus* Fryer and Paggi, originally recorded in the Paraná River, has been found in Mexico and Central America; in contrast, the distribution of *Ilyocryptus elegans* Paggi, until the present, seems to be limited to the waters of the Paraná River main channel in its middle reach (Paggi 2004).

9.2 Zooplankton of Rivers and Floodplain Shallow Lakes: Similarities and Differences

Although aquatic environments of the Paraná River could be basically classified as rivers or lakes, there is a gradient between and within each group. Among the rivers, we should consider the main channel, with rigorous physical conditions, and the secondary channels and tributaries that run in the floodplain and generally have a comparatively lower flow and current velocity. Among the floodplain lakes, in addition to variations in shape, size, and degree of marginal cover, there are differences in the degree of connection to the river: there are lakes that are directly connected to it through a short channel of reversible flux, and others that are connected through intermediate, generally temporary water bodies, activated by the flood pulse (Drago 1981). The former are almost permanently connected to the river and suffer rapid changes at the hydrometric level, sometimes behaving as semi-lotic systems; the latter, in contrast, have a higher water residence time.

In lotic environments, rotifers and nauplii dominate, and low density of individuals is observed. These characteristics are more notorious in the main channel than in the secondary channels (José de Paggi 1981, 1983, 1984). Zooplankton in the main channel is comparatively less abundant and shows a higher predominance of rotifers than in the secondary channels and tributaries of the Paraná River middle reach. Some of these differences were observed during a sampling in medium and high waters at 25 stations belonging to different channels along a reach of 440 km (José de Paggi 1983, 1988, Table 9.1). Annual studies carried out in secondary channels near the Resistencia City-Corrientes City and Paraná City-Santa Fe City cross sections also showed that maximum densities were higher than those of the main channel (José de Paggi 1981; Martínez and Frutos 1986; Frutos 1998).

Among the lentic environments, indirectly connected lakes show a comparatively higher mean abundance than those with direct connection to the river, as observed between the El Tigre and Los Matadores lakes, located near the city of Santa Fe (José de Paggi and Paggi, unpublished). El Tigre Lake, indirectly connected to the river, showed an annual mean abundance of 605.5 ind./L (CV 144) for 1987–1988 (samplings every fortnight). The Los Matadores Lake, directly connected to the river, showed annual means of 169.2 and 306 ind./L (CV 177 and 132), for 1974–1975 and 1977–1978, respectively (samplings every fortnight).

Generally, lakes show a comparatively higher zooplankton density than that of the river Fig. 9.1; they also have a comparatively higher proportion of adult copepods, which increases the mean size. On the other hand, larger rotifers, as *Asplanchna* spp. and *Brachionus calyciflorus*, are more frequent in lakes than in the main channel (José de Paggi 1980; Paggi 1980).

Table 9.1 Mean values of environmental factors, richness, and zooplancton density in medium water (*1*) and high water (*2*) section km 971–531 (José de Paggi 1988)

	Discharge (m^3/s)		Current flow (m/sc)		Conductivity (μS/cm)		Richness		Zooplancton density (Ind./L)	
	1	2	1	2	1	2	1	2	1	2
Main channel (6)	8.733.3	12.000	0.87	1.16	74.19	80.00	20	27	54.37	10.72
Scondary channels (8)	248	475	0.52	0.55	445.00	203.00	45	45	122.34	15.19
Tributaries on right margin (7)	13.9	9	0.33	0.08	3.960.00	3.082.00	30	39	153.77	387.89
Tributaries on left margin (4)	19.25	46	0.26	0.27	1.030.00	841.20	32	41	147.38	252.98

( ) Study localities number

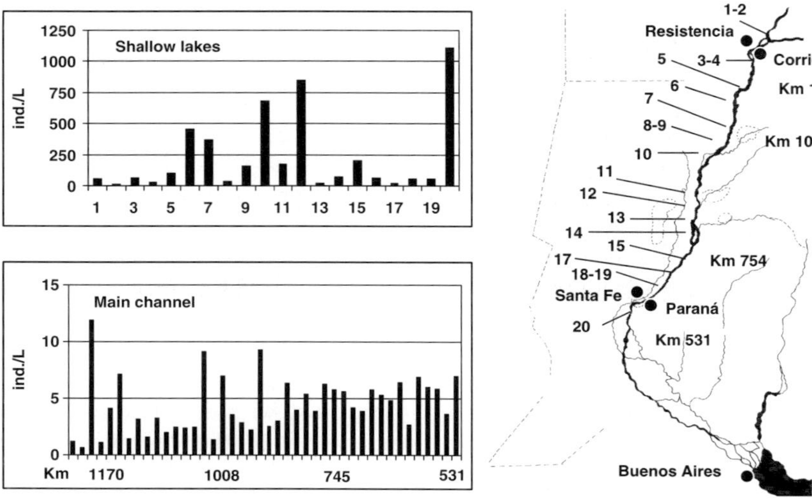

Fig. 9.1 Zooplankton abundance in 20 floodplain shallow lakes and 42 localities of the main channel (José de Paggi 1980)

Zooplankton composition is similar between rivers and lakes, but some species are more frequent and abundant in a certain type of environment. *Bosminopsis deitersi* is more frequent and abundant in the main channel, while *Diaphanosoma* spp. is more frequent in lakes (Paggi 1980). A higher number of littoral planktonic species is observed in lakes, particularly in the most shallow and vegetated ones (José de Paggi 1984; José de Paggi and Paggi, unpublished). The proportion of limnetic species is directly related to depth of water bodies (Paggi 1980).

Lakes show higher richness than rivers; alpha or local diversity for a given moment can vary from very few species in non-vegetated lakes to more than 50 rotifer species in only one sampling, as in Junco and Concepción lakes, near the mouth of the Corriente River (José de Paggi 1996). The higher instantaneous richness in floodplain lakes can be related to environmental heterogeneity (José de Paggi 1984), as well as to morphometric parameters. Paggi (1980) found a relationship between size of the area and number of species in 21 flooding valley lakes ($r = 0.57$, $p = 0.01$).

There is a transversal gradient of water bodies from the main channel to the floodplain lakes. The water residence time, very low in an extreme (main channel) and high in the other one (more isolated lakes) determines the gradient of abundance, the proportion of crustaceans, and zooplankton species richness (José de Paggi 1988; Paggi and José de Paggi 1990; José de Paggi 2004b, Fig. 9.2). Mean nutrient levels, total organic matter, biomass, and complexity of phytoplankton, benthos and fish communities also increase towards the main channel-secondary channels direction, and

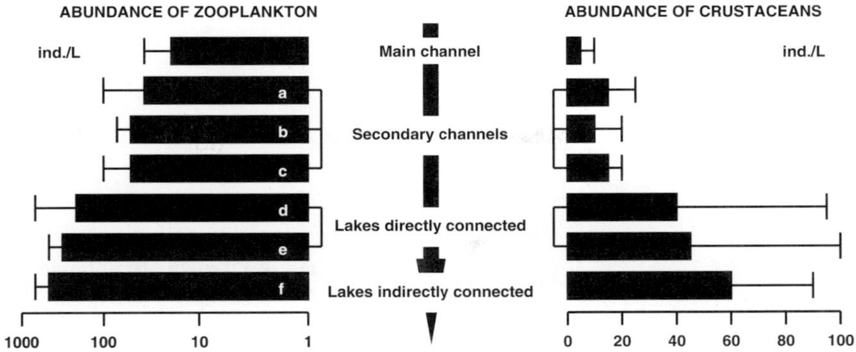

Fig. 9.2 Zooplankton abundance (logarithmic scale) and Crustacean abundance in the main channel (km 603), secondary channels and lakes with different degree of connection. References: *a* Colastiné River, *b* Santa Fe River, *c* Correntoso River, *d* Los Matadores Lake, *e* La Cuarentena Lake, *f* El Tigre Lake (José de Paggi 1981, 1984; Paggi and José de Paggi unpublished)

even more if plain lentic environments are included in the analysis (different authors in Quiros and Cuch 1989).

9.3 Factors Regulating Zooplankton Abundance and Seasonal Distribution in Floodplain Lakes

Seasonal distribution of zooplankton in shallow lakes has been investigated in water bodies close to the beginning and the end of the Paraná River middle reach. In these studies, zooplankton showed marked temporal and spatial variations in abundance.

In Laguna Turbia and Los Pájaros lakes (27°25′S and 58°45′W), indirectly connected to the river, near the Resistencia-Corrientes section, Frutos (1993a 1993b, 1996) found the lowest zooplankton densities in the isolation periods, being microcrustaceans the most abundant group. When the lakes were connected to the river, there was an increase in transparency and depth, and the highest zooplankton abundance was recorded (700 ind./L). In contrast, in the Catay Lake, directly and closely connected to the other lakes, abundance varied inversely to the river hydrometric level (Frutos 1993a). Monthly or fortnightly surface samples were obtained in the limnetic zone of these lakes.

In the lakes near to the Paraná-Santa Fe section, the highest zooplankton abundance was observed in the isolation phases (Bonetto and Martínez de Ferrato 1966; José de Paggi 1993; Paggi unpublished). In the small El Tigre Lake (31°41′S/60°40′W), indirectly connected to the river (3.35 ha in isolation period) and sampled every fortnight taking the whole water column from surface to bottom, there was an increase in conductivity, phytoplankton and

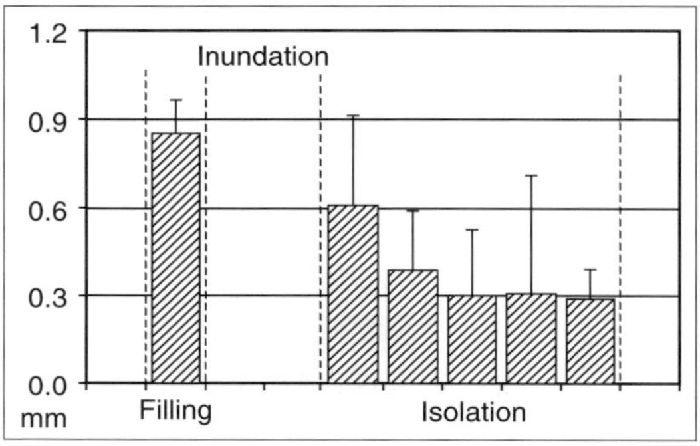

Fig. 9.3 Length variation in *Diaphanosoma fluviatile*, El Tigre Lake (Paggi, unpublished)

zooplankton abundance during the isolation phase (José de Paggi 1993; García de Emiliani 1997). In this period, the most abundant group was rotifers (up to 4,300 ind./L). Calanoid copepods in the limnetic area and cyclopoid copepods in the littoral zone, in contrast, showed lower concentrations than in the flooding period (José de Paggi 1993). In the lake isolation phase, predation of small carnivorous fishes on copepods and cladocerans is intense, as was demonstrated by the analysis of their digestive tract (Oliveros pers. comm.) and the size distribution of *Diaphanosoma fluviatile* (Paggi unpublished, Fig. 9.3). This would generate a lower competition and intrazooplanktonic predation, favoring the dominance of rotifers Fig. 9.4. *Tropocyclops, Mesocyclops* and *Microcyclops,* present in the littoral zone, showed they had consumed small rotifers (José de Paggi 1993).

Although these lakes are shallow environments, changes in the diel vertical distribution of zooplankton have been detected. In two different lakes, Don Felipe and El Tigre, *Diaphanosoma brevirreme* was more abundant at the surface during the night (Martinez de Ferrato 1967; José de Paggi 1995). In the El Tigre Lake, reverse rotifer migrations were observed in the littoral zone, while movements were less conspicuous in the limnetic area. Evidence of horizontal migrations were also detected in this lake; in the littoral zone, rotifers were more abundant during the morning; in contrast, in the limnetic area, they were more abundant during the night. Hourly changes in the vertical distribution of zooplankton difficult the comparison of abundance and temporal variations between surface-sampled lakes and water column-sampled lakes.

Abundance can also vary horizontally in function of the environmental heterogeneity of lakes, determined by the morphometry and the existence of vegetated littoral areas. For example, in the El Tigre Lake, rotifers were significantly more abundant in the limnetic and littoral areas, dominated by *Typha* sp. and *Paspalum repens*, than at a station located in a limnetic-littoral transitional area and at a littoral area station dominated by *Eichhornia crassipes* (José de Paggi 1993). The limnetic area showed a low crustacean density (maximum 33 ind./L) (Paggi, unpublished); the highest crustacean

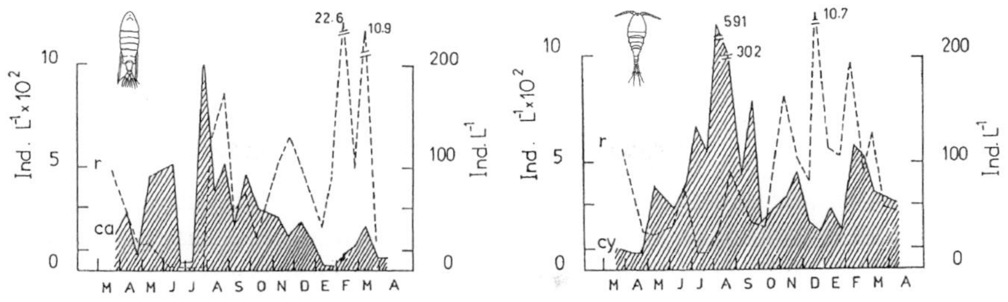

Fig. 9.4 Variations in abundance of rotifers and calanoid copepods (limnetic zone) and rotifers and cyclopoid copepods (littoral zone) in El Tigre Lake (José de Paggi 1993)

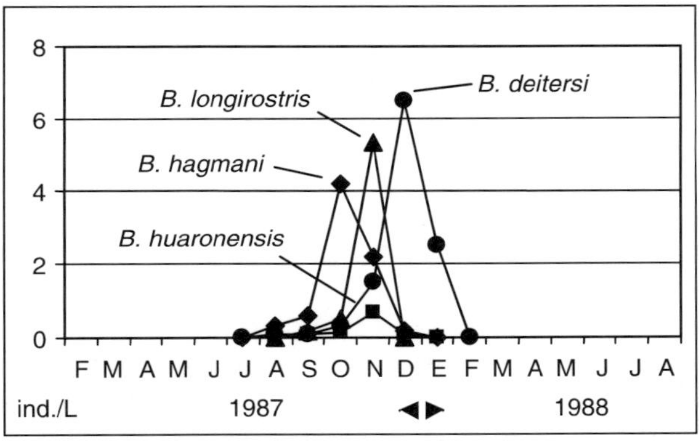

Fig. 9.5 Abundance of different species of *Bosmina* and *Bosminopsis* in the littoral area of the El Tigre Lake (Paggi unpublished).

abundance was found at the littoral stations (maximum 1,350 ind./L). Successions that suggest active competition processes were recorded at these stations (Fig. 9.5).

La Cuarentena (31°42′S–60°37′W) is a dendritic-shaped lake, originated by the junction of various segments of abandoned channels (Drago 1973). Its arms have different depths and plant coverage. At a level of 3.32 m in the Paraná River (at Paraná City), the extension of the non-vegetated area was 274.5 ha (Oldani and Tablado 1985). In its southeast extreme, it is directly communicated to the main channel of the Paraná River by a narrow and shallow channel, while by the north extreme it receives water from flooded plains. Its particular morphometry increases the environmental complexity and generates areas with different water residence time. Zooplankton samples were analyzed in 1980–1981 at several times of the year, integrating surface to bottom samples (José de Paggi 2004b; Paggi and José de Paggi unpublished). Eleven sampling stations were considered along a water residence time gradient that began at station 1, the most isolated from the river influence, and concluded at station 11, in the Paraná River. During the flooding, water went into the lake by the north extreme of the island, producing a deep sweeping action. Zooplankton density was very low, even lower than in the river (14 ind./L), with dominance of small sizes: rotifers, Chydoridae and Moinidae.

During the lake isolation, in which there only was a contact with the river through a short channel, the areas were differentiated in function of their water residence time. The most isolated ones showed up to 700 ind./L of total zooplankton. Among microcrustaceans, *Boeckella bergi* and *Notodiaptomus incompositus* were dominant, with sizes of up to 1,500 μm. In the river and in the connective channel, *Bosmina* spp. dominated, showing sizes smaller than 500 μm. The proportion of microcrustaceans was higher at all stations during the low water period Fig. 9.6. Although the highest water residence time at some stations can influence the dominance of copepods due to their K-strategist character, this size distribution can also indicate a higher predation by fish in the center of the lake.

9.4 Flood Pulse, Physical Variables and Potamozooplankton Dynamics

Zooplankton abundance in the main channel of the Paraná River middle section varies temporarily from less than 1 ind./L to 195 ind./L (Paggi and José de Paggi 1974; José de Paggi 1984; Bonetto and Corrales 1985). The pattern of temporal distribution of zooplankton abundance showed a relationship with the hydrological cycle. It can be observed in Fig. 9.7 where the results of simultaneous studies at two section of river are shown. The highest density of organisms was observed in low waters and the lowest density in high waters; a similar behavior was observed for phytoplankton in the same period (Perotti de Jorda 1984). A more or less prolonged period of low waters implies a higher water residence time in the channel, which, together with a comparatively lower flow velocity, are favorable conditions for zooplankton development. The stability of phases in terms of low hydrometric variability can be important for zooplankton, determining favorable conditions for community re-organization. The amplitude of phases could influence phytoplankton structure. In the most prolonged low water phase (September to February), occurring in 1971–1973 in the Paraná-Santa Fe section, an increase in adult cladocerans and copepods was recorded (Paggi and José de Paggi 1974). In 1980, when the river level was always high, with a low range of hydrometric variability, a higher proportion of crustaceans than in 1977, 1978 and 1979 was recorded (José de Paggi 1984).

The flood pulse is also a hydrosedimentological pulse. Paraná River carries away an important load of suspended sediments, with values of up to 600 mg/L (at km 603). The time-lag of this pulse is about 22–39 days in relation to flooding

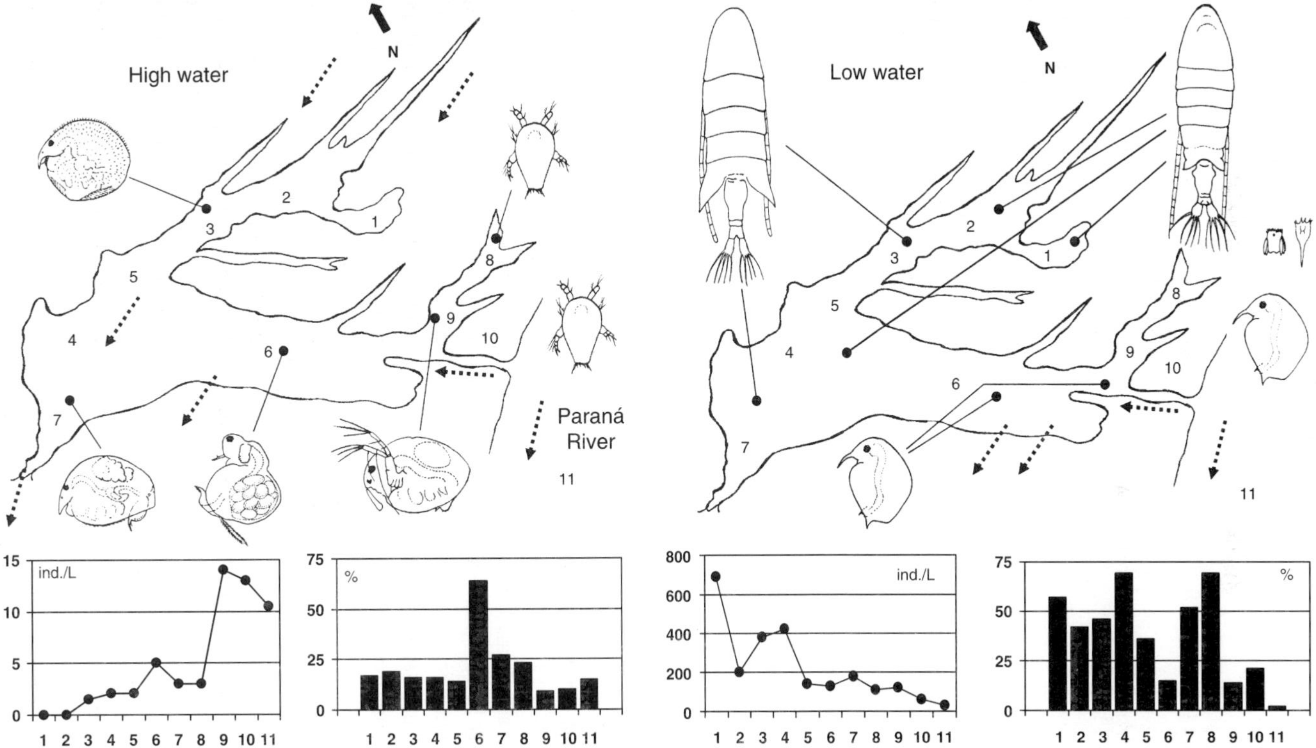

Fig. 9.6 Abundance of zooplankton and percentage of crustaceans in La Cuarentena Lake during high and low waters. Dominant microcrustaceans are indicated (Paggi and José de Paggi unpublished). *Dashed lines* indicate the direction of water fluxes.

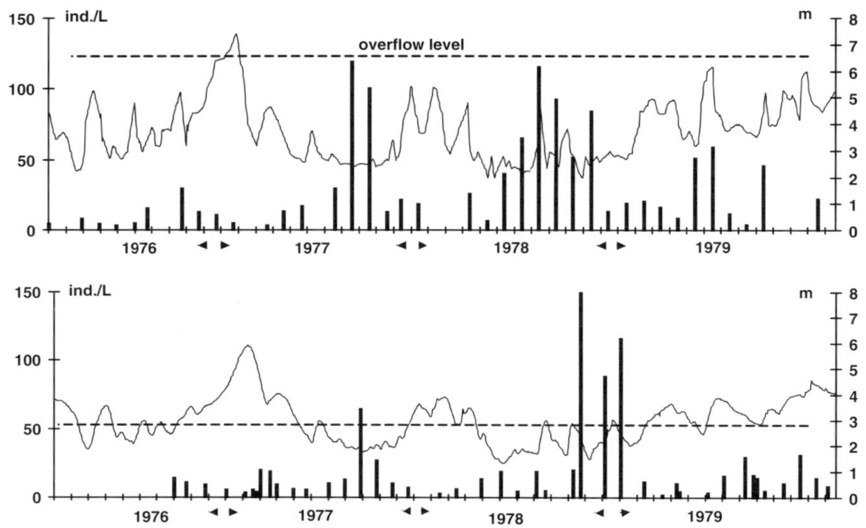

Fig. 9.7 Hydrometric level and zooplankton abundance in the Resistencia City-Corrientes City cross section, km 1,208 (*upper panel*) and Paraná City-Santa Fe City cross section, km 603 (*lower panel*). The overflow level is indicated for the ports of Corrientes City (Neiff 1999) and Paraná City, respectively. The sampling period in the second cross section began in September 1979 (Bonetto and Corrales 1985; José de Paggi 1984)

(Drago and Amsler 1988). In the Paraná-Santa Fe section, km 603 (1976–1980), there was an inverse relationship between abundance and suspended sediments (r: –0.35 $P<0.009$, $n=54$, José de Paggi unpublished). The highest zooplankton abundance was observed when suspended sediment concentrations were lower than 200 mg/L and, inversely, the lowest zooplankton abundance was found with suspended sediment concentrations of 600 mg/L. Only *Lecane proiecta* was relatively abundant, with high concentrations of suspended sediments, and *Bosminopsis deitersi* was the only crustacean present with 500 mg/L of suspended sediments (José de Paggi 1984). In filter feeders species with food size selection, the inorganic suspended sediment can be incorporated into feeding with the consequent disadvantages. In the case of the Paraná River, suspended sediments are constituted by the wash load in more than 60%, with a high percentage of particles of sand and clay lower than 16 μm, (Drago and Amsler 1981); therefore, they are found within the size range usually filtered by plankters (Lampert and Sommer 1997).

The ability of some cladocerans to discriminate between particles can have strong implications in the plankton composition of the Paraná River. Experimental studies showed that reproduction rate of *Moina micrura*, a frequent species in this river, is not affected by mean concentrations (250 mg/l) of inorganic particles lower than 16 μm, but, in contrast, growth and reproduction of *Diaphanosoma birgei* are strongly affected. This genus is very frequent in plain lakes but is uncommon in lotic environments (José de Paggi 1994).

Temperature and flow velocity of water can also influence zooplankton. The lack of seasonal regularity in the hydrometric behavior of the Paraná River, contrarily that recorded in highly regular rivers as the Orinoco (Vásquez and Wilbert 1992), interferes with the analysis on the impact of temperature. However, a peak in zooplankton abundance has been observed in spring (September to November), even when the river remained in high waters (José de Paggi 1984). Temperature could increase the positive effects of falling waters or reduce the impact of high waters on the egg development rates and fecundity of organisms.

The water flow has an advective mechanical effects on planktonic populations. In the main channel of the Paraná River, annual mean values of current velocity between 1977 and 1980 varied between 1.20 and 1.30 m/s (CV 21–25) (José de Paggi 1984). A significant inverse relationship was found between current velocity and abundance, not only in the main channel but also in less important rivers of lower current velocities (Paggi and José de Paggi 1990). Water turbulence and current velocity favor a uniform distribution of zooplankton along the water column, even in areas of more than 20 m in depth (José de Paggi 1985).

There are no clear evidences that biological mechanisms controlling zooplankton in lakes have a similar effect zooplankton populations in the main channel. Although phytoplankton biomass shows low values (Schiaffino 1981; Perotti de Jorda 1984; Bonetto 1983; Bonetto et al. 1983), other feeding resources would be available, as organic matter absorbed in sediments, detritus, bacteria or ciliates. It is probable that phytoplankton would not be a limiting factor in large rivers (greater than five orders), as suggested by other authors (Pace et al. 1992). Inversely, there are no evidences of an important control of phytoplankton by zooplankton, as has been recorded in small rivers. Only in an secondary channel of Paraná River, García de Emiliani (1997) found an increase in the proportion of net phytoplankton in relation to nanoplankton in low waters, which could be related to the consumption of small algae by zooplankton, since at that moment zooplankton was more abundant.

In the main channel there are not evidences of a heavy predation pressure upon zooplankton by invertebrates. *Asplancha* and adult cyclopoid copepods are not abundant (José de Paggi 1984; Bonetto and Corrales 1985), and larvae of *Chaoborus* are restricted to lakes. Regarding to fish predation, the main zooplankton consumers in the main channel are larvae: there are no frequent or abundant planktivorous adult fishes. The Paraná River, as other large tropical and temperate rivers of South America, is characterized by the high abundance of detritivorous fishes, Prochiolodontidae, Curimatidae and Loricariidae, and the presence of great predators, as Pimelodidae of genus *Pseudoplatystoma* (Quiros 1988). Many species are totally or partially spawners, and zooplanktophagous larvae are abundant in the Paraná River only during the end of spring and summer (from November to January), since species reproduce during high waters. There are records of up to 7,300 larvae in 1,000 m^3 in the main channel (Oldani 1990). *Prochilodus lineatus* represents 50% of total fish biomass (Oldani 1990); its zooplanktophagous larvae mainly feed on cladocerans, up to a size of 8 mm in length. Afterwards, when larvae reach around 14 mm, rotifers and algae are their most important foods (Rossi 1992). Larvae of other frequent species in the river consume cladocerans and rotifers (Rossi 2001), or cladocerans and copepods (Rossi 1989). This means that fish larvae can only affect zooplankton abundance during a short period of the year. The negative effect of flooding, mainly during the summer, can be boosted due to the higher abundance of zooplanktophagous fish larvae in the main channel and secondary channels.

Other factors of possible biological control of zooplankton are currently beginning to be analyzed. At the end of the 90s, the presence of an invasive mollusk, *Limnoperna fortunei* Dunker (Darrigan and Ezcurra de Drago 2000), was recorded for the first time in the middle Paraná River. Diverse studies have demonstrated the negative impact of the invasion of filtering bivalve mollusks on river zooplankton in Europe and North America, for example, *Dreissena polymorpha* (Jack and Thorp 2000, among others). We still do not know the effects of *L. fortunei* on zooplankton of the Paraná River main channel. Undoubtedly, studies previous to the invasion will be very useful in future analyses. *L. fortunei* shares with *D. polymorpha* a similar size, fast growth, rapid dispersion of its larvae and high filtering rate (Sylvester et al. 2004; Boltovskoy et al. 2006). In contrast, some characteristics of the rivers studied for *D. polymorpha* are less comparable with the Middle Paraná River; for example, its high discharge, that reaches a mean of 16,400 m^3/s (Depetris and Pasquini 2007) in the Middle Paraná River.

9.5 Zooplankton of the Main Channel and the Contribution of High Reaches and the Floodplain

As was previously pointed out, zooplankton of the middle Paraná River was studied simultaneously at two sections located at 1,208 and 603 km from its outlet in the La Plata River (José de Paggi 1984; Bonetto and Corrales 1985). The abundance of organisms was similar in both sections, although there was a distance of 600 km between them. The average abundance was 31 ind./L in the former and 21 ind./L in the latter.

In a longitudinal distribution study, carried out in the main channel along the whole middle section, a downstream increase in the zooplankton abundance was not observed though this trend was detected for the biomass (José de Paggi 1980). Frutos (2004), in a study that included the Paraguay River, even observed a decrease in zooplankton density downstream. In medium and high waters, zooplankton distribution was also analyzed along a shorter section of 440 km. In none of the cases was there an increase in density downstream (José de Paggi 1983, 1988).

In spite that the time of residence of the water in the Middle Paraná River can be approximately 20 days (Drago 1990) and that ovigerous females of various species are not infrequent in spring (José de Paggi unpublished) successful reproduction and recruitment reproduction in the channel is probably very low due to its harsh environmental conditions.

If there is an active in transit reproduction, a significant increase in density in the final sections of the Middle Paraná River could be expected, but this was not observed. On the contrary, in other rivers like Apure River main stem, in the Orinoco River system (Saunders and Lewis 1989) found that densities and species number increase markedly over 600-km reaches.

It is probable that flow velocity and transported sediment load regulate strongly the development of populations in the channel. Water flow velocity can be the main cause of mortality in rivers in which there is not an intense predation effect by vertebrates and invertebrates. If we compare abundance in the Upper Paraná with that in the Middle Paraná, the last one shows a higher density of organisms Table 9.2. This could be related to the Paraguay River inputs and to the higher development of the middle section floodplain (Sendacz 1993).

Richness of the middle section is higher than that of the Upper Paraná River, but there are no important differences between richness recorded at the beginning of the middle section (km 1,208) and that found downstream, after more than 600 km. However, the low number of species in common between both sections shows that the persistence of species in the channel is low.

Table 9.2 Range of variation of environmental factors, chlorophyll *a* and zooplankton at different transversal sections of Paraná River (Corrales 1979; Bonetto 1983; Bonetto et al. 1983; Jose de Paggi 1984; Perotti de Jorda 1984; Bonetto and Corrales 1985)

	Upper Paraná	Middle Paraná	
Km	1,583–1,380	1,208	603
Sampling period	1976–1978	1976–1979	1976–1980
Temperature (°C)	16–30	16.6–29.5	14.9–28
Transparency (m)	0.08–1.28	0.01–0.90	0.08–0.49
Conductivity (μS/cm)	29–98	35–630	60–118
Suspended sediment (mg/L)	10–120	0.5–1,221	58–609
Chlorophyll -a (mg/m^3)		2–18	3–17
Zooplankton density (ind/L)*	2–60 (18)	1–195 (31)	1–148(21)
Species number	58	127	113

*Parentheses indicates mean value, n=22, 44, 51 (Km 1,583–1,380, 1,208 and 603, respectively)

The analysis of the zooplankton composition along the Middle Paraná River, considering nine different cross sections, including samples from the center and at both banks, (José de Paggi 1980) shows that only five species from a total of 57 species recorded in all the river reach, were common to the nine sections: *Keratella tropica, K. americana, Polyarthra trigla, Bosminopsis deitersi* and *Ceriodaphnia cornuta*, i.e., only 8.7% of species remained throughout the 707-km section. Only eight species were present in 65–90% of cross sections, i.e., less than 20% of species were recorded in 65–100% of all stations in 707 km. There is a high number of species that did not come from upstream and that probably have been incorporated from the floodplain, as a contribution of secondary channels and lentic environments, and tributary streams. It is possible that the contribution of the floodplain or 'lateral dimension' sensu Ward (1989), would be more important for zooplankton richness of the Paraná River than the longitudinal dimension.

Regarding richness, although the local contribution from the floodplain to the channel evidences to be higher than the input from upstream, there have been 225 rotifer species recorded in the floodplain, while only 113 species were recorded in the main channel, along more than 7 years of study (José de Paggi 2004b). Reduced hydraulic washout in floodplains as a consequence of the flow attenuation in forested areas may also be an important factor reducing the likelihood of zooplankton export, as suggested by Vásquez and Rey (1992). This 'biological rugosity' can determine that the final number of species entering the in channel is low. Hamilton et al. (1990) found that macrophytes act as a true biological filters of transported zooplankton. Moreover, Depetris and Kempe (1990) suggested that the wooded floodplain is an effective filter to retain sediments and particulate organic debris. Then, the richness of reaches species could be an indicator of the potential diversity or biotic complexity of the floodplain along the reaches, the characteristics of the flow inside them, and transport conditions between floodplain and river.

9. 6 Connectivity and Species Richness

Hydrological connectivity is the exchange of water, nutrients, and organisms between environments of the river floodplains, activated by the flood pulse (Amoros and Roux 1988). Connectivity and movements between water bodies should act as a homogenizing force, decreasing beta diversity (Amoros and Bornette 1999). Beta diversity is a measure of turnover of species between habitats and it could be considered as a measure of the degree of connectivity between habitats (Ward et al. 1999). The richness of 11 lakes located at the cross section Santa Fe City to Paraná City was analyzed. Beta diversity decreased slightly in high water, but decreasing were higher considering rotifers separately (José de Paggi 2004 b). Homogenizing by the flooding water

Table 9.3 Gamma, alpha and beta diversity for different groups of waterbodies at different water level

	Gamma diversity	Alpha diversity	Beta diversity	Habitat diversity
Between shallow lakes				
Zooplankton LW	84	17,7	37	11
Zooplankton HW	87	19,8	33	11
Rotifers LW	50	11	35,4	11
Rotifers HW	41	15	17	11
Between floodplain and main channel				
Zooplankton RW	62	16,6	11	25
Zooplankton HW	72	13,2	18	25

LW low water, *HW* high water, *RW* rising water
Habitat diversity: number of localities sampled
Data from Jose de Paggi 1983, 1988, 2004b

would probably be more effective for this group because of its small size and limited capacity of movements. Beta diversity between fluvial environments of the floodplain was low in rising waters and high waters, particularly in the first one when connectivity begins to act (Table 9.3).

The highest annual cumulative richness for 1977–1980 in the main channel (km 603) was produced when the hydrometric level showed the highest variation range (José de Paggi 1984). The increase in the river-floodplain connectivity can increase the recruitment of the floodplain species and consequently the diversity in the channel.

The temporal presence of some species in the channel would be a valuable indicator of the flows of water between floodplain and main channel and the contribution of different source areas. The Chydoridae, with littoral and benthic taxa, present a high diversity in floodplain shallow lakes (Paggi 1980), but, in the channel, several species were recorded whenever the river level was above 4 meters (Paggi and José de Paggi 1974). Most chydorids are associated to a substrate. Their presence during maximum river overflows can indicate the resuspension of floodplain sediments and vegetable particles, and later washout. It is also possible that hatching of resting eggs in sediments of areas surrounding the lakes or the own channel might also be contributing to the populations.

Some species of genera *Collotheca* and *Sinantherina* were found in the main channel (km 603) only when the periods of water movements from the floodplain to the channel were concluded. The higher diversity of genus *Brachionus* in the channel was associated to the largest and longest flood phases in 1977 (14 taxa) and 1980 (10 taxa), while they were comparatively low in 1978 and 1979 (eight and seven taxa, respectively). In these last two years, the hydrometric conditions suggest an ineffective exchange between main channel and floodplain, considering that *Brachionus* presented a high diversity in the floodplain-lakes at the middle Paraná River (José de Paggi 1984, 2004).

9.7 Conclusions

Zooplankton of Middle Paraná River lotic environments show similar characteristics to that from other rivers of the world, i.e., dominance of small organisms, such as rotifers, Bosminidae and nauplii (Shiel et al. 1982; Kobayashi et al. 1996). Density is low comparable to that of other South American rivers without a strong anthropic impact (Vasquez and Rey 1989).

Factors controlling abundance and distribution of fluvial zooplankton are mostly physical and related to the pulse regime, which affects other biotic communities of temperate and tropical floodplain rivers (Junk et al. 1989; Neiff 1999). In the main channel of the Paraná River, biological mechanisms of control zooplankton populations would be less important than physical mechanisms of control. The influence of the flow regime on lotic zooplankton was stressed by several authors (Shiel et al. 1982; Paggi 1984; Saunders and Lewis 1988; Pace et al. 1992; Basu and Pick 1996). The particular negative influence of current velocity and turbidity on the Paraná River zooplankton was also recorded in other rivers (Thorp et al. 1994). One of the causes of the rotifer dominance in the Paraná River main channel, in addition to its short generation time, which compensates losses by dragging, can be the higher tolerance of this group to suspended sediments. Rotifers are less affected than cladocerans and copepods, and this can modify the patterns of intraspecific competition (Kirk and Gilbert 1990; Jack et al. 1993).

Zooplankton is more abundant in lakes and it is very taxonomically diverse, particularly when there is a higher environmental heterogeneity, as that shown by heavily vegetated lakes. The higher water residence time determines higher plankton abundance; this factor has been analyzed by several authors (Pace et al. 1992; Basu and Pick 1996, 1997; Pourriot et al. 1997), and it would be even more important for zooplankton than for phytoplankton, since the latter presents a shorter generation time (Basu and Pick 1996). Directly connected lakes show a comparatively lower abundance than indirectly connected lakes with a higher isolation time. These results are contradictory to those found in the Upper Paraná River (Anderson et al. 2004). As suggested by Amoros and Bornette (2002), there are probably multiple factors interacting in the connections between environments.

The influence of the flood pulse on zooplankton of floodplain lakes would be related to their degree of connection with the river. In the directly

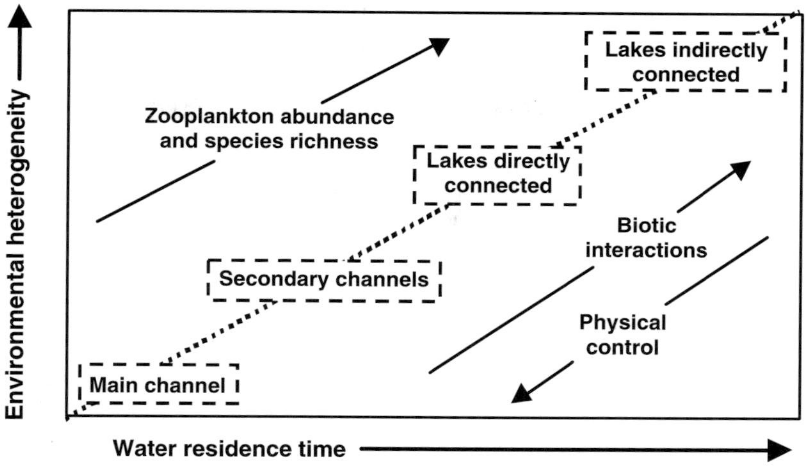

Fig. 9.8 Illustrative diagram of the factors that control abundance and richness of zooplankton in the main channel and the plain in the Paraná City-Santa Fe City cross section

connected lakes, physical factors associated to the flood pulse would be more important. In more isolated lakes, population control by the flood pulse would be lower, and biotic interactions, competition and predation would be more important (Fig. 9.8). Another important gradient for zooplankton is environmental heterogeneity, which becomes more complex along the environmental mosaic of the floodplain. Vegetated lakes offer a great variety of microhabitats with which they increase diversity (José de Paggi 1993, 2004b). The gradient of relative influence of the physical and biological control that extends from the main channel, dominated by physical factors, to the lakes, where biotic interactions show a higher influence, has been observed for zooplankton in the Danube River floodplain system (Baranyi et al. 2002), and for other communities in large rivers (Pecharsky et al. 1990; Statzner 1987; Johnson et al. 1995).

References

Anderson S, Aoyagui M, Bonecker C (2004) Rotifers in different environments of the Upper Paraná River floodplain (Brazil): richness, abundance and the relationship with connectivity. Hydrobiologia 522:281–290

Amoros C, Bornette G (1999) Connectivity and biocomplexity in waterbodies of riverine floodplains. Freshw Biol 47:761–776

Amoros C, Bornette G (2002) Connectivity and biocomplexity in waterbodies of riverine floodplains. Freshw Biol 47:761–776

Amoros C, Roux AL (1988) Interactions between water bodies within the floodplains of large rivers: functions and development of connectivity. In: Schreiber KF (ed) Fluvial hydrosystems. Chapman and Hall, London, pp 125–133

Baranyi C, Hein T, Holarek C, Keckeis S, Schiemer F (2002) Zooplankton biomass and community structure in a Danube River floodplain systems: effects of hydrology. Freshw Biol 47:473–482

Basu BK, Pick FR (1996) Factors regulating phytoplankton and zooplankton biomass in temperate lakes. Limnol Oceanogr 41:1572–1577

Basu BK, Pick PR (1997) Phytoplankton and zooplankton development in a lowland, temperate river. J Plankton Res 19:237–253

Boltovskoy D, Correa N, Cataldo D, Sylvester F (2006) Dispersion and ecological impact of the invasive freshwater bivalve *Limnoperna fortunei* in the Río de la Plata Watershed and beyond. Biol Invasions 8(4):947–963

Bonetto AA, Corrales M (1985) Zooplancton del Rio Paraná Medio: variaciones temporales y distribucionales en el área de confluencia con el río Paraguay. Ecosur 12/13:1–23

Bonetto AA, Martinez de Ferrato A (1966) Introducción al estudio del zooplancton de las cuencas isleñas del Paraná Medio. Physis 26(72):385–396

Bonetto CA (1983) Fitoplancton y producción primaria del Paraná Medio. Ecosur 10(19/20):79–102

Bonetto CA, Zalocar de Domitrovic Y, Vallejos E (1983) Fitoplancton y Producción primaria del Rio Alto Paraná (Argentina). Physis 41(101):81–93

Corrales MA (1979) Contribución al conocimiento del zooplancton del Alto Paraná. Ecosur 6:185–205

Darrigan G, Ezcurra de Drago E (2000) Distribución de *Limnoperna fortunei* (Dunker 1857) (Mytilidae), en la cuenca del Plata, Región Neotropical. Medio Ambiente 13:75–79

Depetris P, Kempe S (1993) Carbon dynamics and sources in the Paraná River. Limnol Oceanogr 38:382–395

Depetris P, Pasquini AI (2007) The geochemistry of the Paraná River: an overview. In: Iriondo M, Paggi JC, Parma J (eds) The Middle Paraná River: limnology of a subtropical wetland. Springer, Berlin Heidelberg New York (in press)

Drago EC (1973) Caracterización de la llanura aluvial del Paraná medio y de sus cuerpos de agua. Boletim Paranaense de Geociencias 31:31–44

Drago EC (1981) Grados de conexión y fases hidrológicas en ambientes leníticos de la llanura aluvial del río Paraná (Argentina). Ecología 6:27–33

Drago EC (1990) Geomorphology of large alluvial rivers: Lower Paraguay and Middle Paraná. Interciencia 15:378–387

Drago EC, Amsler, ML (1981) Sedimentos suspendidos en el tramo medio del río Paraná: variaciones temporales e influencia de los principales tributarios. Rev Asoc Cienc Nat Litoral 12:28–43

Drago EC, Amsler M (1988) Suspended sediment at a cross section of the Middle Paraná River: concentration, granulometry and influence on the main tributaries. Sediment Budgets 174:381–385

Frutos SM (1993a) Zooplancton en cuerpos de agua isleños del Bajo Paraná. Ambientes subtropical 3:87–121

Frutos SM (1993b) Zooplancton de la Laguna Turbia (Isla del Cerrito) en la confluencia de los ríos Paraná y Paraguay (Argentina). Rev Brasil Biol 56:569–580

Frutos SM (2004) Abundancia y riqueza de especies del zooplancton en el eje fluvial Paraguay-Paraná Com. Cient. Tecn. Univ. Nac. Nordeste B-046:4 p

Frutos SM (1996) Zooplancton de la laguna Turbia (Isla del Cerrito) en la confluencia de los ríos Paraná y Paraguay (Argentina). Rev Brasil Biol 56(3):569–580

Frutos SM (1998) Densidad y diversidad del zooplancton en los ríos Salado y Negro–Planicie del río Paraná Argentina. Rev Brasil Biol 58(3):431–444

Garcia de Emiliani MO (1997) Effects of water level fluctuations on phytoplankton in a river-floodplain lake system (Paraná River, Argentina). Hydrobiologia 357:1–15

Hamilton K, Sippel S, Lewis WM, Saunders III J (1990) Zooplankton abundance and evidence for its reduction by macrophyte mats in two Orinoco flood plain lakes. J Plankton Res 12:345–363

Jack JD, Thorp JH (2000) Effects of the benthic suspension feeder *Dreissena polymorpha* on zooplankton in a large river. Freshw Biol 44:569–579
Jack JD, Wickham SA, Toalson S, Gilbert JJ (1993) The effect of clays on a freshwater plankton community: an enclosure experiment. Arch Hydrobiol 127:257–270
Johnson B, Richardson W, Naimo TJ (1995) Past, present, and future concepts in large river ecology. Bioscience 45(3):134–141
Junk WJ, Bayley PB, Sparks RE (1989) The flood pulse concept in river-floodplain systems. In: Dodge DP (ed) Proceedings of the International Large River Symposium. Can Spec Publ Fish Aquat Sci 106:110–127
José de Paggi SB (1978) Observaciones sobre algunos rotíferos nuevos para la Fauna Argentina. Neotropica 24:99–104
José de Paggi SB (1979) Contribución al conocimiento de la fauna argentina de Rotíferos. II Algunas especies de los géneros *Lecane* (Nitzch) y *Lepadella* Bory de St.Vincent. Neotropica 25:37–44
José de Paggi SB (1980) Campaña limnológica Keratella I en el río Paraná Medio: zooplancton de ambientes lóticos. Ecología 4:69–75
José de Paggi SB (1982) Contribución al conocimiento de la fauna argentina de Rotíferos. III Nuevos registros específicos en ambientes del río Paraná medio. Neotropica 28:117–124
José de Paggi SB (1981) Variaciones temporales y distribución horizontal del zooplancton en algunos cauces secundarios del río Paraná Medio. Stud Neotrop Fauna Environ 16:185–199
José de Paggi SB (1983) Estudio sinóptico del zooplancton de los principales cauces y tributarios del valle aluvial del río Paraná: tramo Goya-Diamante. I Parte. Rev Asoc Cienc Nat Litoral 14:163–178
José de Paggi SB (1984) Estudios limnológicos en una sección transversal del tramo medio del río Paraná: distribución estacional del zooplancton. Rev Asoc Cienc Nat Litoral 15:135–155
José de Paggi S (1985) Distribución vertical del zooplancton en el río Paraná Medio (km. 603). Rev Hydrobiol Trop 18(1):13–21
José de Paggi SB (1988) Estudio sinóptico del zooplancton de cauces y tributarios del valle aluvial del Río Paraná, Tramo Goya-Diamante (II Parte). Stud Neotrop Fauna Environ 23:149–163
José de Paggi SB (1990) Ecological and biogeographical remarks on the rotifer fauna of Argentina. Rev Hydrobiol Trop 23(4):297–311
José de Paggi SB (1993) Composition and seasonality of planktonic rotifers in limnetic and littoral regions of a floodplain lake (Paraná river system). Rev Hydrobiol trop 26:53–63
Jose de Paggi S (1994) Zooplancton del Río Paraná: microcrustáceos y material inorgánico en suspension. Tesis Magister en Ecologia Acuática Continental, Universidad Nacional del Litoral 66 pp
José de Paggi S (1995) Vertical distribution and diel migration of rotifers in a Paraná River, floodplain lake. Hydrobiologia 310:87–94
José de Paggi SB (1996) Rotifera (Monogononta) diversity in subtropical waters of Argentina. Annales d Limnologie 32:209–220
José de Paggi SB (2004a) Diversidad de Rotiferos Monogononta en el Bajo Paraná. Tesis Doctoral, Universidad Nacional de La Plata, 200 pp
José de Paggi SB (2004b) Diversidad de Rotíferos Monogononta del Litoral Fluvial Argentino. INSUGEO 12:185–194
Junk WJ, Bayley PB, Sparks RE (1989) The flood pulse concept in river floodplain systems. Can Spec Fish Aquatic Sci 106:110–127
Kirk KL, Gilbert JJ (1990) Suspended clay and the population dynamics of planktonic rotifers and cladocerans. Ecology 71:1741–1755
Kobayashi T, Gibbs P, Dixon P, Shiel RJ (1996) Grazing by a river zooplankton community: importance of microzooplankton. Mar Freshw Res 47:1025–36
Lampert W, Somner U (1997) Limnoecology: the ecology of lakes and streams. Oxford University Press, New York

Martínez CC, Frutos SM (1986) Fluctuación temporal del zooplancton en Arroyos y Esteros del Chaco Oriental (Argentina). Ambiente Subtrop 1:112–133

Martínez CC, José de Paggi SB (1988) Especies de Lecane Nitzch (Rotifera Monogononta) en ambientes acuáticos del Chaco oriental y del valle aluvial del río Paraná (Argentina). Rev Hydrobiol Trop 21(4):279–295

Martinez de Ferrato A (1966) Nuevos cladóceros para las aguas argentinas. Physis 26:397–403

Martinez de Ferrato A (1967) Nuevos cladóceros para las aguas argentinas. II. Acta Zool Lillo 23:325–330

Morello J (1984) Perfil ecológico de Sudamérica. Características estructurales de Sudamérica y su relación con espacios semejantes del planeta. Ediciones Cultura Hispana, 93 pp

Neiff JJ (1999) El régimen de pulsos en ríos y grandes humedales de Sudamérica. In: Malvarez AI, Kandus P (eds) Tópicos sobre grandes humedales sudamericanos. ORCYT-MAB (UNESCO) pp 99–146

Oldani N (1990) Variaciones de la abundancia de peces del valle del río Paraná (Argentina) Rev Hydrobiol Trop 23:67–76

Oldani N, Tablado A (1985) Dinámica temporal de pequeños peces de agua libre en la laguna "La Cuarentena" (Isla Carabajal, río Paraná medio). Stud Neotrop Fauna Environ 20(1):49–58

Pace ML, Findlay SEG, Lints D (1992) Zooplankton in advective environments: the Hudson river community and a comparative analysis. Can J Fish Aquat Sci 49:1060–1069

Paggi JC (1972) Nota sistemática acerca del algunos Cladóceros del género *Chydorus* Leach 1843 de la Rep. Argentina. Physis 31:223–236

Paggi JC (1973a) Contribución al conocimiento de los Cladóceros dulceacuícolas argentinos. Physis 32:105–114

Paggi JC (1973b) Contribución al conocimiento de los Rotíferos dulceacuícolas planctónicos de la República Argentina. Physis 32:321–330

Paggi JC (1973c) Acerca de algunas especies de la familia Moinidae (Crustacea, Cladocera) de la República Argentina. Physis 32:269–277

Paggi JC (1978) Revisión de las especies argentinas del género *Diaphanosoma* Fisher (Crustacea, Cladocera). Acta Zool Lillo 33:43–65

Paggi JC (1978) Sobre la presencia de *Trochosphaera aequatorialis* Semper y *Horaella thomassoni* Koste (Rotatoria, Testudinellidae) en lagunas del valle de inundación del río Paraná. Rev Asoc Cienc Nat del Litoral 9:77–82

Paggi JC (1979) Revisión de las especies argentinas del género *Bosmina* Baird agrupadas en el subgénero *Neobosmina* Lieder (Crustacea, Cladocera). Acta Zool Lillo 35:137–162

Paggi JC (1980) Campaña limnológica Keratella I en el río Paraná medio (Argentina). Zooplancton de ambientes leníticos. Ecología 4:77–88

Paggi JC (2004) Importancia de la fauna de "Cladoceros" (Crustacea, Brachiopoda) del Litoral Fluvial Argentino. INSUGEO 12:5–12

Paggi JC, José de Paggi SB (1973) Sobre algunos rotíferos nuevos para la Fauna Argentina. Rev Asoc Cienc Nat Litoral 4:49–60

Paggi JC, José de Paggi SB (1974) Primeros estudios sobre el zooplancton de aguas lóticas del Paraná Medio. Physis B 86(33):91–114

Paggi JC, José de Paggi SB (1990) Zooplankton of lotic and lenitic environments of the Middle Paraná River. Acta Limnol Brasil 3:685–719

Pourriot R, Rougier C, Miquelis A (1997) Origin and development of river zooplankton: example of the Marne. Hydrobiologia 345:143–148

Perotti de Jorda NM (1984) Estudios limnológicos en una sección transversal del tramo medio del río Paraná: Biomasa y productividad del fitoplancton. Rev Asoc Cienc Nat Litoral 15:117–133

Quirós R (1988) Resultados del Simposio Internacional sobre grandes ríos y su aplicabilidad a los grandes ríos de América latina. COPESCAL Doc Ocas 5, 70 pp

Quirós R, Cuch S (1989) The fisheries and limnology of the Lower Plata basin. In: Dodge DP (ed) Proceedings of the International Large River Symposium. Can Spec Publ Fish Aquat Sci 106:429–443

Rossi LM (1989) Alimentación de larvas de *Salminus maxillosus* Val 1840 (Pisces Characidae). Iheringia Ser Zool 69:49–59
Rossi LM (1992) Evolución morfológica del aparato digestivo de postlarvas y prejuveniles de *Prochilodus lineatus* (Val., 1847) (Pises, Curimatidae) y su relación con la dieta. Rev Hydrobiol Trop 25:159–167
Rossi LM (2001) Ontogenetic diet shifts in a neotropical catfish, Sorubim lima (Schneider) from the River Paraná system. Fish Manage Ecol 8:141–152
Saunders JF, Lewis WM Jr (1989) Zooplankton abundance in the lower Orinoco River, Venezuela. Limnol Oceanogr 34:397–409
Schiaffino M (1981) Campaña limnológica Keratella I en el río Paraná Medio: XIII Fitoplancton de ambientes lóticos. Rev Asoc Cienc Nat Litoral 12:140–17
Sendacz S (1993) A study on the zooplankton community of the upper Paraná and floodplain lagoons. PhD Thesis. University of São Paulo, São Paulo
Shiel RJ, Walker KF, Williams WD (1982) Plankton of the lower River Murray, South Australia. Aust J Mar Freshw Res 33:210–227
Sylvester F, Dorado J, Boltovskoy; Juárez A, Cataldo A (2004) Filtration rates of the invasive pest bivalve *Limnoperna fortunei* as a function of size and temperature. Hydrobiologia 534:71–80
Thorp JH, Black AR, Haag KH (1994) Zooplankton assemblages in the Ohio River: seasonal, tributary, and navigation dam effects. Can J Fish Aquat Sci 51:1634–1643
Vasquez E, Rey J (1989) A longitudinal study of zooplankton along the Lower Orinoco River and its Delta (Venezuela). Ann Limnol 25:107–120
Vásquez E, Rey J (1992) Composition, abundance and biomass of zooplankton in Orinoco floodplain lakes, Venezuela. Ann Limnol 28(1):3–18
Vasquez E, Wilbert W (1992) The Orinoco: physical, biological and cultural diversity of a major tropical alluvial river. In: Calow P, Petts G (ed) The river handbook: hydrological and ecological principles, vol 1. Blackwell Scientific Publications, London, pp 448–470
Ward JV (1989) The four-dimensional nature of lotic ecosystem. J North Am Benthol Soc 8:2–8
Ward JV, Tockner K, Schiemer F (1999) Biodiversity of floodplain river ecosystems: ecotones and connectivity. Regul River: Res Manage 15:125–139

10 Benthic Invertebrates

Inés Ezcurra de Drago[1], Mercedes Marchese[2], and Luciana Montalto[2]

10.1 Introduction

The Middle Paraná River in contrast to the upper and lower segments, is less affected by the anthropogenic action, since there are no dams in this sector and the river runs through a very little industrialized region of Argentina. Therefore, it is an interesting area upon which to apply ecological theories and explain the dynamics and functioning of a highly complex system, even in quite natural conditions. The most applied concepts in the Paraná River system are the flood pulse (Junk et al. 1989), reconsidered and widened by Neiff (1990, 1999), Tockner et al. (2000) and Junk and Wantzen (2002), the connectivity among environments (Amoros and Roux 1988; Amoros and Bornette 2002), the analysis of patterns and processes in relation to the four dimensions (Ward 1989), and the analysis at different hierarchical scales (Frissell et al. 1986). On the other hand, although stream hydraulics (Statzner and Higler 1986) and the Hyporheic Corridor Concept (Stanford and Ward 1993) are applicable, the information on these topics is still very scarce.

The Middle Paraná River floodplain constitutes an environmental mosaic that includes secondary channels of different flow and discharge (from < 50 to > 2,000 $m^3.s^{-1}$), lakes of different origin and morphometry (i.e., irregular shape with several round-oval, elongated embayments, etc., according to Paira and Drago Chap. 3), different successional stages, according to the degree of connectivity and location in the floodplain, as well as a wide range of temporary wetlands. Therefore, there are representatives of almost all invertebrate taxonomic groups in the benthos of this system (Fig. 10.1). The geomorphological, hydrological and chemical characterization of the Paraná River is given by Iriondo and Paira (Chap. 1), Iriondo (Chap. 2), Paira and Drago (Chap. 3), Drago (Chap. 4), Amsler et al. (Chap. 5) and Depetris and Pasquini (Chap. 6). The river-floodplain system is very dynamic due to its hydrosedimentological regime, with an amplitude, duration and frequency

[1]Instituto Nacional de Limnología, INALI (CONICET-UNL). José Maciá 1933, 3016 Santo Tomé, Argentina, e-mail: inesezcurra@arnet.com.ar

[2]INALI (CONICET-UNL) and Facultad de Humanidades y Ciencias (FHUC-UNL), Ciudad Universitaria 3000, Santa Fe, Argentina

M.H. Iriondo, J.C. Paggi, and M.J. Parma (Eds.)
The Middle Paraná River: Limnology of a Subtropical Wetland

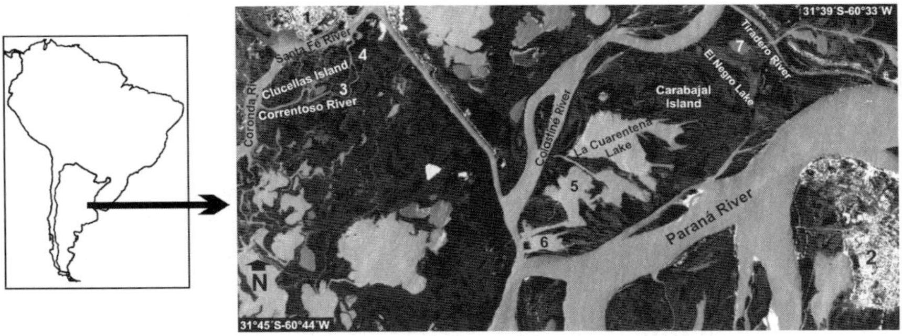

Fig. 10.1 Satellite image showing the environments analyzed in the alluvial plain of the Middle Paraná River. *1* Santa Fe City. *2* Paraná City. *3* El Tigre Lake. *4* Los Matadores Lake. *5* Vuelta de Irigoyen Lake. *6* El Puesto Lake. *7* La Mira Lake. Scale 1:50000, May 2004

upon which the offers of habitats, resources, refuges, distribution, abundance and complexity of benthic species assemblages depend.

The stream benthos is widely known worldwide since various decades ago, while that of large rivers has been more recently analyzed (Monakov 1969; Mordukhai-Boltovskoi 1979; Castella et al. 1991; Dudgeon 1995; Tockner et al. 1998; Rempel et al. 2000; Buffagni et al. 2000; Malmqvist 2002). Benthos of the Neotropical region, characterized by large rivers, was mainly analyzed in the Paraná River since more than two decades ago (summarized in Bonetto and Wais 1995; Takeda et al. 1997; Marchese et al. 2002; Takeda and Fujita 2004), as well as in the Uruguay River (Di Persia and Olazarri 1986; Ezcurra de Drago and Bonetto 1969) and, more recently, in the Paraguay River (Takeda et al. 2000; Barbosa et al. 2001; Ezcurra de Drago et al. 2004; Marchese et al. 2005).

On the other hand, knowledge on invertebrate diversity and dynamics in marginal wetlands of the Paraná River system is still very scarce, limited to studies on invertebrate colonization of different decomposed plant species (Neiff and Poi de Neiff 1990; Bruquetas de Zozaya and Neiff 1991; Poi de Neiff 1991; Poi de Neiff and Casco 2003; Capello et al. 2004). Studies on some taxa present in Middle Paraná River temporal wetlands have been recently carried out (Montalto and Marchese 2005; Montalto and Paggi 2006). There are studies on some aquatic and terrestrial arthropods of the Amazonian floodplain (Irlmer 1975; Reiss 1976; Nessimian et al. 1998; Walker 1986; Adis 1992; Junk 1997), and on invertebrates of temporary wetlands in the United States (Wissinger 1999) and Africa (Pacini and Harper 2001).

In this chapter, we summarize studies carried out on benthic organisms (≥ 100 μm) since 1969 in floodplain water bodies, since 1980 in the main channel and secondary channels of the Middle Paraná River, and since 2001 in marginal wetlands. Results on adult stages of indigenous Bivalvia (Hyriidae and Mutelidae) are not included.

10.2 Spatiotemporal Longitudinal Patterns

Benthos structure of the main channel and floodplain secondary channels with permanent flow (mean annual discharge > 500 $m^3 s^{-1}$, Drago et al. 2003) from the Middle Paraná River is very different in the central strip and the banks (Marchese and Ezcurra de Drago 1992; Marchese et al. 2002; Drago et al. 2003). A decrease in density, biomass, species richness and diversity in the longitudinal dimension of the Middle Paraná River was mainly observed in the main channel, while the Shannon index increased in the secondary channels (Fig. 10.2).

The central strip with sandy beds, mobile dunes and scarce organic matter content (Table 10.1) shows the highest density, lowest biomass, low species richness, and diversity of the Middle Paraná River mesohabitats. *Narapa bonettoi* (Oligochaeta) is dominant and often the only benthic species (maximum up to 112,000 ind. m^{-2}, Varela et al. 1983), and it is associated with lower densities of *Myoretronectes paranaensis, Itaspiella parana* (Turbellaria), *Haplotaxis aedeochaeta* (Oligochaeta), and *Tobrilus* sp. (Nematoda) (Fig. 10.3). *Potamocaris* spp. (Crustacea Harpacticoida) and *Parachironomus* sp. (Diptera, Chironomidae) are occasionally recorded. The main channel banks are silt-clayed, with high organic matter content in the bottom sediments and similar species assemblages to those of the floodplain channels. Oligochaeta (*Paranadrilus descolei, Limnodrilus hoffmeisteri,*

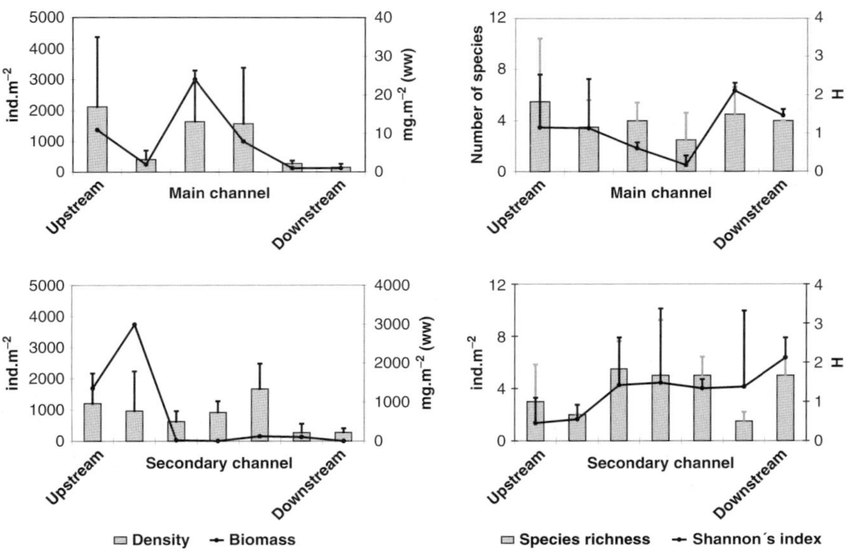

Fig. 10.2 Mean values of species richness, Shannon's index, density (ind. m^{-2}) and biomass (mg. m^{-2}) in longitudinal dimension of the main channel and permanent flow secondary channel. The *bars* indicate the standard deviation of three values

Table 10.1 Physical and chemical values (minimum-maximum) of the main channel and floodplains habitats of the Middle Paraná River

	Main channel (central strip)	Permanent flow secondary channels	Intermittent flow secondary channels	Floodplain lakes connected	Floodplain lakes with indirect connection	Floodplain lakes isolates	Temporary marginal wetlands
Discharge ($m^3.s^{-1}$)	4,000–60,000*	75–2,279	29–279				
Current velocity ($m.s^{-1}$)	0.57–1.77	0.21–0.78	n/r–0.79				
Depth (m)	9–50	4–11	4.5–7	1.10–5.70	1.50–4.60	0.50–5.80	0–1
Sand (%)	94–99	81–100	6–99	15–75	2–85	2–75	22–90
Silt (%)	0–6	0–18	0.1–65	5–80	10–80	10–85	4–32
Clay (%)	0–1	0–52	0.1–40	4–60	5–75	6–75	2–58
Organic matter (%)	0.003–0.021	0.012–0.105	0.403–1.355	0.59–1.98	1.3–10.6	1.89–6.5	0.3–18.5
Temperature (°C)	14–28	14–28	13–27	13–27	12–27	12–27	12.5–25.5
Oxygen ($mg.l^{-1}$)	6.5–9.0	5.0–9.8	0–8.9	4.6–8.2	0–8.5	0–8.9	0–7
pH	7.0–7.4	6.8–7.6	6.2–7.3	6.9–7.3	6.5–7.4	6.0–7.5	7.2–7.6
Conductivity ($\mu S.cm-^{1}$)	65–110	72–580	75–540	85–230	95–380	82–450	110–480
Secchi disc (m)	0.17–0.42	0.07–0.48	0.07–0.65	0.11–0.45	0.11–1.10	0.15–1.40	0.08–0.31
Vegetation cover (%)	0	0–10	4–90	7–20	6–30	10–100	50–100

*maximum discharge during El Niño events in 1983
n/r no recorded by the flow meter

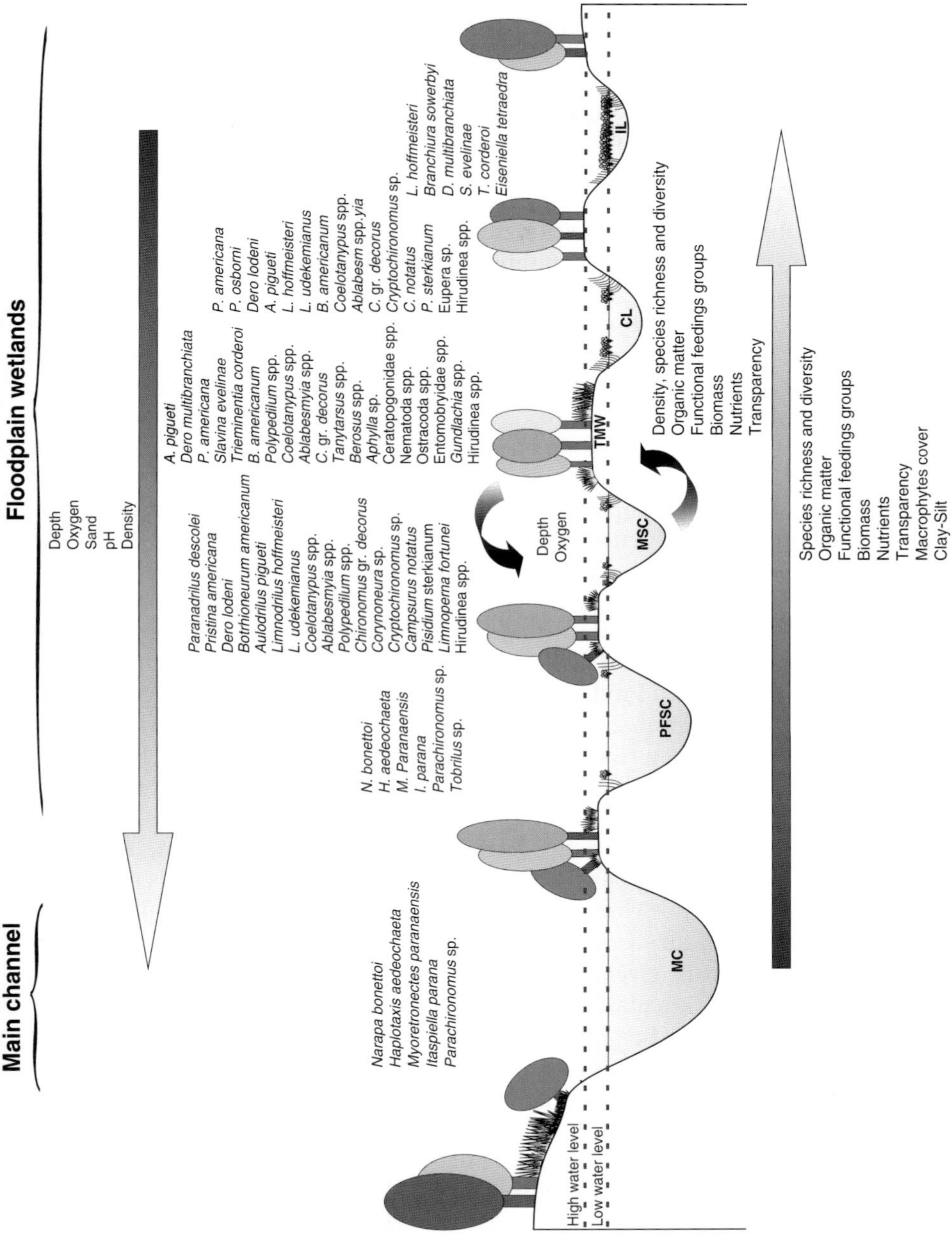

Fig. 10.3 Lateral gradients in physical, chemical, communities parameters and benthic species assemblages in the river-floodplain transect (indicated by *straight arrows*). The *small arrows* show the marginal wetland-river dynamic. *MC* main channel, *PFSC* permanent flow secondary channel, *MSC* major secondary channel, *TMW* temporal marginal wetland, *CL* connected lake, *IL* isolated lake. Modified from Marchese and Ezcurra de Drago (1992). With kind permission of Springer Science and Business Media

Aulodrilus pigueti, Bothrioneurum americanum, Pristina americana) and Chironomidae (*Cryptochironomus* sp., *Coelotanypus* sp., *Ablabesmyia* spp., *Axarus* sp.) are dominant taxa (Fig. 10.3). These mesohabitats showed higher benthic biomass, species richness, species diversity and lower density than the central strip (Marchese and Ezcurra de Drago 1992; Marchese et al. 2002). Along the Middle Paraná River banks, some species of Porifera, as *Trochospongilla repens, T. paulula* and *Oncosclera navicella*, developed on artificial substratum, are occasionally detected (Marchese et al. 2002). The central strip of rivers, as was established by Thienemann (1913) for lakes, is the best descriptive and diagnostic area for rivers because of the constant benthic species assemblages.

The only remarkable longitudinal change was observed in basaltic rock beds in the Apipé rapids and Paso de la Patria (Corrientes Province), with great populations of Porifera, as *Oncosclera navicella, O. schubarti, Trochospongilla repens, T. ringueleti* and *Corvospongilla seckti* (Ezcurra de Drago 1993), the epifaunal Bivalvia *Byssanodonta paranensis, Limnoperna fortunei* and the Gasteropoda *Potamolithus* sp., only recorded during low water, owing to the difficult sampling.

Similar benthic species assemblages were also registered in the Upper and Lower Paraná River from Tres Lagoas (Mato Grosso do Sul, Brazil, 20°47′S–51°42′W) to the mouth of the Río de la Plata (Varela et al. 1983; Montanholi-Martins and Takeda 1999; Peso and Bechara 1999; Marchese et al. 2002; Takeda and Fujita 2004), in the Upper and Lower Paraguay River (Ezcurra de Drago et al. 2004; Marchese et al. 2005), and in the Upper and Middle Uruguay River (Ezcurra de Drago pers. obs.). Basically, there is a stable benthic composition along approximately 3,500 km and 18° of latitude (from Cáceres, Brazil, 16°03′S–57°42′W, to the mouth of the Río de La Plata, 34°27′S–58°23′W) in the Paraguay-Paraná hydrosystem, and also along 1,200 km in the upper reach of the Paraná River, between Três Lagoas and Corrientes City. These species-poor, sand-adapted benthic assemblages, with small-size organisms, are similar to the bottom fauna from mobile sand-bottom rivers worldwide (e.g., White Nile (Monakov 1969); Athabasca River, Canada (Barton and Lock 1979); Malaysian rivers (Lim 1987); Amur River (Bogatov et al. 1995), and Ganges River (Dudgeon 1995)). There is a coincidence between the endemic dominant psammophilic species of Crustacea (gammarids) in the Volga River (Mordukhai-Boltovskoi 1979) and the endemic dominant psammophilic species of Oligochaeta (*Narapa bonettoi*) and Turbellaria (*Myoretronectes paranaensis* and *Itaspiella parana*) in the Neotropical large rivers (Paraná, Paraguay, Iguazú and Uruguay). We hypothesize that this benthic species assemblage may also be found in sandy sediments of other large Neotropical systems, such as the Magdalena, the Amazon, and the Orinoco Rivers (Ezcurra de Drago et al. 2004). We emphasize that benthos of large rivers' main channels, with sandy-mobile beds, has a high similarity world-wide, with high density, low species richness and very low biomass. However, there are variable species assemblages due to different biogeographical regions and ecological characteristics, mainly related with the chemical water quality (Marchese and Ezcurra de Drago 1992; Marchese et al. 2002; Ezcurra de Drago et al. 2004; Marchese et al. 2005).

On the other hand, there are noticeable variations in benthos composition and structure at the longitudinal dimension in the floodplain secondary channels with intermittent flow (mean annual discharge <170 $m^3 s^{-1}$, Drago et al. 2003), located in sectors with anastomosing patterns, which are very common in the floodplain network. Temporal variations in benthos structure and composition are influenced by the life cycles of dominant species, the hydroperiod, according to its duration and amplitude, and the environment successional stages. In the main channel, the greatest changes are given by the life cycle and reproductive strategies of *N. bonettoi*. In intermittent flow secondary channels (e.g., Correntoso River, 94 $m^3 s^{-1}$), changes are not only given by the life cycle of organisms but also by the influence of the flood pulse, that affects benthos succession. In these channels, benthic spatio-temporal changes produced by geomorphological, hydrological and sedimentological variations in the longitudinal dimension between the straight reaches (SR) and the confluence scour holes (CSH) were observed by Ezcurra de Drago et al. (unpublished) (Fig. 10.4a–c). During the drought phase, a lentification process is produced along the river, with hypoxia and/or anoxia mainly in the bottom of the CSH. There are an increase in the conductivity and percentages of silt, clay and organic matter, with a species assemblage similar to those of the SR, characterized by *Paranadrilus descolei, Limnodrilus hoffmeisteri, L. udekemianus, Pristina americana, Bothrioneurum americanus* and *Aulodrilus pigueti*, although with different relative densities (Fig. 10.4c). In the high water phase, current velocity, dissolved oxygen, and percentages of sand in the bottom sediments increase, while the conductivity, silt and clay percentage, and organic matter values decrease (Fig. 10.4c). During this phase, the greatest changes in the structure and benthic composition between SR and CSH are produced. There is a decrease in density in SR and a high increase in CSH, where the benthic assemblage changes drastically, with a very low density of *L. hoffmeisteri, L. udekemianus* and *P. americana*, and high densities of *N. bonettoi* and *M. paranaensis*. These are typical species of the main channel central assemblage (Figs. 10.3, 10.4c). In these microhabitats, the flow pulse (sensu Tockner et al. 1998) produces the washout of fine sediments in the shallow upstream and downstream borders, as in the bottom of the CSH, by a process of turbulent currents (Fig. 10.4a–b), creating the conditions required for the development of both species. Moreover, the influence of hydraulic parameters is also registered (Statzner and Higler 1986), as was observed by Blettler (pers. com.).

The highest benthic composition and structure changes between SR and CSH were recorded during La Niña phase (from October to December 1988) and El Niño phases (from May to August 1992 and from November 1997 to July 1998) (Fig. 10.5). The drought phase produces a benthos remarkable

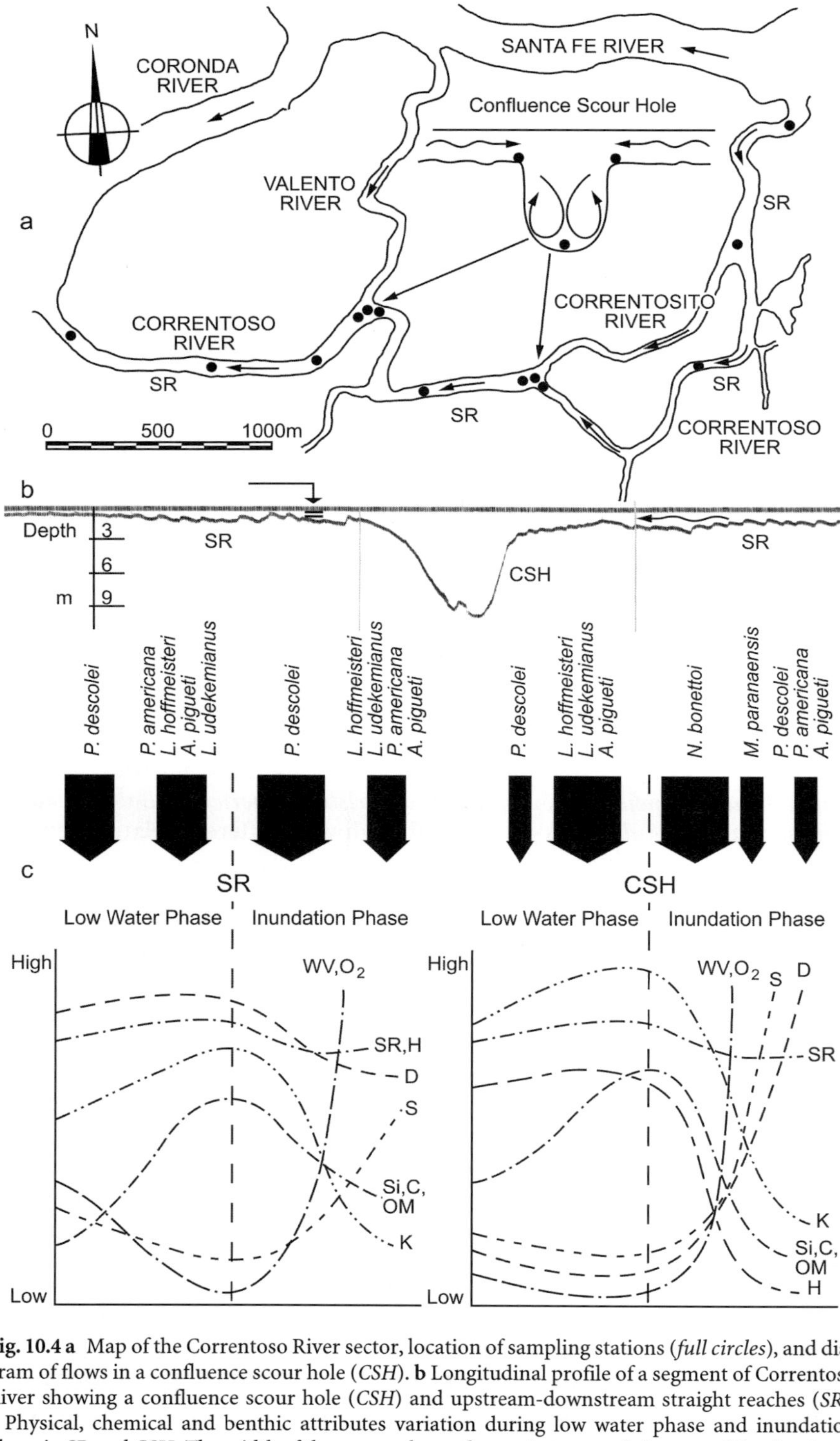

Fig. 10.4 a Map of the Correntoso River sector, location of sampling stations (*full circles*), and diagram of flows in a confluence scour hole (*CSH*). **b** Longitudinal profile of a segment of Correntoso River showing a confluence scour hole (*CSH*) and upstream-downstream straight reaches (*SR*). **c** Physical, chemical and benthic attributes variation during low water phase and inundation phase in SR and CSH. The *width of the arrows* shows the spatiotemporal changes of the dominant species

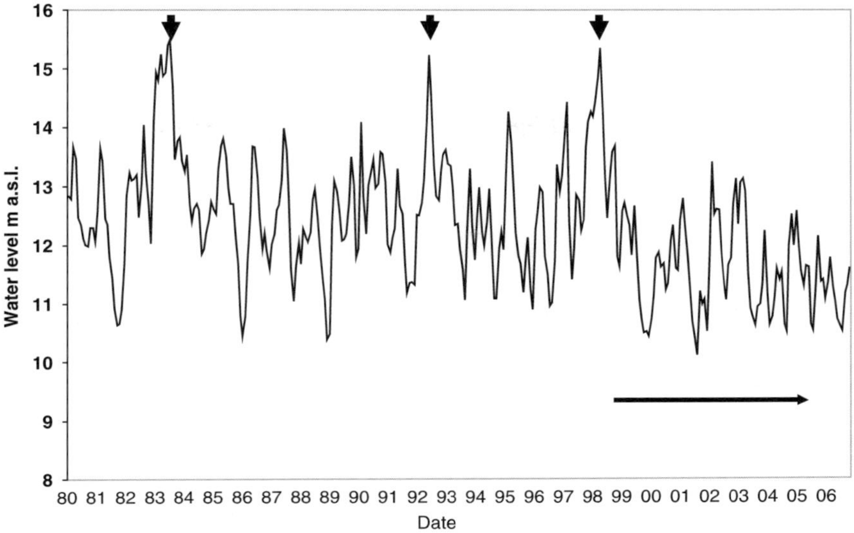

Fig. 10.5 Daily water levels of the Middle Paraná River (at Santa Fe Harbour staff gauge). *Vertical arrows* show the ENSO events. *Horizontal arrows* show the low water period from 1999 to 2006

stress, as a consequence of the changes produced by the advance of successional stages. The studies carried out on the Correntoso River benthos in SR and CSH in 1988–2006 show an evolution from an active channel to an environment, beginning a long-term lentification process (Drago et al. 2003), with predominance of sedimentation periods over erosion periods, from 1988 to 1992 (El Niño events). These conditions favored a marked increase in the rooted aquatic vegetation (mainly *Panicum elephantipes, Paspalum repens* and *Echinochloa polystachya*) and floating aquatic vegetation (*Eichhornia crassipes* and *E. azurea* as dominant species). During El Niño 1992, a reset of the system was produced and, afterwards, the succession advanced until the following El Niño events in 1997–1998, where the system was not able to reset again. The low hydrometric levels reached by the ordinary flooding and other hydrometric levels of lower magnitude than ordinary floods that occurred from 1999 to 2006 (Fig. 10.5) have allowed a great vegetation cover that reached 90% in the Correntoso River. These conditions allowed only scarce periods of a very low water movement, with a notable increase in fine bottom sediments (silt 56.93%, clay 41.41% and sand 1.66%) and OM (6.3%), especially CPOM. At this stage of the successional process, intermittent flow secondary channels benthos showed the lowest values of species richness (1–2 tolerant species), density and Shannon's index.

Although the El Niño events produce a reset of the floodplain environments, as mentioned before, such a process is not produced with the same intensity in all habitats, depending on the location of the alluvial plain channel. The benthic fauna begins to be similar to that of older floodplain lakes, completely covered by floating grasses and rooted vegetation when the channel is located in the old floodplain area, far away from the main channel, with a complex array of hydrosedimentological, topographic and vegetation influences and more complex connectivity processes (Drago et al. 2003).

10.3 Spatiotemporal Lateral and Vertical Patterns

In the lateral and temporal dimensions (Ward 1989), it is essential to consider the concept of the flood pulse (Junk et al. 1989) when analyzing the relationships between the structure of the fluvial environment and the dynamics of its particular biota. The Paraná River floodplain, as those of other large South-American rivers (e.g., Amazon, Orinoco, Magdalena, etc.), is characterized by a high macrophyte richness and production, that provide a great content of organic matter to the bottom, producing hypoxia and/or anoxia conditions in many environments, mainly in the low water period and at the beginning of the flooding phase. In the Carabajal Island (where El Puesto, La Cuarentena, El Negro and La Mira lakes are found, Fig. 10.1), mean values of dry organic matter for *Eichhornia crassipes* were 119 t × 100 ha of water body, and in the Clucellas Island (where Los Matadores and El Tigre lakes are located, Fig. 10.1), 624 t of dry organic matter (Bayo et al. 1981) were registered. This produced a decrease in richness and diversity of benthic species, and an increase in more tolerant species, that reach a larger size and biomass due to a lower interspecific competition. Nutrient recycling in the floodplain produces direct effects on the biota, and the highest fluvial animal biomass is directly or indirectly derived from production within the alluvial plain and not from organic matter produced upstream, as proposed by the River Continuum Concept (Vannote et al. 1980). The increase in nutrients is related to the interaction between aquatic environments and their margins, since decomposed aquatic macrophyte biomass and accumulated detritus are partially drifted to the terrestrial system, and the same situation occurs with the terrestrial vegetation that contributes to aquatic productivity. Studies on nutrient transference between aquatic and terrestrial zones are still scarce, but it is clear that both zones are interrelated and depend on each other on the environmental productivity and the hydrosedimentological regime (Carignan and Neiff 1992; Neiff et al. 2001).

The large river floodplain, in regions with seasonal precipitations, shows an annual hydrometric regime according to the generally predictable high and low water phases. Such fluctuations (floods and droughts), as described by Spark et al. (1990), are not disturbances but predictable changes, except if they show a high magnitude of maximum and/or minimum level, as the ENSO events, or if they are out of time. The Paraná River is subjected to an annual regime, with a low water phase produced in August–September and a high water phase approximately produced in March–April, although sometimes considerable temporal variations are observed in different years (Marchese et al. 2002). The results of alterations in the hydrological regime are associated to changes in the balance of production and respiration processes and in nutrient recycling patterns of the different floodplain habitats (Bonetto et al. 1986; Junk 1984; Camargo and Esteves 1996) and in the main river channel (Bonetto 1976; Neiff 1990).

In the river–floodplain lateral dimension, a gradient from the main channel to the temporary marginal wetlands is observed, with an increase in species richness, diversity and biomass, and a decrease in density (Figs. 10.3, 10.6). In this way, the typical species assemblage for each type of environment increase their composition in the transversal zonation (Fig. 10.3), where some common species are found (mainly oligochaetes and chironomids) among the different floodplain environments, including the temporary wetlands. In the river-floodplain transect, we find more than 150 taxa (without including Turbellaria, Nematoda and Crustacea) with a higher representativeness of oligochaetes in the channels and lakes, and insects in the temporary wetlands.

10.4 Temporary Marginal Wetlands

In the literature, marginal wetlands have been denominated as aquatic-terrestrial transitional zones or ATTZ (Junk et al. 1989), marginal fluvial wetlands (Neiff et al. 1994, 1999), or riparian wetlands (Mitsch 1996). Migration from the channels towards the alluvial plain is a common mechanism of invertebrate colonization in these systems (Irmler 1989; Montalto and Marchese 2005). In this way, marginal wetlands always show a higher number of species, some of them being characteristic of these habitats and temporarily absent in the overflowed river (Montalto and Marchese 2005). The high richness registered in these environments (up to 110 taxa), can be explained by the contribution of invertebrates associated to vegetation inputs in the overflowing phase. Moreover, the great diversity of species in temporary floodplain wetlands can be due to the insufficient time to allow a competitive exclusion. If we also consider the approximate total number of taxa (150) in the river–floodplain system, the evidence demonstrates that a physical factor (the flood pulse) produces and maintains a very diverse and dynamic habitat structure, with which a great biodiversity is generated, despite the stress of these transitional zones. The drastic change between aquatic and terrestrial phases gives as a result of high seasonal losses for most animal and plant populations, but these losses tend to recover by the rapid growth, early maturity, high rates of reproduction in *r*-strategist organisms and rapid dispersion, showing a high resilience. However, in other moments, *K*-strategists can be favored by their competitive capacity.

Pacini and Harper (2001) stated that it is often difficult to clearly distinguish wetland invertebrates from generically aquatic ones, or even from terrestrial ones. However, some invertebrates may be considered as good descriptors of these habitats. The main features that distinguish these animals are their adaptive, physiological and behavioral strategies to survive in habitats with wide changes during the hydroperiod, including their tolerance to extreme drought conditions. Aquatic invertebrates in these habitats must either survive to non-flooded periods with desiccation resistant life-cycle stages or persist in nearby water bodies and recolonize the temporal wetlands again after overbank flooding. Many aquatic species possess traits that enhance their resistance to water loss, such as diapause, cysts, and physiological tolerance to the drying phase (Laddle and Bass 1981; Junk 1997; Pacini and Harper 2001; Montalto and Marchese 2005).

10.5 Floodplain Lakes

Lakes show a high heterogeneity in the spatiotemporal benthos distribution, with variations related to the origin, age, morphometry, area, degree of plant cover, granulometric composition and amount of bottom sediment organic matter. The distance from the lake to the channels and the topography of the island in which the lake is located directly affect its degree of connectivity with the river (Amoros and Roux 1988), which influence the benthos temporal variations, produced by the flood pulse (FPTV) or by advanced successional stages (SS). Benthos composition and structure show higher spatial variations in irregular lakes with several embayments (e.g., La Cuarentena and El Puesto, Chap. 3) than in round-oval lakes (e.g., El Tigre, El Espinillo, Los Matadores and La Mira), since those present a higher heterogeneity, while elongated lakes (e.g., El Negro) are more uniform (Fig. 10.1). FPTV were registered with a higher intensity in the elongated lakes and in the directly or indirectly connected round-oval lakes, while they were only detected in the isolated lakes during ENSO events.

In the isolated round-oval lakes (e.g., El Tigre), marked spatiotemporal variations were registered due to differences in the temporal water quality, with an increase in conductivity (450 $\mu S.cm^{-1}$) during low water due to a seepage by groundwater, sand patches (70–90% and 0.86–1.89% OM), and silt clayed patches (90–98% and 3.4–6.0% OM), with bottom hypoxia conditions. Chironomidae (*Polypedilum* spp., *Aedokritus* sp., *Axarus* sp.) were dominant in sand patches and Chaoboridae (*Chaoborus* sp.), Tubificinae (*A. pigueti*, *Limnodrilus hoffmeisteri*), Ephemeroptera (*Campsurus* cf. *notatus)*, Hirudinae (*Helobdella adiastola*), Bivalvia Sphaeriidae (*Pisidium sterkianum*) and Chironomidae (*Chironomus* gr. *decorus)* were more abundant in silt-clayed patchiness. When the water level increases, the conductivity decreases and this in turn produces a decrease in density of *A. pigueti* and an increase of *Pristina americana, Dero lodeni, Trieminentia corderoi, Slavina evelinae* and *Branchiura sowerbyi*, always showing low biomass values (Fig. 10.6). In this

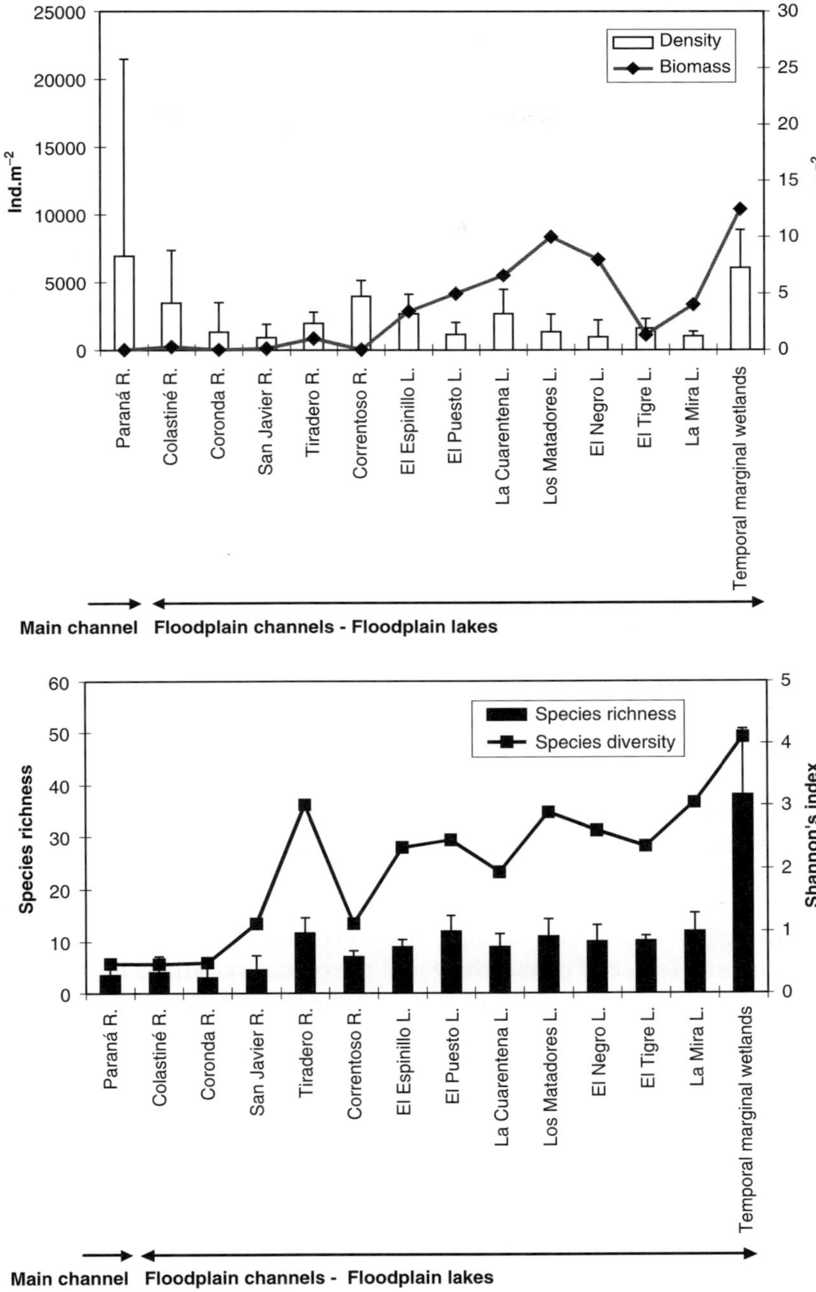

Fig. 10.6 Species richness, Shannon's index, density (ind. m^{-2}) and biomass (mg. m^{-2}) in lateral dimension from the main channel to temporary marginal wetlands

type of lake, which shows a long low water period (1999–2006), a very advanced successional state is observed, with 100% plant cover and great necromass in the bottom, creating adverse conditions for benthic invertebrates.

In an isolated (but directly connected) round-oval lake (e.g., Los Matadores), most differences were registered in the sector directly influenced by the river–lakes relationship, with lower organic matter content in the bottom sediments and without periods of hypoxia in the center of the lake (Table 10.1). These conditions determined a different benthic composition and structure, with an increase in density and biomass of *Campsurus* cf. *notatus*, mainly in the higher water movement sector (river–lake connection), a decrease in *Aulodrilus piqueti*, and an absence of *Branchiura sowerbyi* and *Chaoborus* sp.

The irregular lakes, with high shore development, several large embayments and an indirect connection to the main channel (e.g., La Cuarentena) show a main basin connected to the river and a smaller secondary basin with lower water movement, higher transparency, plant cover, and organic matter in bottom sediments (Table 10.1). These differences determine noticeable spatiotemporal variations in the benthic composition and structure. The benthic composition of the major basin is similar to that registered in the Los Matadores Lake for the sector that is connected to the river, with dominance of *Campsurus* cf. *notatus* in density and biomass. In the sector where the lake is connected with the minor basin, there is a great development of Sphaeriidae bivalves (mainly *Pisidium sterkianum* and *Eupera* cf. *klappenbachi*), that increases biomass, and in the minor basin, Oligochaeta and Chironomidae predominate, showing a similar composition to that of isolated round oval lakes during the high water level. In the embayments, there are hypoxia and/or anoxia conditions, and only tolerant species are registered (*Limnodrilus hoffmeisteri, Chironomus* gr. *decorus* and *Chaoborus* sp.). At the end of spring, the appearance of adults of *C. notatus* and *C.* gr. *decorus* begins, producing a sudden decrease in benthic biomass, while at the beginning of summer there is an increase in density because of egg hatching; high biomass is reached again in autumn. Because of the high area and shore development, the origin and degree of connectivity and the spatial variations in the benthic composition and structure, La Cuarentena Lake shows benthic species assemblages characteristic of the different successional stages of the Middle Paraná River floodplain lakes: major basin → minor basin → vegetated embayments.

10.6 Vertical Dimension

The vertical dimension also plays an important role in these systems, since it can provide refuge for egg incubation, juvenile hatching and habitat for the first stages, constituting a faunistic reserve capable of recolonizing the system after adverse conditions (Marmonier et al. 1992; Stanford and Ward 1993; Brunke and Gonser 1999). In the Paraná River marginal wetlands, many invertebrate species (chironomids, bivalves, oligochaetes, etc.) can move towards into the bottom sediments during the drought phases until favorable conditions are re-established (Montalto and Paggi 2006). Specimens of *Aulodrilus pigueti* have also been registered in sand banks 30 m from the Paraná River main channel (Marchese pers. obs.).

10.7 Factors Influencing the Benthic Structure

Physical and biological processes depend on the scale at which the observations are made. For example, physical processes, such as the discharge, and hydraulic parameters, substratum stability, erosion or deposition processes, temperature, chemical conditions (conductivity, pH), and macrophyte cover strongly affect distributional patterns reflected in differences between rivers or catchments. Grain size of bottom sediments, organic matter content, oxygen, food availability and biotic interactions, including predation and competition, create localized effects (Fig. 10.7). The benthic species assemblages composition in the Paraná system exhibits more variation in large spatial scales, where geophysical processes occur, than in smaller scales, as also reported by Li et al. (2001).

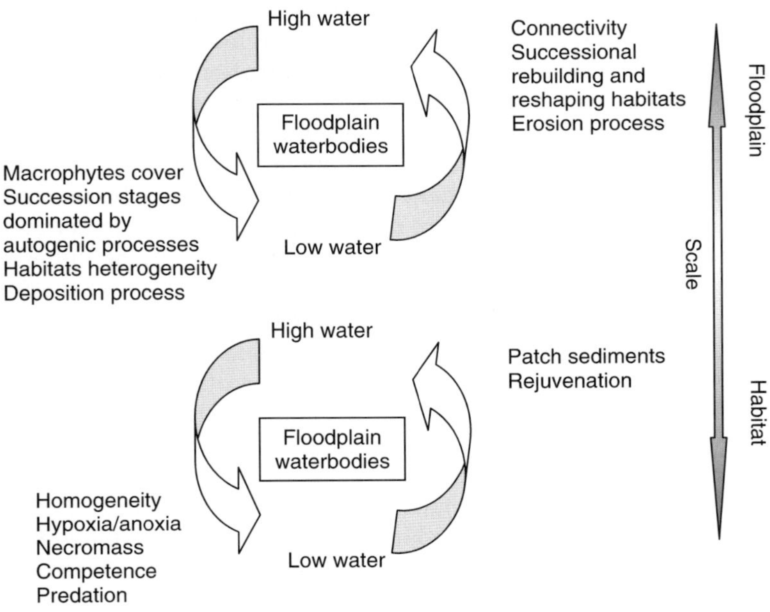

Fig. 10.7 The theoretical framework of the hydrosedimentological pulses effects on the floodplain water bodies at different scales

Shifts in the composition of benthic assemblages and total density are determined temporally in a two-time scale, the annual flood pulse and the connectivity degree among floodplain environments, and the long term phenomena involving ecological succession (Amoros and Bornette 2002). The differences are more evident among environments during low water (including low discharge secondary channels, indirectly connected floodplain isolates, and temporary wetlands), and within-streams at high water level, when the system is more hydrologically controlled. Low water is a more stressful factor than high water for benthic communities, in coincidence with Neiff (1996) and Junk et al. (1989) for vegetation and fishes. Wind, precipitations and temperature are also important factors with local effects, mainly in temporary wetlands.

The life cycles of many floodplain species are related to the flood pulse, its annual regulation, amplitude and duration, which, in turn, produce a selective pressure on aquatic organisms that colonize the temporary wetlands during the high water level. The invertebrate aquatic assemblages are replaced by terrestrial assemblages in marginal wetlands during the drought period.

10.8 Functional Feeding Groups and Food Webs

Organic matter provided by macrophytes is an important source of energy for floodplain habitats; the available particulate organic matter (POM) is not rapidly consumed by benthic macroinvertebrates and it is mainly accumulated in isolated floodplain lakes and temporary marginal wetlands, producing hypoxia or anoxia conditions. Riparian vegetation provides, in general, a lower amount of organic matter than aquatic plants. The paucity or absence of typical shredders in tropical streams was reported by many authors, and Wantzen and Wagner (2006) summarized various hypotheses for this situation. In the Middle Paraná River, we found many invertebrates that may be considered as shredders, for example *Polypedilum* spp., *Phaenosepctra* sp., *Endotribelos* sp. and *Hyalella curvispina*, and small crabs that chew conditioned litter or live vascular plant tissue, as defined by Cummins et al. (2005). These species contribute to the breakdown and reduce the amount of available coarse particulate organic matter (CPOM). However, the floodplain environments are a high necromass storage zone, mainly represented by roots and stems of macrophytes and contributions from riparian vegetation. The gathering collectors (Oligochaeta, Chironominae, *Campsurus* cf. *notatus*, etc.) are the most important feeding group, including many species in all the environments. The filtering collectors (*Limnoperna fortunei, Byssanodonta paranensis, Pisidium* spp. *Corbicula* spp., *Eupera* spp., and species of Mycetopodidae, Hyriidae, Hydropsychidae and Porifera) are recorded in several habitats from the main channel banks and across the floodplain. Predators (Hirudinea, Tanypodinae, Turbellaria, *Chaoborus* sp., Odonata, Tipulidae, Coleoptera, Decapoda Brachyura) are always represented, reaching the highest percentages in the temporary wetlands and floodplain lakes. The scrapers (*Heleobia parchappei, H. guaranitica, Asolene puelchella, Gundlachia* sp., *Marisa planogyra, Biomphalaria* sp. and *Pomacea* spp.) reach the highest density in isolated floodplain lakes, or in those indirectly connected ones with the river. High density of *Potamolithus* spp. is detected in the main channel in basaltic rocks of the Apipé rapids bed and Paso de la Patria (Corrientes Province).

There is an increasing gradient in functional group complexity from the main channel and secondary channels of high discharge and permanent flow to isolated floodplain lakes and marginal temporary wetlands represented by gathering-collectors, filtering-collectors and predators to gathering-collectors, filtering-collectors, scrapers, shredders, and predators. The main channel is characterized by the dominance of fine particulate organic matter in suspended loads and very scarce POM deposited in the bottom sediments, while floodplain environments are dominated by coarse particulate organic matter. The heterotrophy with leaf litter is the dominant base of benthic web foods in the floodplain environments. We also suggest that the central strip of the main channel and major secondary channels are autotrophic, with *Aulacoseira granulata* (Bacillariophyceae) as the dominant basis of simple benthic food webs, typical of these habitats, since *A. granulata* (dominant algae in the phytoplankton of the main channel, see Chap. 7) is the main food for *N. bonettoi*. Heterotrophy, with POM derived from the riparian zone and the floodplain during the high water level, is the basis of food webs in the banks of the main channel and high discharge secondary channels. However, large floodplain rivers have P/R values higher than 1 (Neiff 1996). It is important to state that the highest organic matter is autochthonous, produced in the same system.

The emphasis on longitudinal patterns and processes has long been considered in river ecology and several fundamental insights suggest changes in the food web along river networks (Vannote et al. 1980). Thompson and Townsend (2005) and Romanuk et al. (2006) found that fish species richness and connectance increase from mountain to foothill to prairie streams, and suggested that the upstream-downstream linkages and the longitudinal dimension structure change along the river through different physiographic regions. In contrast, macroinvertebrates and fish richness and connectance in large river floodplains increase in the lateral dimension from the main channels to floodplain lakes, with different connectivity degree, that may also be related to lateral changes in productivity. Predation by higher trophic levels (fishes, amphibians, birds, mammals) on benthic invertebrates was mainly observed in floodplain lakes, secondary channels, and marginal wetlands (Oliveros 1980; Beltzer 1991; Lajmanovich 2000; Peltzer and Lajmnovich 2004) (Fig. 10.8). Benthic communities of the main channel central strip and high discharge secondary channels are characterized by typical species assemblages with very scarce biomass, providing poor resources for higher trophic levels and, thus, the food-web structure seems to be limited with both prey and predators being benthic invertebrates (Fig. 10.8). However, fish larvae and juveniles feed on drifting chironomids and Naidinae

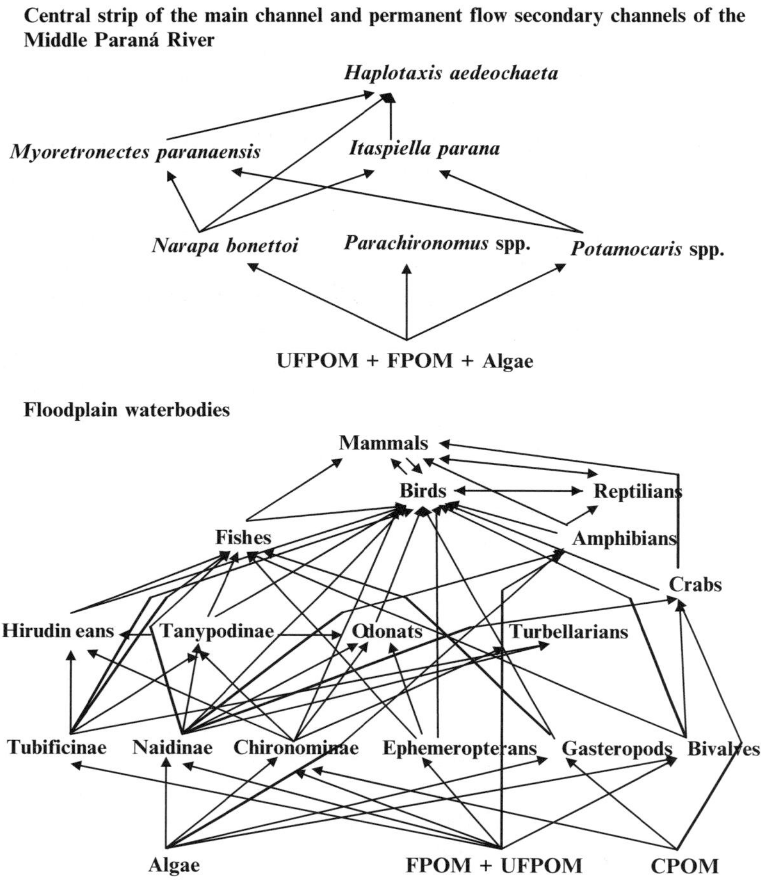

Fig. 10.8 Hypothetical pathways of matter flow in the river-floodplain of the Middle Paraná River

that could be resuspended by ascent currents or detached from the marginal vegetation (see Rossi et al. chapter 12). In floodplain environments, the predation pressure of benthivorous fishes appears to be much stronger during the low water level than during the high water level.

10.9 Bivalvia Invasive Species in Benthic Communities

Limnoperna fortunei (Dunker 1857) is an Asiatic invasive bivalve registered in the Middle Paraná River alluvial plain periphyton and benthos since 1996, reaching high population densities (Darrigran and Ezcurra de Drago 2000). The floodplain during the high water phase provides to *L. fortunei* a great diversity of colonizing substrates, in the river banks and in the river overflowing zones or in the marginal wetlands. The species colonization is favored when there is a big and long flooding period, as was observed in El Niño during 1998, where *L. fortunei* reached dense populations (Montalto et al. 1999). The extensive distribution of this species is due to its larval development, its extensive reproductive period, prolonged for almost all the year (Darrigran et al. 1999; Cataldo and Boltovskoy 2000; Ezcurra de Drago et al. 2006), and to the presence of the byssus from the last planktonic larval stage that allows it to attach to any type of substrate. Thus, *L. fortunei* is planktonic in its larval stages and benthic and associated to aquatic and riparian vegetation of all environments in its juvenile and adult stages in the Paraná River system.

Other invasive species are *Corbicula fluminea* (Muller 1774) and *C. largillierti* (Philippi 1811), native of Southeast Asia, north and east Australia and Africa, were introduced in Argentina in 1981 (Ituarte 1981). These species are not very abundant and are only occasionally registered in the environments of the Middle Paraná River, while they reach high densities in the Upper and Lower Paraná River segments (Takeda and Fujita 2004; Rodrigues Capítulo et al. 1998).

10.10 Conclusions

The Middle Paraná River system is arrayed as a mosaic of habitats undergoing succession. Key factors controlling benthic structure and composition are hydrosedimentological flood pulses, hydraulic conditions, hydrological connectivity degree, aquatic plants cover and water quality. Thus, benthic communities from aquatic habitats of the Paraná River system show the following patterns:

- The central strip is the most descriptive and diagnostic area of the rivers due to a higher stability of the benthic structure, with lower riparian and land-use influence.
- Benthic structure and composition in the main and secondary channels of higher discharge and permanent flow are similar along the longitudinal dimension, with a slight decrease in density and biomass downstream and very little seasonal variations.
- Benthic structure and composition in secondary channels of low discharge and intermittent flow show marked spatiotemporal variations between the straight reaches (SR) and the confluence scour holes (CSH). During the low water phase, the composition is similar between both sectors, but in the high water phase it changes abruptly.
- Benthic structure and composition in floodplain lakes vary according to origin, morphometry, floodplain location, degree of connectivity with the channels and successional stages. The greatest variations are given between the irregular

lakes connected with several embayments, and the isolated round-oval lakes with a decreasing gradient in density, biomass, species richness and diversity.
- The increase in the water level phase produces a reset in the successional stages, mainly in the directly or indirectly connected lakes, while in secondary channel with intermittent flow and in the isolated lakes, far away from the main channel, such reset is partially verified in the El Niño events.
- The spatiotemporal variations in the temporary marginal wetlands are very pronounced in relation to the flood pulse and the river overflow, with a gradual change in the composition of aquatic and terrestrial invertebrates. However, many aquatic invertebrates remain at these sites even in the drought period because of the adaptive strategies (cysts, eggs, refuge in hyporheos, etc.) that allow them to recolonize the habitats when conditions are favorable again.
- A gradient from the main channel to permanent flow secondary channels, to isolated lakes and temporary marginal wetlands is observed, with a decrease in density and an increase in biomass, richness, and diversity of species. In the same direction, an increase in the complexity of functional groups and trophic webs, and in the connectance, is produced.
- The shredders functional group, although scarce, is more represented in the floodplain environments where CPOM is very high. However, a great amount of necromass accumulates in the bottom of floodplain habitats, mainly in those of advanced successional stages.
- The typical benthic species assemblage of the main channel central strip and higher secondary channels is characterized by endemic species of the Neotropical region, as *Narapa bonettoi, Myoretronectes paranaensis, Itaspiella parana, Haplotaxis aedeochaeta,* while *Potamocaris* spp., *Tobrilus* sp. and *Parachironomus* spp. are occasionally registered.
- The typical benthic species assemblage of the main channel banks and higher secondary channels is similar to that of the secondary channels that show intermittent flow, with dominance of *Paranadrilus descolei, Botrhioneurum americanum, Campsurus* cf. *notatus* and *Pisidium sterkianum.*
- In lakes connected to the river, the typical benthic species assemblages is similar to that of secondary channels banks with permanent flow and that of the center of intermittent flow, except for *P. descolei*, which is only registered in rivers. In isolated lakes or in embayments of irregular lakes with more advanced successional stages and a higher content of organic matter in bottom sediments, the typical assemblage is given by the dominance of *Branchiura sowerbyi, Chaoborus* sp. and *Chironomus* gr. *decorus.*
- In temporary wetlands, in addition to many oligochaete species common to other floodplain environments, there is a higher representativeness of Insecta (mainly Coleoptera *Berosus* spp., Chironomidae *Polypedilum* spp., Ceratopogonidae spp., Dolychopodidae spp., Tipulidae spp., Odonata *Aphylla* sp.) and Ostracoda.
- Spatial heterogeneity, habitat structure, temporal instability and high productivity determine the higher importance of the lateral dimension than the longitudinal dimension in large floodplain rivers.

Acknowledgements. We thank the field and laboratory technical staff of the National Limnological Institute (INALI-CONICET-UNL) for their outstanding collaboration, and the National Council of Scientific and Technical Research for the grant to carry out this study. We thank Dr. C. Ituarte for Sphaeriidae identifications.

References

Adis J (1992) How to survive six months in a flooded soil: strategies in Chilopoda and Symphyla from central Amazonian floodplains. Stud Neotrop Fauna Environ 27:117–129

Amsler ML, Drago EC, Paira AR (2007) Fluvial sediments. Main channel and alluvial plain interrelationship. In: Iriondo M, Paggi JJ, Parma MJ (eds) The Middle Paraná River: limnology of a subtropical wetland. Springer, Berlin Heidelberg New York, pp 123–142

Amoros C, Bornette G (2002) Connectivity and biocomplexity in waterbodies of riverine floodplains. Freshw Biol 47:761–776

Amoros C, Roux AL (1988) Interaction between water bodies within the floodplains of large rivers: function and development of connectivity. In: Schreiber KF (ed) Connectivity in landscape ecology. Münsterche Geographische Arbeiten 29:125–130

Barbosa FAR, Faria Pereira MC, Avreu Vianna J (2001) Water quality, phytoplankton and benthic invertebrates of the Upper and Lower Río Paraguay Basin, Paraguay. In: Chernoff B, Willink PW, Montalbault JR (eds) A biological assessment of the aquatic ecosystem of the Río Paraguay basin. Conservation International, Washington, DC, pp 61–67

Barton DR, Lock MA (1979) Numerical abundance and biomasa of bacteria algae and macrobenthos of a large northern river, the Athabasca. Internat Rev Ges Hydrobiol 64:345–369

Bayo V, Lallana VH, Lorenzatti E, Marta MC (1981) Evaluación cuantitativa de la vegetación acuática en islas del vale aluvial del río Paraná Medio. Parte1. Ecología 6:67–72

Beltzer A (1991) Aspects of the foraging of the waders *Tringa flavipes, Calidris fuscicolollis* and *Charadrius collaris* (Aves: Scolopcidae, Charadriidae) in Del Cristal Pond (Santa Fe, Argentine). Stud Neotrop Fauna Environ 26:63–73

Bonetto AA (1976) Calidad de las aguas del río Paraná. Introducción a su estudio ecológico. Dirección Nacional de Construcciones Portuarias y Vías Navegables, INCYTH PNUD ONU Argentina

Bonetto AA, Varela ME, Bechara JA (1986) El bentos del Paraná Medio en el tramo Corrientes-Esquina. Ecosur 12/13:37–57

Bonetto AA, Wais IR (1995) Southern South American streams and rivers. In: Cushing CE, Cummins KW, Minshall GW (eds) River and streams ecosystems. Elsevier, Amsterdam, pp 257–293

Bogatov V, Sirostky S, Yuriev D (1995) The ecosystem of the Amur River. In: Cushing CE, Cummins KW, Minshall GW (eds) River and streams ecosystems. Elsevier, Amsterdam, pp 601–614

Brunke AC, Gosner T (1999) Hyporheic invertebrates—the clinal nature of interstitial communities structured by hydrological exchange and environmental gradients. J N Am Benthol Soc 18:344–362

Bruquetas de Zozaya I, Neiff JJ (1991) Decomposition and colonization by invertebrates of *Typha latifolia* L. litter in Chaco cattail swamp (Argentina). Aquat Bot 40:185–193

Buffagni A, Crosa GA, Harper DM, Kemp JL (2000) Using macroinvertebrates species assemblages to identify river channel habitat units: an application of the functional habitats concept to a large, unpolluted Italian river (River Ticino, northern Italy). Hydrobiologia 435:213–225

Camargo AFM, Esteves FA (1996) Influence of water level variation on biomass and chemical composition of the aquatic macrophyte *Eichhornia azurea* (Kunth) in an oxbow lake of the rio Mogui-Guaçu (Sâo Paulo, Brasil). Arch Hydrobiol 135(3):423–432

Capello S, Marchese M, Ezcurra de Drago I (2004) Descomposición y colonización por invertebrados de hojas de *Salix hundboldtiana* en la llanura aluvial del río Paraná Medio. Amazoniana 18:125–143

Carignan R, Neiff JJ (1992) Nutrient dynamics in the floodplain ponds of the Paraná River (Argentina) dominated by *Eichhornia crassipes*. Biogeochemistry 17:85–121

Castella E, Richardot-Coulet M, Roux C, Richoux P (1991) Aquatic macroinvertebrates assemblages of two contrasting floodplains: the Rhône and Ain Rivers, France. Regul River Res Manag 6:289–300

Cataldo D, Boltovskoy D (2000) Yearly reproductive activity of *Limnoperna fortunei* (Bivalvia) as inferred from the occurrence of its larvae in the plankton of the lower Paraná River and the Río de la Plata estuary (Argentina). Aquat Ecol 34:307–317

Cummins KW, Merritt RW, Andrade PCN (2005) The use of invertebrate functional groups to characterize ecosystem attributes in selected streams and rivers in south Brazil. Stud Neotrop Fauna Environ 40(1):69–89

Darrigran G, Ezcurra de Drago I (2000) Distribución de *Limnoperna fortunei* (Danker, 1857) (Mytilidae) en la Cuenca del Plata, Región Neotropical. Medio Ambiente 13(2):75–79

Darrigran G, Penchaszadeh P, Damborenea C (1999) The life cycle of *Limnoperna fortunei* (Danker, 1857) (Bivalvia: Mytilidae) from a neotropical temperate locality. Shell Res 18(2):361–365

Depetris PJ, Pasquini AI (2007) The geochemistry of the Paraná River: an overview. In: Iriondo M, Paggi JJ, Parma MJ (eds) The Middle Paraná River: limnology of a subtropical wetland. Springer, Berlin Heidelberg New York, pp 143–174

Di Persia DH, Olazarri J (1986) Zoobenthos of the Uruguay system. In: Davies BR, Walker KF (eds) The ecology of river Systems. Junk Publishers, Dordrecht, pp 623–630

Drago EC (2007) The physical dynamics of the river–lake floodplain system. In: Iriondo M, Paggi JJ, Parma MJ (eds) The Middle Paraná River: limnology of a subtropical wetland. Springer, Berlin Heidelberg New York, pp 83–122

Drago E, Ezcurrra de Drago I, Oliveros O, Paira A (2003) Aquatic habitats, fishes and benthic assemblages of the Middle Paraná River. Amazoniana 17:291–341

Dudgeon D (1995) The ecology of river and streams in tropical Asia. In: Cushing CE, Cummins KW, Minshall GW (eds) River and streams ecosystems. Elsevier, Amsterdam, pp 615–657

Ezcurra de Drago I (1993) Distribución geográfica de las esponjas argentinas (Porifera: Sponguillidae, Potamolepidae, Metariidae). Relaciones zoogeográficas, vías de doblamiento. In: Boltovskoy A, López HL (eds) Conferencias de Limnología. La Plata, Argentina, pp 115–125

Ezcurra de Drago I, Bonetto AA (1969) Algunas características del bentos en los saltos del río Uruguay, con especial referencia a la ecología de los poríferos. Physis 28:359–369

Ezcurra de Drago I, Marchese M, Wantzen KM (2004) Benthos of a large neotropical river: spatial patterns and species assemblages in the Lower Paraguay and its floodplains. Arch Hydrobiol 160:347–374

Ezcurra de Drago I, Montalto L, Oliveros OB (2006) Desarrollo y ecología larval de *Limnoperna fortunei*. In: Darrigran G, Damboronea C (eds) Bioinvasión del mejillón dorado en el continente americano. La Plata, Agentina, pp 83–91

Frissell CA, Liss WJ, Warren CE, Hurley MD (1986) A hierarchical framework for stream habitat classification: viewing streams in a watershed context. Environ Manage 10:19–214

Iriondo MH (2007) Geomorphology. In: Iriondo M, Paggi JJ, Parma MJ (eds) The Middle Paraná River: limnology of a subtropical wetland. Springer, Berlin Heidelberg New York, pp 33–52

Iriondo MH, Paira AR (2007) Physical geography of the basin. In: Iriondo M, Paggi JJ, Parma MJ (eds) The Middle Paraná River: limnology of a subtropical wetland. Springer, Berlin Heidelberg New York, pp 7–31

Irlmer U (1975) Ecological studies of the aquatic soil invertebrates in tree inundation forest of Central Amazon. Amazoniana 3:337–409

Irlmer U (1989) Population ecology and migration of *Dero multibranchiata* STIEREN, 1892 (Naididae, Oligochaeta) in Central Amazon Inundation Forest. Amazoniana 9:31–52

Ituarte C (1981) Primera noticia acerca de la introducción de pelecípodos asiáticos en el área rioplatense (Mollusca: Corbiculidae). Neotropica 27:79–83

Junk WJ (1984) Ecology of the várzea, floodplain of Amazonian whitewater rivers. In: Sioli H (ed) The Amazon: limnology and landscape ecology of a mighty tropical rivers and its basin. WJ Junk Publishers, Dordrecht, pp 215–243
Junk WJ (1997) The central Amazon floodplain, ecology of a pulsing system. Springer, Berlin Heidelberg New York
Junk WJ, Bayley PB, Sparks RE (1989) The flood pulse concept in river–floodplains systems. Can Spec Publ Fish Aquat Sci 106:110–127
Junk WJ, Wantzen KM (2002) The flood concept: new aspect, approaches, and applications: an update. Proceedings of the 2nd large River Symposium (LARS), Pnom Phen, Cambodia, pp 117–140
Ladle M, Bass JAB (1981) The ecology of a small chalk stream and its responses to drying during drought condition. Arch Hydrobiol 90:448–466
Lajmanovich RC (2000) Interpretación ecológica de una comunidad larvaria de anfibios anuros. Interciencia 25(2):71–79
Li J, Herlihy A, Gerth W, Kaufmann P, Gregory S, Urquhart S, Larsen DP (2001) Variability in stream macroinvertebrates at multiple spatialscales. Freshw Biol 46:87–97
Lim RP (1987) Water quality and faunal composition in the streams and rivers of the Ulu Endau area, Johore, Malaysia. Malay Nat J 41:337–347
Marchese M, Ezcurra de Drago I (1992) Benthos of the lotic environments in the middle Paraná River system: transverse zonation. Hydrobiologia 237:1–13
Marchese M, Ezcurra de Drago I, Drago E (2002) Benthos invertebrates and physical habitat relationships in the Paraná River flood–plain system. In: Mc Clain M (ed) The ecohydrology of South American rivers and wetlands. IAHS Spec Publ 6:111–132
Malmqvist B (2002) Aquatic invertebrates in riverine landscapes. Freshw Biol 47:679–694
Marchese MR, Wantzen KM, Ezcurra de Drago I (2005) Benthic invertebrate assemblages and species diversity patters of the upper Paraguay River. River Res Appl 21:485–499
Marmonier P, Dole-Olivier MJ, Creuzé Des Chǎtelliers M (1992) Spatial distribution of interstitial assemblages in the floodplain of the Rhône River. Reg Rivers 7:75–82
Mistch WJ (1996) Managing the world's wetlands—preserving and enhancing their ecological functions. Verh Int Ver Limnol 26:139–147
Monakov AV (1969) The zooplankton and zoobenthos of the white Nile and adjoining waters in the Republic of the Sudam. Hydrobiologia 33:161–185
Montanholi-Martins MC, Takeda AM (1999) Communities of benthic Oligochaetes in relation to sediment structure in the upper Paraná River, Brazil. Stud Neotrop Fauna Environ 34:52–58
Montalto L, Oliveros OB, Ezcurra de Drago I, Demonte LD (1999) Peces del río Paraná Medio predadores de una especie invasora: *Limnoperna fortunei* (Bivalvia, Mytilidae). Rev FABICIB 3:85–101
Montalto L, Marchese M (2005) Cyst formation of Naididae and Opistocystidae (Oligochaeta) as adaptive strategy for tolerating drought in fluvial wetland of Paraná River. Wetlands 25(2):488–494
Montalto L, Paggi A (2006) Diversity of chironomid larvae in a marginal fluvial wetland of the Middle Paraná River floodplain, Argentina. Int J Lim 42:289–300
Mordukhai-Boltovskoi D (1979) Zoobenthos and other invertebrates living on substrata. In: Mordukhai-Boltovskoi D (ed) The river Volga and its life. WJ Junk Publishers, Dordrecht, pp 235–268
Neiff JJ (1990) Ideas para la interpretación ecológica del Paraná. Interciencia 15(6):424–441
Neiff JJ (1996) Large rivers of South America: toward the new approach. Verh Int Ver Limnol 26:167–180
Neiff JJ (1999) El régimen de pulsos en ríos y grandes humedales de Sudamérica. In: Malvarez I (ed) Tópicos sobre humedales subtropicales y templados de Sudamérica. UNESCO, Montevideo Uruguay, pp 97–146
Neiff JJ, Poi de Neiff A (1990) Litterfall, leaf decomposition and litter colonization of *Tessaria integrifolia* (Compositae) in Paraná River floodplain. Hydrobiologia 203:45–52

Neiff JJ, Iriondo MH, Carignan R (1994) Large tropical South American wetlands: a review. In: Neimann RJ, Decamps H (eds) The ecology and management of aquatic-terrestrial ecotones. Unesco Washington, DC, pp 156–165

Neiff JJ, Poi de Neiff AG, Casco S (2001) The effect of prolonged floods on *Eichhornia crassipes* growth in the Paraná River floodplain lakes. Acta Limnol Bras 13(1):51–60

Nessimian JL, Dorvillé AM, Sanseverino AM, Baptista DF (1998) Relation between flood pulse and functional composition of the macroinvertebrates benthic fauna in the lower Rio Negro, Amazonas, Brasil. Amazoniana 15(1/2):35–50

Oliveros OB (1980) Campaña limnológica "Keratella I" en el río Paraná Medio: Aspectos tróficos de los peces de ambientes lenítIcos. Ecología 4:115–126

Pacini N, Harper M (2001) Biodiversity and conservation of afrotropical wetland invertebrates. In: Gopal B, Junk WJ, Davis JA Biodiversity in wetlands: assessment function and conservation, vol II. Backhuys Publishers, Leiden, The Netherlands, pp 133–156

Paira AR, Drago EC (2007) Origin, evolution and types of floodplain waterbodies. In: Iriondo M, Paggi JJ, Parma MJ (eds) The Middle Paraná River: limnology of a subtropical wetland. Springer, Berlin Heidelberg New York, pp 53–81

Peltzer PM, Lajmanovich RC (2004) Anuran tadpole assemblages in riparian areas of Paraná River (Argentina). Biodiv Conserv 13(10):1833–1842

Peso J, Bechara JA (1999) Estructura del zoobentos del Embalse Yacyretá en dos estaciones de muestreo, antes y después del llenado a cota 76 s.n.m. Rev Ictiol 7:37–47

Poi de Neiff A (1991) Caracterización funcional de los invertebrados en ríos de llanura del Chaco Oriental. Rev Bras Biol 50:875–882

Poi de Neiff A, Casco SL (2003) Biological agents that accelerate winter decay of *Eichhornia crassipes* Mart. Solms. in northeastern Argentina. In: Thomaz SM, Bini LM (eds) Ecologia e Manejo de Macrófitas Aquáticas. EDUEM, Maringá, Brasil, pp 127–144

Reiss F (1976) Charaktersierung zentralamazonischer Seen und Überschwemmungs-wälder. Amazoniana 5(1):3–23

Rempel LL, Richard JS, Healey MC (2000) Macroinvertebrates community structure along gradients of hydraulic and sedimentary conditions in a large gravel-bed river. Freshw Biol 45:57–73

Rodrigues Capìtulo A, César II, Tassara MP, Paggi AC, Remes Lenicov M (1998) Distribution of the macrobenthic fauna of the South Coastal Fringe of the Río de La Plata (Argentina). Impact of the urban contamination. Ver. Int. Verein. Limnol 26:1260–1265

Romanuk TN, Jackson LJ, Post JR, McCauley E, Martinez ND (2006) The structure of food webs along river networks. Ecography 29:3–10

Rossi L, Cordiviola de Yuan E, Parma MJ (2007) Fishes. In: Iriondo M, Paggi JJ, Parma MJ (eds) The Middle Paraná River: limnology of a subtropical wetland. Springer, Berlin Heidelberg New York, pp 305–325

Sparks R, Bayley P, Koholer S, Osborne L (1990) Disturbance and recovery of large floodplain rivers. Environ Manage 14(5):699–709

Standford JA, Ward JV (1993) An ecosystem perspective of alluvial rivers: connectivity and the hyporheic corridor. J N Am Benthol Soc 12:48–60

Statzner B, Higler B (1986) Stream hydraulics as a major determinant of benthic invertebrate zonation patterns. Freshwater Biol 16:1217–139

Takeda AM, Shimitzu G, Higuti J (1997) Variações espaço temporais da comunidade zoobentica. In: Vazzoler AE, Agostinho AA, Hahn NS (eds) A planície de inundação dọ Alto Rio Paraná: Aspectos físicos, químicos, biológicos y sócio econômicos. Ed UEM, Maringá Brasil, pp 155–175

Takeda AM, Callisto M, Barbosa F (2000) Zoobenthos survey of the Pantanal, Mato Grosso do Sul, Brasil. In: Willink PW, Chernoff B, Alonso LE, Montanbault JR, Lourival R (eds) A biological assessment of the aquatic ecosystems of the Pantanal, Mato Grosso do Sul, Brasil. Bull Biol Assess 18:103–106

Takeda AM, Fujita DS (2004) Benthic invertebrates. In: Thomaz SM, Agostinho AA, Hanh NS (eds) The Upper Paraná River and its floodplain: physical aspect, ecology and conservation. Backhuys Publ, Leiden The Netherlands, pp 191–208

Thienemann A (1913) Der Zusammenhang zwischen dem Sauertoffgehalt des Tiefenwassers und der Zusammensetzung der Tierfauna unserer Seen. Int Revue ges Hydrobiol Hidrogr 6:243–249

Tockner K, Schiemer F, Ward JV (1998) Conservation by restoration: the management concept for a river–floodplain system on the Danube River in Austria. Aquat Conserv 8:71–86

Tockner K, Malard F, Ward JV (2000) An extension of the flood pulse concept. Hidrol Process 14:2861–2883

Thompson RM, Townsend CR (2005) Energy availability, spatial heterogeneity and ecosystem size predict food-web structure in streams. Oikos 108:137–148

Vannote RL, Minshall GW, Cummins KW, Sedell JR, Cushing CE (1980) The river continuum concept. Can Fish Aquat Sci 37:130–137

Varela ME, Bechara J, Andreani NL (1983) Introducción al estudio del bentos del Alto Paraná. Ecosur (Argentina) 19(19/20):103–126

Walker I (1986) Experiments on colonization of small waterbodies by Culicidae and Chironomidae as a function of decomposing plants substrates and their implications for natural Amazonian ecosystem. Amazoniana 10(1):113–125

Wantzen KM, Wagner R (2006) Detritus processing by invertebrate shredders: a neotropical-temperate comparison. J N Am Benthol Soc 25(1):216–232

Ward JV (1989) The four-dimensional nature of lotic ecosystems. J N Am Benthol Soc 8:2–8

Wissinger SA (1999) Ecology of wetland invertebrates: synthesis and application to wetlands management. In: Batzer DP, Rader RB, Wissinger SA (eds) Invertebrates in freshwater wetlands of North America: ecology and management. Wiley, New York, pp 1043–1086

Zalocar de Domitrovic Y, Devercelli M, García de Emiliani MO (2007) Phytoplankton. In: Iriondo M, Paggi JJ, Parma MJ (eds) The Middle Paraná River: limnology of subtropical wetland. Springer, Berlin Heidelberg New York, pp 177–203

11 Littoral Communities. Macrocrustaceans

Pablo Collins[1,2,3,4], Veronica Williner[1,2], and Federico Giri[1,3]

11.1 Introduction

The littoral community of lotic and lentic environments of the Middle Paraná River is complex and dynamic, providing shelter to a high biological diversity and abundant populations. A numerically important group and with active participation in the community structure is that of macrocrustaceans, and, specifically, the order Decapoda, superorder Eucarida Calman 1904, subphylum Crustacea Brünnich 1772. The taxonomic unit Decapoda records more than 8,500 species, most of them restricted to marine areas (Brusca and Brusca 1990); however, some of them have successfully conquered freshwater and brackish environments. Among the latter, mangrove swamps and lenitic environments of large river floodplains provide the highest diversity and density of individuals due to their environmental heterogeneity (Bliss 1989).

South American freshwater decapods are grouped into seven families (Manning and Hobbs 1977; Rodríguez 1981, 1992; Magalhães and Türkay 1996; Morrone and Lopretto 2001). Their number in the Middle Paraná River environments decreases to four, systematically ordered according to Martin and Davis (2001) in Sergestidae, Palaemonidae, Aeglidae and Trichodactylidae (Table 11.1) (Lopretto 1995; Collins et al. 2002).

The analysis of this fauna entails to interpret macrocrustacean life in these unstable environments. The first crustacean fossil records correspond to the early Cambrian (550 million years ago), although they could also be placed in the Precambrian era (Scholtz 2004). Penaeoidea prawns, most of them marine, except for *A. paraguayensis*, with the most ancestral characters. The other prawns (Stenopodidea and Caridea) would come from independent evolutionary lines, together with Astacidea and Palinurea. The first Anomuran record appeared in the Jurassic, so they would be closer to Caridea and Brachyura, while many crab species appeared in the Eocene (Schweitzer and Feldmann 2005).

[1]Instituto Nacional de Limnología (CONICET-UNL), José Maciá 1933, 3016 Santo Tomé, Argentina
[2]FByCB-FHUC, Universidad Nacional del Litoral, Argentina
[3]FCyT, Universidad Autónoma Entre Ríos, Argentina
[4]e-mail: pcollins@arnet.com.ar

M.H. Iriondo, J.C. Paggi, and M.J. Parma (Eds.)
The Middle Paraná River: Limnology of a Subtropical Wetland

Table 11.1 List of Decapod taxa found in the Middle Paraná River and its floodplain ordered systematically according to Martin and Davis (2001). Some subdivisions in the Infraorder Brachyura were suppressed for an easier compilation of records

Suborder	Infraorder	Family	Species	Vulgar name
Drendrobranchiata		Sergestidae	*Acetes paraguayensis*	Planctonic shrimp
Pleocyemata	Caridea	Palaemonidae	*Macrobrachium borellii*	Prawn
			Palaemonetes argentinus	Prawn
			Pseudopalaemon bouvieri	Prawn
	Brachyura	Trichodactylidae	*Trichodactylus borellianus*	Crab
			Dilocarcinus pagei	Crab
			Sylviocarcinus australis	Crab
			Zilchiopsis collastinensis	Crab
			Z. oronensis	Crab
			Poppiana argentiniana	Crab
			Valdivia camerani	Crab
	Anomura	Aeglidae	*Aegla uruguayana*	Pancora or Crab
			A. platensis	Pancora or Crab

Since the radiation occurring after the effects of the Cretaceous/Paleozoic event (K/P) (Schweitzer and Feldmann 2005), some groups and species invaded freshwater environments, developing strategies that allowed them to survive in unstable conditions. The beginning of radiation could have begun in the Amazon River area for the family Trichodactylidae; in prawns, the information is confusing, indicating a group in Central America and another one in the Amazon. In Aeglidae, ancestors in marine sediments from Mexico (*Protaegla minuscula*) (approximately 110 million years) and New Zealand (*Haumuriaegla glaessneri*) were observed (Feldmann 1984; Feldmann et al. 1998). The hypotheses of their entrance to South America refer to a radiation in the Indo-Pacific region associated to the continental drift (Oligocene). The entrance would have been from the Río de la Plata, from Chile, during the glaciations, or from the Pacific Ocean, between Chile and Peru (Ortmann 1902; Schmitt 1942; Ringuelet 1949a, 1949b, 1949c; Perez-Lozada et al. 2004) (Fig. 11.1).

The evolution and radiation of freshwater decapod fauna coincides temporarily with the formation of the Paraná River, taking part in the dynamics showed from the beginning of this hydrosystem (see Chap. 2). These decapods came from Atlantic species that entered through the Río de la Plata, the Amazon, through ephemeral connections with the Paraguay River or its tributaries, and the High Paraná River, with specimens originated in the "Mata Atlantica" region (Schmitt 1942; Ringuelet 1949b, 1949c; Lopretto 1980; Collins et al. 2002; Collins 2007).

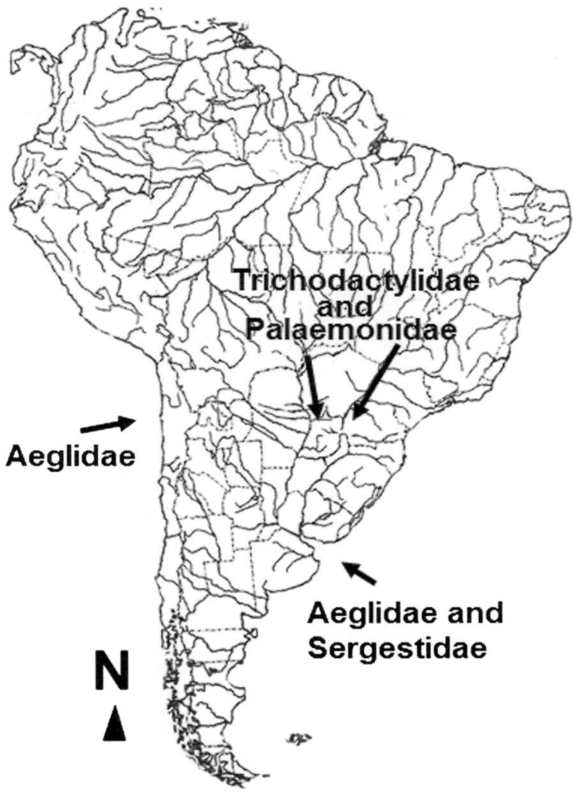

Fig. 11.1 Radiation of freshwater decapods to the Middle Paraná River

11.2 Biological Features

11.2.1 Morphology

The body of decapods can be divided into cephalothorax and abdomen. The first one includes cephalic and thoracic segments, where gills are closed by lateral extensions. Two subgroups can be recognized in the Middle Paraná River according to their type of gills. The first subgroup includes prawns with dendrobranchiate gills and planktonic eggs (only one species, *Acetes paraguayensis*). The other subgroup presents phylobranchiate gills and eggs transported under the abdomen of the female, including Caridean prawns, "páncora" crabs (*Aegla* sp., the only anomuran present in freshwater environments and endemic of South America), and true crabs (Rodríguez 1980) (Fig. 11.2).

The cephalothorax projects a rostrum in prawns and "páncora" crabs, being absent in true crabs. The eyes present a stalk; the first three pairs of

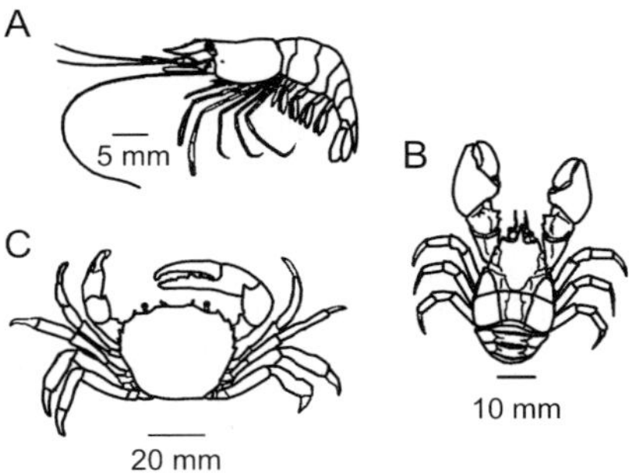

Fig. 11.2 Drawing of the Decapod morpho-types inhabiting the Middle Paraná River. **a** Palaemonidae-Sergestidae (prawn); **b** Aeglidae (pancora); **c** Trichodactylidae (true crab) (modified from Collins et al. 2004)

thoracic appendices are small and modified for feeding. The other five pairs of appendices are pereiopods. The abdomen is extended backwards in prawns, while in "páncora" crabs and true crabs there is a shortening and flexion upwards. Moreover, they have five pleopods and a pair of terminal uropods that can show different degrees of development.

Crustacea generally have the capacity to loose appendices by autotomy or appendotomy at certain risks or physical stress, as occurs generally in arthropods (Maruzzo et al. 2005). This phenomenon has been registered, under laboratory conditions, in decapods of the Middle Paraná River area as a consequence of chemical stress caused by agrotoxics (Montagna and Collins 2006).

Prawns show a small size, reaching a maximum record of 10 cm of total length (*M. borellii*) (Gonzalez-Baro et al. 1990), but being more frequent between 15 and 65 mm of cephalothorax length (Collins 1999a). "Páncora" crabs (Aeglidae) show mean sizes between 15 mm and 40 mm (Giri and Collins 2004), while among true crabs (Trichodactylidae), there are small-size specimens, as the genus *Trichodactylus* (between 10 and 25 mm carapace width (CW) (Collins et al. 2006), and medium-size specimens (between 25 and 40 mm CW) (*Dilocarcinus* sp., *Zilchiopsis* sp.) (Collins et al. 2002; Williner and Collins 2002a).

11.2.2 Growth

Growth is discontinuous, due to the rigid exocuticle, defined by two parameters: the intermolt time and the increase per molt. These parameters vary according to age, sex, temperature, population structure, type of food and environmental quality (Collins 1997a; Collins and Petriella 1999; Renzulli and Collins 2000; Williner and Collins 2000, 2003; Montagna and Collins 2004; Collins and Cappello 2006).

In turn, relative growth of body parts has different patterns of development associated to endogenous and exogenous factors (Collins 2001). This defines undifferentiated phases with isometric growth, corresponding to juveniles, and other differentiated phases, for adults, with different types of allometric growth. The transition between phases occurs through critical molts, indicating the beginning of gonadal maturation (Collins and Petriella 1999). The differentiation between sexes becomes important in the territorial defense, combats, displacements and reproductive courtships (Collins 2001). Variations in growth, as in the geometrical shape of the exoskeleton, are observed at population, sexual and specific levels (Giri and Collins 2004).

11.2.3 Internal Medium

Hyposaline environments, as the Paraná River, guided the appearance of mechanisms that allowed the decrease in permeability and salt losses, assuring gaseous exchange. During intermolt, the tegument has an effective impermeability (Fig. 11.3), being this stage longer than in marine crustaceans. Gills, branchiostegites and branchial chambers contribute directly in

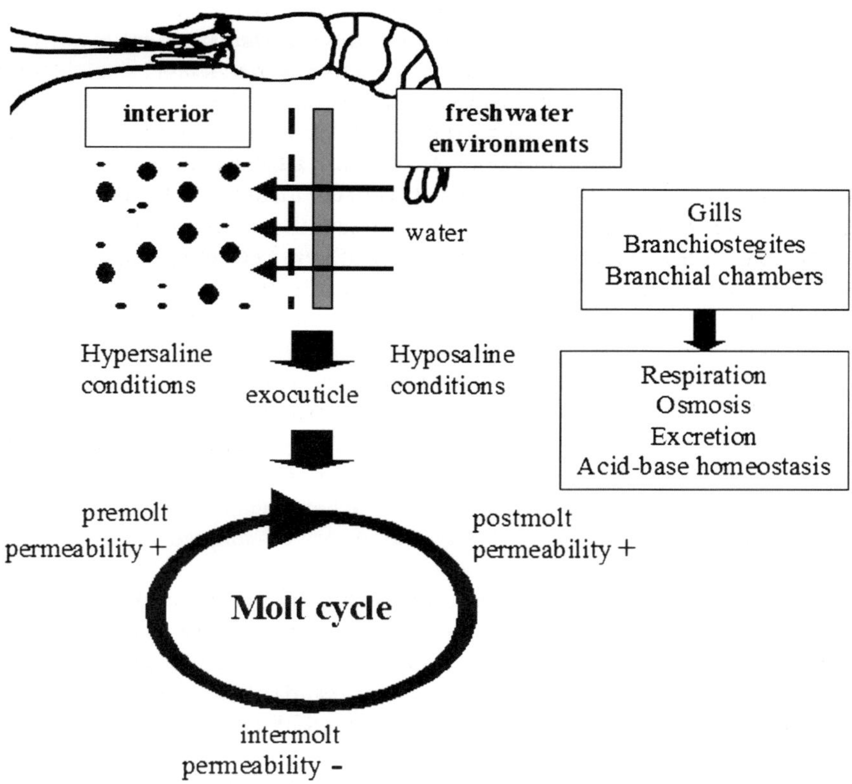

Fig. 11.3 Relation between freshwater environments and internal medium of decapods indicated the some critical region and permeability condition during the molt cycle

respiration, osmosis, excretion and acid-base homeostasis (via membrane carriers such as Na+/NH4+ or Cl–/HCO3–) (Schmidt-Nielsen 1997). The branchial chamber can be shut, becoming independent of the aquatic medium for a certain time, allowing the animals to move extensive sections through the mainland, as observed in some crabs (Fernández and Collins 2002).

Excretion is mainly done through a gland with an excretion pore in the antennal base. The primary function of this organ is to regulate the volume of fluids and salt concentration. This gland has a long nephridial channel, probably an adjustment to freshwater life, since it allows reabsorption of salts, producing a fluid that is hypoosmotic compared to haemolymph (Fingerman 1992).

Reproduction is carried out through unflagellated sperms carried inside spermatophores. These are eliminated by a gonopore placed at the base of the coxa of the fifth pair of pereiopods. Spermatophores are transferred to the female by the two first pleopods. The greater specialization occurs in crabs, in which the posterior pair acts as a piston. These crustaceans show external fertilization and spawning.

The stomach is composed of a cardiac and a pyloric chamber. The first one has slightly sclerotized supporting elements in its anterior wall, and the ventral surface of the pyloric chamber presents hardened borders covered by setae of different length. Setae transport and select the ingested material towards the stomach (Boschi 1981; Collins 2000), varying in number, shape, size, degree of suture and structure calcification in the different species. In the group Dendrobranchiata, represented by the genus *Acetes*, a certain uniformity and scarce stomach development are observed. Other adaptive mechanisms are the reduction in size and sclerotized structures (Caridea), or the increase in size and number of sclerotized structures (Aeglidea and Trichodactylidae). Within the Decapoda, crabs show the highest development in size and organ architecture (Icelly and Nott 1992).

11.3 Intra–Interspecific Population Interaction

Macrocrustacean populations show an ordering and spatial distribution ruled by physical and biological factors that mark some heterogeneity within a hydrosystem (Walker 1994; Carter et al. 1996). Species are co-adapted to the dynamics imposed in the Paraná River from its formation, habitat characteristics and stability degree, being the product of a joint organic evolution (Collins 2000).

Intra and/or interspecific interactions in prawns and crabs can modify some population parameters, as survival, growth, fecundity and/or local migration. In turn, populations are disturbed by extrinsic and intrinsic pressures, changing their densities along the year, as occurs in other crustaceans (Palmer et al. 1996). The elimination of individuals in an area occurs frequently or rarely, for prolonged or short periods of time, due to the intensity of the disturbance (e.g., flooding, drought, thermal extremes, loss of refuge) (Collins 2000).

On the other hand, the presence of prawns and crabs in an environment depends on the habitat-vegetation (e.g., species and plant age) - sediment (e.g., type of sediment and granulometry) relationship, physicochemical properties, food abundance, presence of preys, among others (Poi de Neiff and Carignan 1997; Lercari and Defeo 1999; Pothoven et al. 2004; Collins et al. 2006). Most lenitic environments are temporarily affected by the input and renovation of river waters (see Chap. 3). This determines that animals be found in a difficult situation between remaining in an environment that is drying, with the possibility of dying, or migrating to search for a place with suitable conditions to survive. When water level decreases, refuges are lost and populations of each species are concentrated, increasing the contact between individuals (Fernández and Collins 2002). These environmental modifications show their effect through movements of prawns and crabs among vegetated areas, non-vegetated areas, bottom sediments and water column, joining a daily activity rhythm (Renzulli and Collins 2001; Williner and Collins 2002a, 2005b).

11.3.1 Spatial Distribution and Agonistic Behaviors

In nature, there is a higher proportion of species with contagious spatial arrangement, as in prawns (*M. borellii*) and crabs (*T. borellianus*) (Williner and Collins 2000; Collins et al. 2006). An exception was observed in *P. argentinus*, since the location of an individual does not affect the presence of another one (Williner and Collins 2000). In natural environments, the spatial pattern of this species can be determined by characteristics of the habitat (presence or absence of vegetation), presence of competing species and/or predators.

There is no dominant morpho in species that present random aggregation (e.g., *P. argentinus*), while the prawn *M. borellii* changes growth rhythm together with an increase in aggressiveness of some specimens, indicating an agonistic behavior (Williner and Collins 2000). The determination of the dominant morpho is demonstrated by changes in the chela and/or body size (Karplus et al. 1991; Collins 2001). These ways of interaction involve behavioral, population dynamics and evolutionary responses (Collins 1995, 1997a, 2000; Renzulli and Collins 2000; Williner and Collins 2002a).

11.3.2 Densities

Densities of Palaemonidae in the Middle Paraná River show four sequential processes that coincide with the hydric and thermal cycles (Fig. 11.4). In certain moments, densities can exceed 500 prawns/m^2 (*P. argentinus* and *M. borellii*) and 170 crabs/m^2 (*T. borellianus*).

In summer, a decrease in densities is observed, together with an increase in the hydric level in the lake area, and a higher availability of the mobile

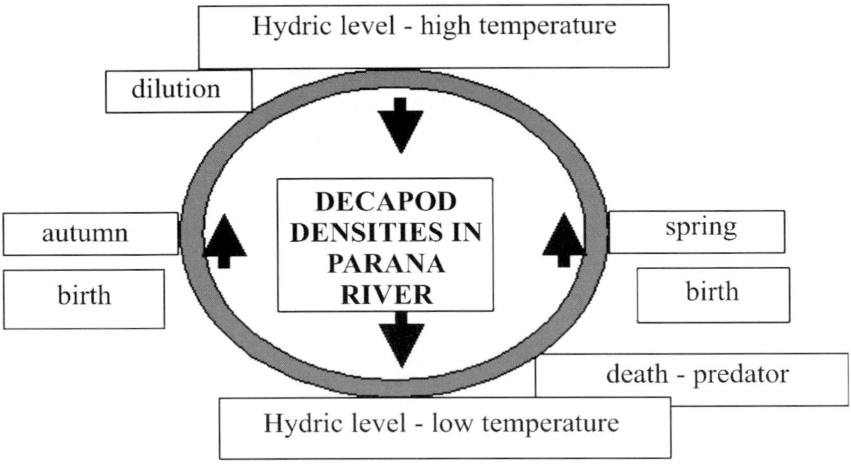

Fig. 11.4 Annual variations in Decapod density in the Middle Paraná River, indicating some factors that affect them. *Arrows* show increasing or decreasing moments

littoral area. This occurs by a process of dilution, together with a migratory event towards new flooded areas or other channels.

Massive migratory processes are common in species that must find brackish water to reproduce (see Sect. 11.5). In the Paraná River, there is only one prawn species that has massive movements: *A. paraguayensis* (Collins and Williner 2003). Mainland movements are specially observed after extreme droughts in crabs (*D. pagei*) (Fernández and Collins 2002). In the Amazon basin, the numerical decrease in the prawn *M. amazonicum* is associated to migrations during floods (Walker and Ferreira 1985). In other environments, factors that set off migratory processes are precipitations, temperature, salinity and nutritional requirements (Bamber and Henderson 1994; Cartaxana 1994; Sogard and Able 1994). There is not enough information yet on prawn and crab migrations in the Paraná River.

In autumn, the number of individuals increases due to the birth of new cohorts and the concentration of specimens, since the littoral area decreases. In winter, densities decrease due to several factors, for example, death of old individuals, as occurs in higher latitudes (Rodríguez Capitulo and Freyre 1989; Spivak 1997), and/or predatory phenomena, coinciding with the decrease in aquatic vegetation refuges. Conditions in this season favor aggressive interactions, producing interspecific competitions (Collins 2000), as occurs in crayfishes (*Procambarus* sp.) (Blank and Fingler 1996).

In spring, an increase in populations occurs due to the input of new cohorts and migrating specimens of other subpopulations. At this stage, reproductive aggregations, a decrease in predation, increase in available refuges in the vegetation and in trophic diversity, and the presence of alternative preys in the littoral community are observed (Collins 2000).

The crab *T. borellianus*, closely related to the floating aquatic vegetation (*Eichhornia* sp.), varies its densities along the year according to a bimodal frequency, coexisting adults and juveniles, coupled with the hydrosedimentological cycle (Collins et al. 2006). This is reflected in different mortality rates for males and females, recruitment pulses and differential movements, according to sex and habitat preferences for each sex (Díaz and Conde 1989). In other tropical and subtropical crabs, simple modes in the size distribution are observed, reflecting differences with populations of temperate regions (Alarcón et al. 2002).

11.3.3 Microspatial Distribution

Species that coexist and have similar requirements can show displacement of characters in some population parameters. This occurs in the most elastic species due to the capacity to vary its needs avoiding the risk of lesions or death. The change can refer to endogenous rhythms or, as it is observed in palaemonids, to variations in the microspatial distribution between the lake-side and the center (Collins 2000). Most abundant Palaemonidae in the Middle Paraná River (*M. borellii* and *P. argentinus*) have a similar geographical distribution, but temporal and evolutionary distances in the conquest of freshwater environments (Boschi 1981). However, they show great similarity in their ecological requirements (Lopretto 1995; Collins and Paggi 1998; Collins 1999b, 2005). Therefore, the microdistribution of these prawns from the lake-side towards the interior reflects seasonal variations (Fig. 11.5).

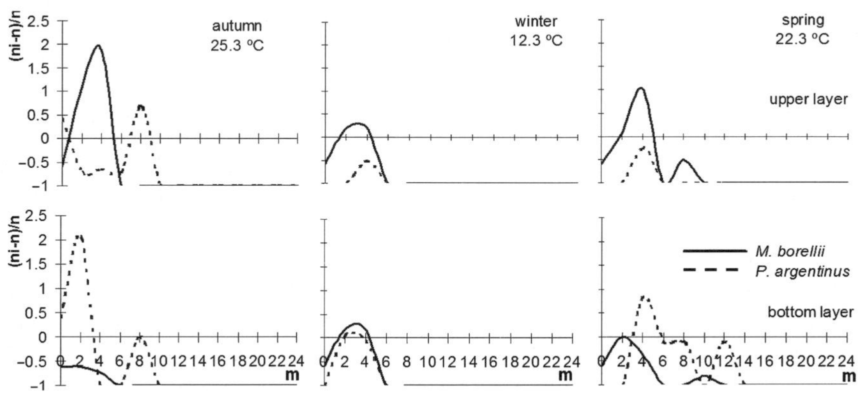

Fig. 11.5 Distribution of prawns *P. argentinus* and *M. borellii* in the micro-space between the lake-side and the center of the lakes in areas with aquatic vegetation. The distribution is measured by a variation rate (*(ni-n)/n*), where *ni* is the densities of a species in a point and *n* is the densities mean of species. Abobe sequences: upper layer of the water, below sequences: bottom layer of the water body. Summer is not represented because limits are not defined

When the most aggressive species (*M. borellii*) is found in high densities, the other one (*P. argentinus*) is found in a lower number. This situation represents a type of adjustment to a competitive pressure (Durret and Levin 1998) during a stable period (autumn), avoiding risks by attacks due to a higher vulnerability (higher ecdysis frequency) (Kneib 1987). In non-vegetated areas, the spatial arrangement is different for the two species. The displacement of *P. argentinus* towards more internal areas of the lake is forced by the presence of the most aggressive species (*M. borellii*), that remains in the area closest to the lake-side, where food abundance is higher.

In winter, there is a spatial selection by behavior according to the thermal tolerance of these species. In this sense, *M. borellii* is more frequent in the area near the lake-side, while *P. argentinus* shows a higher tolerance to cold waters. In spring, modifications are dynamic, with active changes, caused by reproductive movements of sexually active specimens, post-larvae, juveniles or different age groups. In turn, the increase in aquatic vegetation favors a higher quantity of possible trophic sites, determining an increase in areas free of competition (Collins 2000).

In summer, when the river rises and the limits of the water bodies are not defined, new habitats are formed due to the widening of the water-land transition zone (Junk et al. 1989). The incorporation of new patches, with unexploited trophic offers and unused refuges, provokes the rupture in the microspatial distribution established previously (Collins 2000).

11.4 Trophic Relationships of Freshwater Decapods

Trophic relationships are strongly influenced by interactions of abiotic and biotic factors (food availability, competition, risk of predation, annual cycles, daily cycles, flood pulses, temperature, reproduction, development, among others). Freshwater decapods are not only predators but also preys molded by these factors that regulate and invigorate the hydrosystem, taking part in trophic chains that involve aquatic, semiterrestrial and terrestrial environments. Prawns (Sergestidae and Palaemonidae) and crabs (Trichodactylidae and Aeglidae) of the Paraná River are omnivorous and use different trophic levels (Table 11.2) (Collins et al. 2007).

11.4.1 Potential Predators and Preys

The importance of decapods in the trophic webs of the Middle Paraná River lies in their high densities and in the nutritional quality that is transferred energetically towards different levels (Fig. 11.6). Firstly, there is cannibalism on recently molted or smaller specimens. On the other hand, several authors consider decapods as important components of the trophic spectrum of the

Table 11.2 List and frequency of items found in the stomach of prawns and crabs inhabiting the Middle Paraná River grouped according to families (* low frequency; ** medium frequency; *** high frequency). *UA* unicellular algae; *FA* filamentous algae; *C* Cyanophyceae; *B* Bacillariophyceae; *PR* plant rests; *F* Fungii; *PROT* Protozoa; *ROTIF* Rotifera; *N* Nematoda; *OS* Ostracoda; *CLAD* Cladocera; *COPCAL* Copepoda Calanoidea; *COPCICL* Copepoda Cyclopoidea; *D* Decapoda; *CH L* Chironomid larvae; *I L* other insect larvae; *A* Acari; *T* Tardigrada; *OLIG* Oligochaeta; *S* sand

	Sergestidae[1]	Palaemonidae[2,3,4,5,6]	Trichodactylidae[7,8,9]	Aeglidae[10]
FA	**	***	**	**
UA	**	***	*	*
C		**	*	
B	**	**	**	**
PR	**	***	***	***
F	*		*	***
PROT	***		**	*
ROTIF	***	*	**	*
N		*		*
OS			*	
CLAD	**	**	*	*
COPCAL	**	**	*	*
COPCICL		**	*	*
D		*		*
CH L		***		
I L	*	***	***	**
A				*
T				***
OLIG	**	***	**	**
S	***	***	***	***

[1]Collins and Williner 2003, [2]Collins and Paggi 1998, [3]Collins 2005, [4]Collins and Williner 2001, [5]Collins and Williner 2002, [6]Collins 1999b, [7]Williner and Collins 2002a, [8]Williner and Collins 1999, [9]Collins and Williner 2005, [10]Williner 2003

aquatic fauna and of terrestrial organisms linked to the hydrosystem. For example, fish (e.g., yellow catfish, "moncholos", catfish in general) (Bonetto et al. 1963; Oliva et al. 1981), amphibians (e.g., frogs), reptiles (e.g., vipers, alligators), birds (e.g., herons, "biguáes") and mammals (e.g., otters, monkeys) (Beltzer 1983a, 1983b; Bond Buckup and Buckup 1994; Williams and Scrochi 1994; Gori et al. 2003; Port-Carvalho et al. 2004; Lopez et al. 2006).

Moreover, decapod potential preys, macropreys and also minor components are included in this section. It is necessary to precise the preys obtained from an active search, as well as those ingested passively, whose capture can be considered as occasional (Fig. 11.6). Natural fauna supplements and satisfies

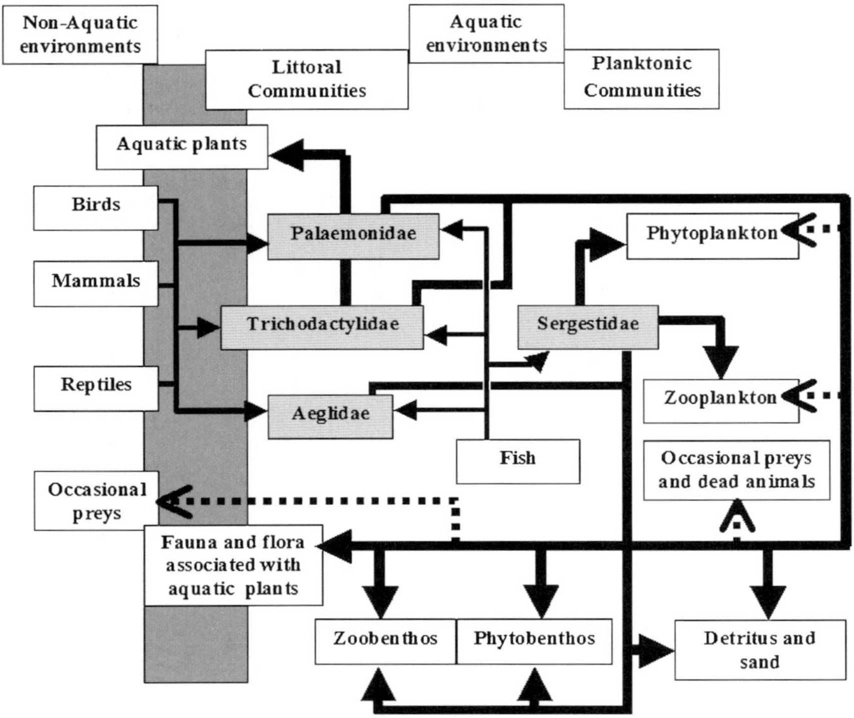

Fig. 11.6 Predators and preys of Decapods in the Middle Paraná River considered non-aquatic and aquatic environments. *Fine line* predators, *thick line* preys (*straight line* more important, *dotted line* occasional uses)

consumers with macronutrients (proteins, lipids, and carbohydrates) and micronutrients, such as vitamins, cholesterol, phospholipids and minerals in an environment where some elements are not quantitatively bioavailable (Collins 2004). Common preys of larger sizes (e.g., insect larvae and oligochaetes) provide high proportions of proteins. Planktonic elements, i.e., phytoplankton (e.g., green algae and diatoms) and zooplankton (e.g., cladocerans, copepods), play an important role in some moments of the ontogenetic and seasonal cycles, providing micronutrients.

11.4.2 Trophic Habits—Trophic Ecology

The feeding habit of many decapods shows changes in relation to larval development. However, prawns and crabs in the Paraná River have a short larval development, except for *A. paraguayensis* and *P. argentinus* (see Sect. 11.5). Although the most frequent habitat is the benthic-littoral area, the first post-hatching stages of the previously mentioned species have a planktonic activity (Boschi 1981; Collins 1999a, 1999b; Collins and Williner 2003).

The most consumed resources are oligochaetes and insect larvae (Table 11.2), except for *D. pagei* and *A. uruguayana* (Williner and Collins 2002a; Collins et al. 2007). The orders Diptera, Tricoptera and Ephemeroptera are the most common ones and the family Chironomidae, the most abundant one (mainly *Chironomus* sp. and *Parachironomus* sp.). Oligochaetes *Dero* sp. and *Prístina* sp. are the most frequent genera. Based on the optimal foraging theory, these species represent an energetic equation favorable to the gain-cost relationship in the obtaining, manipulation and digestion of food (Popchencko 1971; Bouguenec and Giani 1989; Collins and Paggi 1998).

Microcrustaceans are frequent in prawns' diets and somewhat less frequent in those of crabs (Table 11.2), including ostracods (e.g., *Cypridopsis* sp.), cladocerans (e.g., *Macrotrix* sp., *Chydorus* sp., *Bosmina* sp., *Bosminopsis* sp.) and calanoid and cyclopoid copepods (e.g., *Notodiaptomus* sp., *Diaptomus* sp., *Macrocyclops* sp., *Eucyclops* sp). Protozoa (e.g., *Diflugia* sp., *Chlamidaster* sp.) and rotifers (e.g., *Brachionus* sp., *Keratella* sp., *Lecane* sp.) are also registered. These groups have representatives in the planktonic, benthic and pleustonic communities (see Chaps. 9, 10).

Filamentous algae (e.g., *Basicladia* sp., *Oedogonium* sp., *Zignema* sp) and unicellular algae (e.g., *Coelastrum* sp., *Ankistrodemus* sp., *Euastrum* sp) are very common in macrocrustacean diets. Diatoms (e.g., *Gomphonema* sp., *Navicula* sp., *Bacillaria* sp.) are also common in the diet of some crabs and prawns (Devercelli and Williner 2006). Rests of aquatic vegetation are numerous and observed in a high frequency (Fig. 11.7).

The selection is carried out towards large preys of slow movements and not towards those evasive and small ones, which can be explained by the energy optimum balance. In turn, food selection fluctuates according to cycles and movements of preys, e.g., vertical and horizontal movements, availability, circadian cycle, seasonal cycle (Collins and Paggi 1998; Collins 1999b; Collins et al. 2006).

11.4.3 Rhythms, Cycles

The trophic activity in macrocrustaceans occurs during all day but not with the same intensity (Fig. 11.8). Age and development state are also variable, being more irregular in juveniles than in adults (Collins 1995, 1997b). The cycles can be modified according to the elasticity of each species and their capacity to respond to an external pressure (Giri et al. 2002; Collins 2005). The crab *D. pagei* usually eats animals (e.g., oligochaetes, rotifers) at night, associated to the incorporation of sand and sediments from benthos, whereas its daily ingestion is related to aquatic vegetation. This variation indicates feeding and refuge-searching movements (Williner and Collins 2002a).

Seasonal changes in trophic activity are associated to the presence and abundance of potential preys and to the influence of the thermal cycle, coupled to the nutritional need of these macrocrustaceans (Collins et al. 2007).

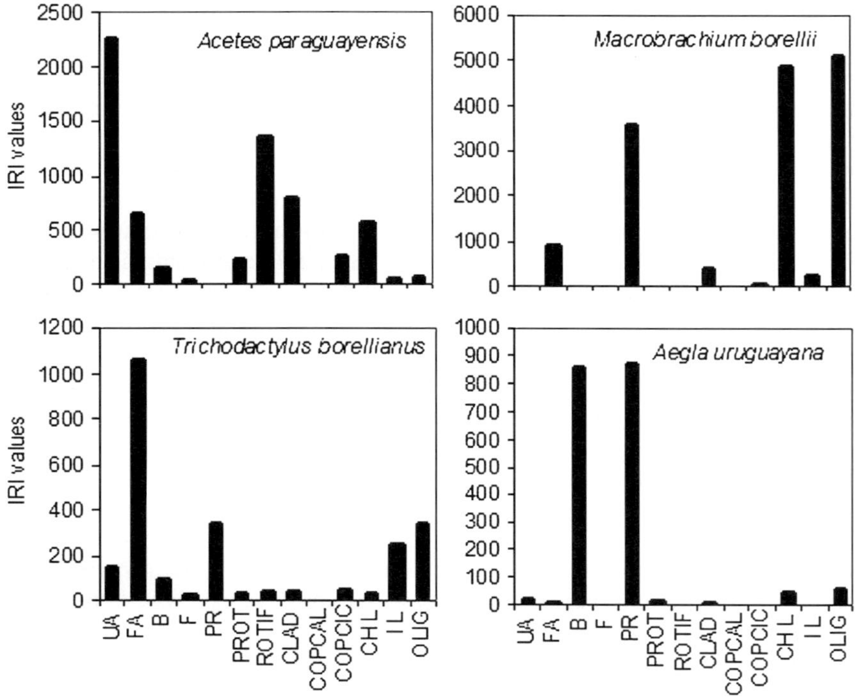

Fig. 11.7 Importance relative index (IRI) of the main food items found in the stomach of prawns and crabs that inhabit the Middle Paraná River (mean values, annual or daily cycles and of various sampled sites). *UA* unicellular algae; *FA* filamentous algae; *B* Bacillariophyceae; *F* Fungii; *PR* plant remains; *PROT* Protozoa; *ROTIF* Rotifera; *CLAD* Cladocera; *COPCAL* Copepoda Calanoidea; *COPCICL* Copepoda Cyclopoidea; *CH L* Chironomid larvae; *I L* other insect larvae; *OLIG* Oligochaeta (modified from Collins and Paggi 1998; Collins 2000, 2005; Williner and Collins 2002a; Collins and Williner 2003; Collins et al. 2007)

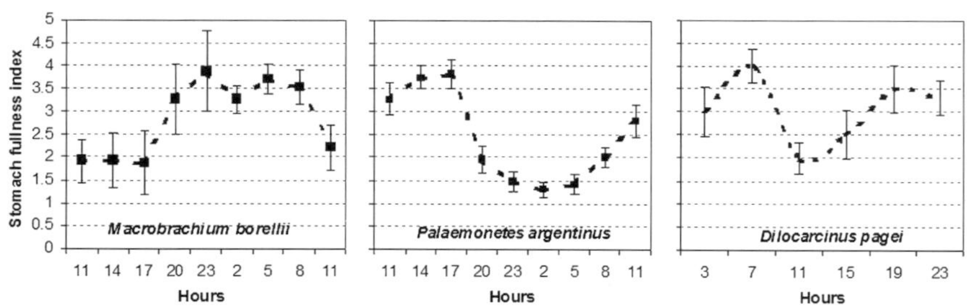

Fig. 11.8 Index of stomach repletion of prawns and crabs during various daily cycles (mean values and standard deviations). Scale: 0 (empty) and 5 (completely full of food stomach) (modified from Collins 2000; Williner and Collins 2002a; Collins 2005; Collins and Williner 2005)

In the Middle Paraná River, higher food consumption occurs at the beginning of spring until autumn, when macrofactors (hydric and thermal cycle) show their maximum values (see Chap. (3)). In this period, growth, development and reproduction are active events, requiring that the trophic activity accompanies successfully (see Sect. 11.5).

In winter, macrophyte cover area decreases due to cold and low river water, so microcrustaceans and other planktonic groups are used as an alternative food source. Meanwhile, in spring, diversity and abundance of algae and plants increase. These ones provide vitamins and essential substances, allowing an optimum and more frequent ecdysis, together with a successful reproduction (Collins 2000).

11.5 Reproduction in Freshwater Decapods

11.5.1 Fertility and Type of Development

From the morphological point of view, decapods are considered as adults when secondary sexual characters are recognizable. However, not all species begin the reproductive season involving the complete adult population. The ovarian maturation of decapods in the Paraná River begins with the increase in temperature, and several spawnings occur during the year. In general, hatched individuals are similar to small adults, except for *A. paraguayensis* and *P. argentinus* that free less developed larvae. The time they are inside the egg is greater than that of marine decapods and the number of eggs is lower (Boschi 1981; Collins 2000; Collins et al. 2004).

In the Middle Paraná River, ovarian maturation firstly begins in prawns (end of July until the beginning of August), and immediately follows in trichodactylid and aeglid crabs. In the former, according to the species, there are differences in the degree of female participation. In *M. borellii*, the reproductive season begins with females of larger sizes (>16 mm CL), whereas all sizes of *P. argentinus* participate from the beginning (8–20 mm CL) (Collins 2000). Earlier gonadal maturation of large specimens of *M. borellii* is another evidence of a hierarchical structure (Williner and Collins 2000).

The following observable manifestation is the presence of ovigerous females, which occurs from August in *P. argentinus* and from September in *M. borellii* (Fig. 11.9). Crab ovigerous females (e.g., *T. borellianus*) appear at the end of winter (August) and are observed until March, with higher occurrence values in November and January (Collins 2000; Collins et al. 2006).

As an adjustment to freshwater environments, prawns and crabs produce a low amount of eggs and of larger sizes than those registered in marine environments. Their number oscillates between 136 and 174 in *P. argentinus*, and between 86 and 107 in *M. borellii* (Collins 2000). The time they take until

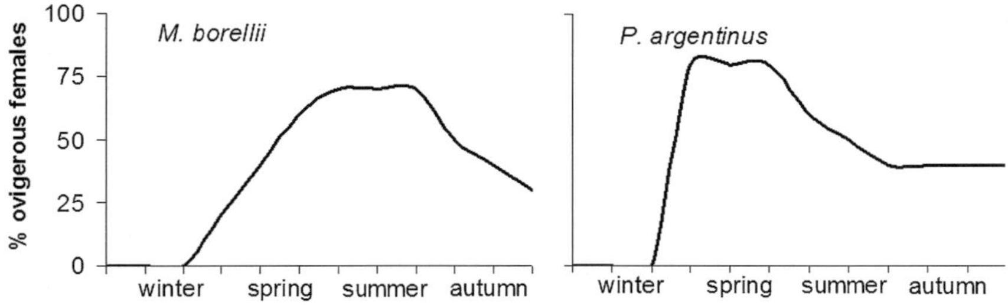

Fig. 11.9 Percentage of prawn adult females that participate in reproduction during the year in Middle Paraná River environments

hatching, according to the type of development, is higher in crabs than in prawns. Among the latter, *P. argentinus* requires the lowest time, since it hatches as a mysis larvae with planktonic habits and positive phototaxism (Boschi 1961, Menu Marque 1973). On the other hand, eggs of *M. borellii* hatch as benthic post-larvae, with similar characteristics to those of adults (Boschi 1961) (see Sect. 11.4). According to Jalihal et al. (1993), they are classified as prawns with partially abbreviated larval development within the category II (see Discussion).

In crabs, the development is complete (Magalhães 2003); organisms are small and morphologically similar to adults, remaining free on their mother or near her. Aeglidae are born as juveniles, very similar to adults, with benthic habits and parental care (Bond-Buckup et al. 1998; López Greco et al. 2004).

There are many types of larval development in Decapods. According to considerations by Jalihal et al. (1993) and Pereira and Garcia (1995), prawns inhabiting the Middle Paraná River should be considered as category IIB (partially abbreviated development). However, pereiopods and pleopods of *M. borellii* are functional, according to category III (completely abbreviated development). In accordance with Jayachandra (2001), palaemonids originated in marine environments but, afterwards, some species evolved, migrated and established in estuary and freshwater environments. The abbreviated larval development is interpreted as an adaptive radiation to freshwater environments, converging independently due to the most recent selective pressure (Pereira and Garcia 1995; Jayachandra 2001). These notions are argued considering that crustaceans would not be monophyletic (Schram and Konemann 2004).

11.5.2 Male:Female Relationship (Sex Ratio) and its Variations

In general, a 1:1 relationship is observed in prawns, with variations related to the environment and moment of the year (Collins 2000). In crabs, a higher presence of females has been observed, except at the end of winter and in spring (Williner and Collins 2002a). The variation in the male: female ratio in *T. borellianus* is negatively correlated with river height; however, there is no relation with other parameters (e.g., temperature and aquatic vegetation). In this case, males show a higher numeric variability, indicating reproductive movements (Collins et al. 2006). Migration models in which males precede females and juveniles in a reproductive displacement are common in some marine crabs and other decapods (Miquel et al. 1985; Fransozo et al. 2003). The higher number of males (*D. pagei* and *T. borellianus*) coincides with a higher proportion of reproductively active females (end of winter-spring) (Williner and Collins 2002a; Collins et al. 2006). On the other hand, variations in the sex ratio can be explained by death of large specimens, as in other decapods (Spivak 1997), whereas the loss of large males could be caused by competition or predation effects (see Sect. 11.4).

11.5.3 Reproduction and Environment

In the two most abundant prawns of the Paraná River there are temporal and spatial differences that allow the reproductive success. The highest proportion of females with gonadal maturation firstly occurs in *P. argentinus* and, afterwards, in *M. borellii* (Collins 2000). In females of the latter, there are qualitative observations that show that largest specimens establish a reproductive territory. This, as well as courtship between males and females, is common in species of *Macrobrachium* (Jayachandra 2001).

The reproductive season is prolonged between 9 and 10 months in this region (from July to May), being shorter in higher latitudes (Schuld and Damborenea 1987). Female gonadal development is induced by the reduction in the level of the gonadal inhibitory hormone, as well as by the mobilization of reserves from the middle intestine gland (hepatopancreas) and the incorporation of nutrients. This important period would be assured by environmental conditions that govern the system since the end of winter and the beginning of spring, favored by the production and diversification of the trophic offer and its optimum nutritional quality (see Chaps. 7, 8, and Sect. 11.4). This situation benefits the first post-hatching stages, generating lower intra-specific competition. In relation to macrofactors that set off the reproductive period, no direct relationship with the increase in river flow or with rains, as occurs in Amazon River prawns (*M. amazonicum*) (Collart 1988) or in prawns from Indian rivers (*M. malalmsonii*) (Ibrahim 1962), has been observed.

The second cohort of palaemonids coincides with the increase in water level, in the flooded area and in refuges, and a greater connection of lakes and rivers. This assures a lower possibility of encounters with competitors and predators.

Particularly, hatching of *P. argentinus* in mysis state occurs before postlarvae of *M. borellii* (Collins 2000). The late appearance of more aggressive

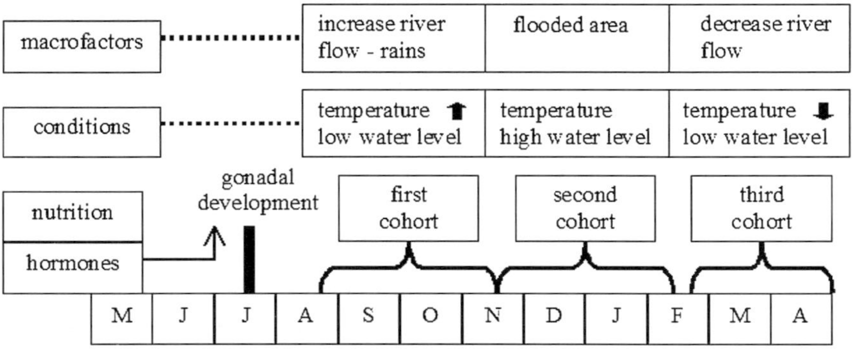

Fig. 11.10 Reproduction and possible adjust to be successful reproductive season. Sequences in reproductive event of freshwater decapods and relation between macrofactors and process in annual cycle

post-larvae favors the development, growth and survival of more larvae of *P. argentinus*. This is added to a higher efficacy in the use of the whole water column and the trophic resources of this last species. As mentioned before, in the first reproductive event, adult females of the *P. argentinus* population participate massively, while in the following events there is a decrease in the number of females involved (Goldstein and Lauria de Cidre 1974; Rodríguez Capitulo and Freyre 1989). On the other hand, not all adult females of *M. borelli* are reproductively active (Fig. 11.10). This could be understood as an action tending to reduce the risks of lesions by attacks and predation of the more aggressive prawns. The post-hatching spatial and trophic competition decreases in the first reproductive event of these sympatric species (Collins 2000). The intimate mother-juvenile relationship at the first stages in trichodactylids (Alarcón et al. 2002; Mansur and Hebling 2002) and aeglids includes parental care, which decreases the risks in juveniles. After the asynchronous hatching, juveniles of *A. uruguayana* remain between 3 and 4 days with the female (López Greco et al. 2004).

11.6 Abiotic Factors Influence in Decapods

Diverse abiotic factors condition the existence of macrocrustacean fauna in the Middle Paraná River, in contrast with other large rivers. Firstly, we should refer to the thermal seasonality of the environment, i.e., there are moments of the year with temperatures higher than 40°C and others that are below 10°C. There is also a wide range of salt concentration in diverse aquatic environments, e.g., there are specimens living in environments with conductivities from 60–7,000 $\mu S\ cm^{-1}$. On the other hand, there is also tolerance to lake desiccation (Fernández and Collins 2002; Collins 2005; Collins unpublished). As has been mentioned previously, these factors condition different aspects of the biology and ecology of this group.

Temperature and hydrometric level vary along the year. For example, the extremes in water level couple with those of temperature, so, in some circumstances, it is complex to determine the influence of each factor. In other decapods, survival is affected by temperature (Chul-Woong and Hartnoll 2000; Carmona-Osalde et al. 2004; Paglianti and Gherardi 2004; Díaz et al. 2002), but there are some species that show some type of tolerance to extreme temperatures.

Growth is also affected by temperature, in the molt increase as in the intermolt periods (Hartnoll 1982; Wu and Dong 2002). In *T. borellianus*, variations in temperature affect the intermolt period, without modifying the increase by molt (Renzulli and Collins 2000). This becomes more complex if we consider the beginning of the ovarian maturation. In *P. varians*, there is a reduction in the intermolt period when temperature increases (Jefferies 1964), whereas in *P. argentinus* (Felix and Petriella 2003) the increase in temperature does not modify the frequency of molt, since ovarian maturation would be producing a higher influence.

During low waters, refuges are lost and populations concentrate. If the water volume decreases drastically, chemical conditions of lakes, swamps or marshes become limiting for the existence of decapods (extreme decrease in oxygen, increase in organic matter decomposition and ionic concentrations, production of harmful gases, changes in pH, among others). The existence of decapods and their permanence suggest the use of diverse strategies as adjustment to these temporary hydrosystems.

As a first strategy, freshwater crustaceans respond to the decrease in oxygen in the water increasing the ventilation rate (Taylor and Taylor 1992; Schmidt-Nielsen 1997). This ventilatory response is necessary but becomes insufficient in certain extreme moments in which they might need other compensatory mechanisms, as the increase in the concentration of haemocyanin, the use of anaerobic processes, or the decrease in the metabolic rate (Schmidt-Nielsen 1997).

11.7 Conclusions

Decapods of the Paraná River have evolved together with the Paraná River and the own characteristics of the littoral community that lives in it. Species show special biological characteristics that allow their development in these hypoosmotic and unstable environments, although some of them still maintain characteristics reminiscent of marine decapods. The Paraná River maintains active and passive exchanges with the Amazon hydrosystem through regular connections, as well as with the "Mata Atlántica" region. The Parapeto River, the Izozog swamp, the Mamore River and the Timane River, which posteriorly drains into the Pilcomayo River, could be active corridors of decapods. Other rivers, as Jauru and Guapare, tributaries of the Paraguay and Amazon rivers, would allow movements of crabs that could walk in the mainland. Despite the low biological diversity of this group, the Middle Paraná River hydrosystem bears high decapod densities with biological and behavioral mechanisms that allow them to coexist. Therefore, it is an important group in the transference of matter and energy within the aquatic system, as well as towards the semi-terrestrial and terrestrial systems.

References

Alarcón DA, Arruda Leme MH, Cobo VJ (2002) Population structure of the freshwater crab *Trichodactylus fluviatilis* Latreille, 1828 (Decapoda, Trichodactylidae) in Ubatuba, northern coast of São Paulo State, Brazil. In: Escobar-Briones E, Alvarez F (eds) Modern approaches to the study of Crustacea. Kluwer, Dordrecht, pp 179–182

Bamber R, Henderson P (1994) Seasonality of caridean decapod and mysid distribution and movements within the Seven estuary and Bristol Channel. Bio J Linn Soc 51:83–91

Beltzer AH (1983a) Alimentación de la garcita azulada (*Butorides striatus*) en el valle aluvial del río Paraná medio (Ciconiiformes: Ardeidae). Rev Hydrobiol Trop 16(2):203–206

Beltzer AH (1983b) Nota sobre fidelidad y participación trófica del "Bigua común" (*Phalacrocorax olivaceus*) en ambientes del río Paraná medio (Pelecaniformes: Phalacrocoracidae). Rev Asoc Cienc Nat Litoral 14(2):111–114

Blank GS, Fingler MH (1996) Interspecific shelter competition between the sympatric crayfish species *Procambarus clarkii* (Girard) and *Procambarus zonangulus* (Hobbs and Hobbs). J Crustac Biol 16(2):300–309

Bliss D (1989) Shrimps, lobsters and crabs: their fascinating life story. Columbia University Press, New York

Bond-Buckup G, Buckup L (1994) A familia Aeglidae (Crustacea, Decapoda, Anomura). Arq Zool (Sao Paulo) 32(4):159–346

Bond-Buckup G, Bueno AP, Kuenecke K (1998) Morphological characteristics of juvenile specimens of *Aegla* (Decapoda, Anomura, Aeglidae). Proc Fourth Int Crustacean Congr 1:371–381

Bonetto AA, Pignalberi C, Cordiviola E (1963) Ecología alimentaria del amarillo y moncholo, *Pimelodus clarias* (Bloch) y *Pimelodus albicans* (Valenciennes) (Pisces, Pimelodidae). Physis 24(67):87–94

Boschi EE (1961) Sobre el primer estadio de dos especies de camarones de agua dulce. (Crustacea, Palaemonidae). Primer Congreso Sudamericano de Zoología 7:69–77

Boschi EE (1981) Decapoda Natantia. Fauna de Agua Dulce de la República Argentina. PROFADU, Buenos Aires

Bouguenec V, Giani N (1989) Les Oligochetes aquatiques en tant que proies des invertrebres et des vertebres: une revue. Acta Oecol Oecol Appl 10 (3):177–196

Brusca R, Brusca G (1990) Invertebrate. Sinauer, Associates, Sunderland Massachusetts

Brünnich MTh (1772) Zoologiae fundamenta praelectionibus academicis accomodata. Grunde i Dyrelaeren. Hafniae et Lipsiae [= Copenhagen and Leipzig]: Apud Fridericus Christianus Pelt

Calman WT (1904) On the classification of the Crustacea Malacostraca. Ann Nat Hist 7(13):144–158

Carmona-Osalde C, Rodriguez-Serna M, Olvera-Novoa M, Gutierrez-Yurrita P (2004) Gonadal development, spawning, growth and survival of the crayfish *Pracambarus llamasi* at three different water temperatures. Aquaculture 232:305–316

Cartaxana A (1994) Distribution and migrations of the prawn *Palaemon longirostris* in the Mira River estuary (southwest Portugal). Estuaries 17(3):685–694

Carter JL, End SV, Kenelly SS (1996) The relationships among three habitat scales and stream benthic invertebrate community structures. Freshw Biol 35:109–124

Chul-Woong O, Hartnoll R (2000) Effects of food supply on the growth and survival of the common shrimp, *Crangon crangon* (Linnaeus, 1758) (Decapoda, Caridea). Crustaceana 73(1):83–99

Collart OO (1988) Aspectos ecológicos do camarão *Macrobrachium amazonicum* (Heller, 1862) no baixo Tocatins (Pa–Brasil). Mem Soc Cs Nat La Salle 48(Suppl):341–353

Collins PA (1995) Variaciones diarias de la actividad trófica en una población de Palaemonetes argentinus (Crustacea Decapoda). Rev Asoc Cienc Nat Litoral 26(1):57–66

Collins PA (1997a) Cultivo del camarón *Macrobrachium borellii* (Crustacea: Decapoda: Palaemonidae), con dietas artificiales. Natura Neotropicales 28(1):39–45

Collins P (1997b) Ritmo diario de alimentación en el camarón *Macrobrachium borellii* (Decapoda, Palaemonidae). Iheringia Sér Zool 82:19–24

Collins PA (1999a) Role of natural productivity and artificial feed in enclosures with the freshwater prawn, *Macrobrachium borellii* (Nobili, 1896). J Aqua Trop 14(1):47–56

Collins PA (1999b) Feeding of *Palaemonetes argentinus* (Nobili) (Decapoda: Palaemonidae) in flood valley of river Paraná Argentina. J Crustac Biol 19(3):485–492

Collins PA (2000) Mecanismos de coexistencia en poblaciones de Palaemónidos diulciacuícolas (Crustacea, Decapoda, Caridea). Tesis Doctoral Universidad Nacional de La Plata, La Plata

Collins PA (2001) Relative growth of the freshwater prawn *Macrobrachium borellii* (Nobili, 1896) (Decapoda: Palaemnidae). Nauplius 9(1):53–60

Collins PA (2004) Cultivo alternativo en el valle aluvial del río Paraná: camarones dulciacuícolas en jaulas flotantes. CIVA2003:427–433

Collins PA (2005) A coexistence mechanism for two freshwater prawns in the Paraná River floodplain. J Crustac Biol 25(2):219–225

Collins PA, Cappello S (2006) Cypermethrin toxicity to aquatic life: bioassays for the freshwater prawn *Palaemonetes argentinus*. Ach Environ Contam Toxicol 51:79–85

Collins PA, Giri F, Williner V (2004) Crustáceos Decápodos del Litoral Fluvial Argentino (Crustacea: Eucarida). INSUGEO Miscelánea 12:253–264

Collins PA, Giri F, Williner V (2006) Population dynamics of *Trichodactylus borellianus* (Crustacea Decapoda Brachyura) and interactions with the aquatic vegetation of the Paraná River (South America, Argentina). Ann Limnol–Int J Lim 42(1):19–25

Collins PA, Paggi JC (1998) Feeding ecology of *Macrobrachium borellii* (Nobili) (Decapoda: Palaemonidae) in the flood valley of the River Paraná, Argentina. Hydrobiologia 362:21–30

Collins PA, Petriella A (1999) Growth pattern of isolated prawns of *Macrobrachium borellii* (Crustacea, Decapoda, Palaemonidae). Invertebr Reprod Dev 36:1–3

Collins PA, Williner V (2001) Espectro trófico del camarón dulciacuícola *Macrobrachium jelskii* en el Parque Nacional Río Pilcomayo. V Congreso Latinoamericano de Ecología, Jujuy, pp 145

Collins PA, Williner V (2002) Espectro trófico natural del camarón *Macrobrachium amazonicum* en el Parque Nacional Río Pilcomayo Argentina. XXIV Congreso Brasileiro de Zoologia, Itajai, Brasil, pp 83

Collins PA, Williner V (2003) Feeding of *Acetes paraguayensis* (Nobili) (Decapoda: Sergestidae) in flood valley of river Paraná Argentina. Hydrobiologia 493:1–6

Collins PA, Williner V (2005) Ecología trófica y ritmo nictimeral del cangrejo *Trichodactylus kensleyi* en Misiones, Argentina. 3° Congreso Argentino de Limnología – CAL III. Chascomús, Argentina, pp 20

Collins PA, Williner V, Giri F (2002) A new distribution record for *Zilchiopsis oronensis* (Miers, 1877) in Argentina. Crustaceana 75(7):931–934

Collins PA, Williner V, Giri F (2007) Trophic relationships in Crustacea Decapoda of a river with floodplain. In: Ashraf MT (ed) Predation in organisms: a distinct phenomenon. Springer, Berlin Heidelberg New York (in press)

Devercelli M, Williner V (2006) Diatom grazing by *Aegla uruguayana* (Decapoda: Anomura: Aeglidae): digestibility and cell viability after gut passage. Ann Limnol – Int J Lim 42(2):73–77
Díaz H, Conde JE (1989) Population dynamics and life history of the mangrove crab *Aratus pisonii* (Brachyura, Grapsidae) in a marine environment. Bull Mar Sci 45:48–163
Díaz F, Sierra E, Re AD, Rodríguez L (2002) Behavioural thermoregulation and critical thermal limits of *Macrobrachium acanthurus* (Wiegman). J Therm Biol 27:423– 428
Durret R, Levin S (1998) Spatial aspects of interspecific competition. Theoretical Pop Biol 53(1):30–43
Feldmann RD (1984) *Haumuriaegla glaessneri* n. gen. and sp. (Decapoda; Anomura; Aeglidae) from Haumurian (Late Cretaceous) rocks near Cheviot, New Zealand. New Z J Geol Geophys 27:379–385
Feldmann RD, Vega FJ, Applegate SP., Bishop GA (1998) Early Cretaceous arthropods from the Tlayúa formation at Tepexi de Rodríguez, Puebla, México. J Paleontol 72(1):79–90
Felix MML, Petriella AM (2003) Molt cycle of the natural population of *Palaemonetes argentinus* (Crustacea, Palaemonidae) from Los Padres lagoon (Buenos Aires, Argentina). Iheringia 93(4):399–411
Fernández D, Collins P (2002) Supervivencia de cangrejos en ambientes dulciacuícolas inestables. Natura Neotropicalis 33(1–2):81–84
Fingerman M (1992) Glands and secretion. In: Harrison FW, Humes AG (eds) Microscopic anatomy of invertebrates, vol 10. Decapoda Crustacea. Wiley, New York, pp 345–394
Fransozo A, Costa RC, Reigada ALD, Nakagaki JM (2003) Population structure of *Aegla castro* Schmitt, 1942 (Crustacea: Anomura: Aeglidae) from Itatinga (SP), Brazil. Acta Limnol Bras 15:13–20
Giri F, Collins P (2004) A geometric morphometric analysis of two sympatric species of family Aeglidae (Crustacea, Decapoda, Anomura). Ital J Zool 71:85–88
Giri F, Williner V, Collins P (2002) Tiempo de evacuación del camarón dulceacuícola *Palaemonetes argentinus* alimentado con larvas de mosquito *Culex pipiens* s.l. FABICIB 6:37–41
Goldstein B, Lauria de Cidre L (1974) Ciclo de maduración sexual y observaciones preliminares sobre el desove del camarón dulciacuícola *Palaemonetes argentinus* (Nobili, 1901) (Crustacea, Decapoda, Palaemonidae) I Hembra. Physis B33(87):165–176
Gonzalez-Baro MR, Irazu C, Pollero R (1990) Palmitoyl-CoA ligase activity in hepatopancreas and gill microsomes of the freshwater shrimp *Macrobrachium borellii*. Comp Biochem Physiol B97:129–133
Gori M, Carpaneto GM, Ottino P (2003) Spatial distribution and diet o the Neotropical otter *Lontra longicaudis* in the Ibera Lake (northern Argentina). Acta Theriol 48(4):495–504
Ibrahim KH (1962) Observations on the fisheries and biology of the freshwater prawn, *Macrobrachium malcolmsonii* H. Mile Edwards of River Godavari. Indian J Fish 9(A)(2):433–467
Icely JD, Nott JA (1992) Digestion and absorption: digestive system and associated organs. In: Harrison FW, Humes AG (eds) Microscopic anatomy of invertebrates, vol 10. Decapoda Crustacea. Wiley, New York, pp 147–202
Hartnoll RG (1982) Growth. In: Bliss DE (ed) The biology of Crustacea vol 2. Academic Press, New York, pp 11–196
Jalihal D, Sankolli K, Shenoy S (1993) Evolution of larval developmental patterns and the process of freshwaterization in the prawn genus *Macrobrachium* Bate, 1868 (Decapoda, Palaemonidae). Crustaceana 65(3):365–376
Jayachandra KV (2001) Palaemonid prawns. Biodiversity, taxonomy, biology and management. Science Publishers, Enfield, NH
Jefferies DJ (1964) The moulting behaviour of *Palaemonetes varians* (Leach) (Decapoda, Palaemonidae). Hydrobiologica 24:457–488
Junk WJ, Bailey PB, Sparks RE (1989) The flood pulse concept in river–floodplain systems. In: Dodge D (ed) Proceedings of the International Larger River Symposiums. Can Spec Fish Aquat Sci, pp 106

Karplus I, Barki A, Israel Y, Cohen S (1991) Social control of growth in *Macrobrachium rosembergii* II. The "leapfrog" growth pattern. Aquaculture 96:325–365

Kneib RT (1987) Seasonal abundance, distribution and growth of postlarval and juvenile grass shrimp (*Palaemonetes pugio*) in a Georgia, USA, salt marsh. Mar Biol 96:215–223

Lercari D, Defeo O (1999) Effects of freshwater discharge in Sandy beach populations: the mole crab *Emerita brasiliensis* in Uruguay. Estuar Coast Mar Sci 49:457–468

López JA, Arias MM, Peltzer PM, Lajmanovich RC (2006) Dieta y variación morfométrica de *Leptodactylus ocellatus* (Linnaeus, 1758) (Anura: Leptodactylidae) en tres localidades del centro-este de Argentina. Bol Asoc Herpetol Esp 16 (1–2):32–39

Lopez Greco LS, Viau V, Lavolpe M, Bond-Buckup G, Rodríguez EM (2004) Juvenile hatching and maternal care in *Aegla uruguayana* (Anomura, Aeglidae). J Crustac Biol 24(2):309–313

Lopretto EC (1980) Análisis de las características del quinto pereiópodo en las especies de *Aegla* del grupo "*platensis*" (Crustacea, Anomura, Aeglidae). Physis B 39(96):37–56

Lopretto EC (1995) Crustacea Eumalacostraca. In: Lopretto EC, Tell G (eds) Ecosistemas de Aguas continentales, Tomo III. Ediciones Sur, La Plata. pp 1001–1039

Magalhães C (2003) Famílias Pseudothelphusidae e Trichodactylidae. In: Melo GAS (ed) Manual de identificaçāo dos Crustacea Decapoda de água doce do Brasil. Loyola, São Paulo, pp 143–287

Magalhães C, Türkay M (1996) Taxonomy of the neotropical freshwater crab family Trichodactilidae. I. The generic system with description of some new genera. Senckenbergiana Biol 75(1/2):63–95

Manning R, Hobbs H (1977) Decapoda. In: Hurlbert SH (ed) Biota Acuática de Sudamérica Austral. San Diego State University, San Diego, pp 157–162

Mansur CB, Hebling NJ (2002) Análise comparativa entre a fecundidade de *Dilocarcinus pagei* Stimpson e *Sylviocarcinus australis* Magalhães & Türkay (Crustacea, Decapoda, Trichodactylidae) no Pantanal do Rio Paraguai, Porto Murtinho, Mato Grosso do Sul. Rev Brasil Zool 19:797–805

Martin JW, Davis GE (2001) An update classification of the recent Crustacea. Nat Hist Mus Los Angeles Co, Contrib Sci 39:1–124

Maruzzo D, Bonato L, Brena C, Fusco G, Minelli A (2005) Appendage loss and regeneration in arthropods: a comparative view. In: Koenemann S, Jenner RA (eds) Crustacea and arthropod relationships. Taylor & Francis Group, Boca Raton, FL, pp 215–245

Menu-Marque SA (1973) Desarrollo larval de *Palaemonetes argentinus* (Nobili, 1901) en el laboratorio (Crustacea, Caridea, Palaemonidae). Physis B 32(85):149–169

Miquel JC, Arnaud PM, Do-Chi T (1985) Population structure and migration of the stone crab *Lithodes murrayi* in the Crozet Islands, Subantarctic Indian Ocean. Mar Biol 89:263–269

Montagna M, Collins P (2004) Efecto de un formulado comercial del herbicida glifosato sobre el cangrejo *Trichodactylus borellianus* (Crustacea, Decapoda: Braquiura). FABICIB 8:227–234

Montagna M, Collins P (2006) Toxicity of glyphosate herbicide formulation upon the freshwater prawn *Palaemonetes argentinus* (Crustacea, Decapoda, Palaemonidae). Nauplius 13(2):149–157

Morrone JJ, Lopretto EC (2001) Trichodactylid biogeographic patterns (Crustacea: Decapoda) and the Neotropical region. Neotrópica 47:49–55

Oliva A, Ubeda C, Vignes EI, Iriondo A (1981) Contribución al conocimiento de la ecología alimentaria del bagre amarillo (*Pimelodus maculatus* Lacépède 1803) del río de la Plata (Pisces, Pimelodidae). Comun Mus Argent Cienc Nat 'Bernardino Rivadavia' Ecol 1(4):31–50

Ortmann AE (1902) The geographical distribution of freshwater decapods and its bearing upon ancient geography. Proc Am Phil Soc 41(171):267–400

Paglianti A, Gherardi F (2004) Combined effects of temperature and diet on growth and survival of young-of-year crayfish: a comparison between indigenous and invasive species. J Crustac Biol 24(1):140–148

Palmer M, Arensburger P, Martín A, Denman W (1996) Disturbance, and path-specific responses: the interactive effects of woody debris and floods on lotic invertebrates. Oecologia 105:247–257

Pereira GA, Garcia JV (1995) Larval development of *Macrobrachium reyesi* Pereira (Decapoda, Palaemonidae), with a discussion on the origin of abbreviated development in Palaemonids. J Crustac Biol 15(1):117–133

Pérez-Lozada M, Bond-Buckup G, Jara C, Crandal K (2004) Molecular systematic and biogeography of the South American freshwater "crabs" *Aegla* (Decapoda: Anomura: Aeglidae) using multiple heuristic tree search approaches. Systematic Biol 53(6):767–780

Poi de Neiff A, Carignan R (1997) Macroinvertebrates on *Eichhornia crassipes* roots in two lakes of the Paraná river floodplain. Hydrobiologia 345:185–196

Popchenko VI (1971) Consumption of Oligochaeta by fishes and invertebrates. J Ichthyol 11(1):75–80

Port-Carvalho M, Ferrari SF, Magalhães (2004) Predation of crabs by Tufted Capuchins (*Cebus apella*) in Eastern Amazonia. Folia Ptimatol 75:154–156

Pothoven S, Fahnenstiel GL, Vanderploeg HA (2004) Spatial distribution, biomass and population dynamics of *Mysis relicta* in Lake Michigan. Hydrobiologia 522:291–299

Renzulli P, Collins P (2000) Influencia de la temperatura en el crecimiento del cangrejo *Trichodactylus borellianus*. FABICIB 4:129–136

Renzulli P, Collins P (2001) Ritmo nictimeral de la actividad locomotora de los cangrejos dulciacuícolas *Dilocarcinus pagei pagei* y *Trichodactylus borellianus*. FABICIB 5:145–153

Ringuelet RA (1949a) Camarones y cangrejos de la zona de Goya (Sergéstidos, Palamonidae y Trichodactylidae) Notas Mus La Plata Zool 14(119):79–109

Ringuelet RA (1949b) Los anomuros del género *Aegla* del noroeste de la República Argentina. Rev Mus La Plata (N.S.) Zool 6:1–45

Ringuelet RA (1949c) Consideraciones sobre las relaciones filogenéticas entre las especies del género *Aegla* Leach (Decapodos Anomuros). Notas Mus La Plata Zool 14:111–118

Rodriguez G (1980) Los Crustáceos Decápodos de Venezuela. Instituto Venezolano de Investigaciones Científicas, Caracas

Rodriguez G (1981) Decapoda. In: Hurlbert SH, Rodríguez G, Santos ND (eds) Aquatic biota of tropical South America, part 1: Arthropoda. San Diego State University, San Diego, pp 51–51

Rodriguez G (1992) The freshwater crabs of America. Family Trichodactylidae and supplement to the family Pseudothelphusidae. Faune Trop 31:1–189

Rodrigues Capitulo A, Freyre L (1989) Demografía de *Palaemonetes (Palaemonetes) argentinus* Nobili (Decapoda Natantia). I Crecimiento. Limnobios 2(10):744–756

Schmidt-Nielsen K (1997) Animal physiology. Adaptation and environment. Cambridge University Press, Cambridge

Schmitt WL (1942) The species of *Aegla*, endemic South American freshwater crustaceans. Proc US Natl Mus 91:431–520

Schram FS, Koenemann S (2004) Developmental genetics and arthropod evolution: on body regions of Crustacea. In: Scholtz G (ed) Evolutionary developmental biology of Crustacea. Balkema Publishers, Lisse, pp 75–92

Scholtz G (2004) Baupläne versus ground patterns, phyla versus monophyla: aspects of patterns and processes in evolutionary developmental biology. In: Scholtz G (ed) Evolutionary developmental biology of Crustacea. Balkema Publishers, Lisse, pp 3–16

Schuld M, Damborenea MC (1987) La fecundidad de *Palaemonetes argentinus* (Crustacea, Palaemonidae) en el canal Villa Elisa (Punta Lara, Provincia de Buenos Aires, Argentina). Ann Mus Hist Nat Valparaíso 18:33–39

Schweitzer CE, Feldmann RM (2005) Decapod crustaceans, the K/P event, and Palaeocene recovery. In: Koenemann S, Jenner RA (eds) Crustacea and arthropod relationships. Taylor & Francis Group, Boca Raton, FL, pp 17–53

Sogard S, Able K (1994) Diel variation in immigration of fishes and decapod crustaceans to artificial seagrass habitat. Estuaries 17(3):622–630

Spivak ED (1997) Life history of a brackish-water population of *Palaemonetes argentinus* (Decapoda: Caridea) in Argentina. Ann Limnol-Int J Lim 33(3):179–190

Taylor HH, Taylor EW (1992) Gills and lungs: the exchange of gases and ions. In: Harrison FW, Humes AG (eds) Microscopic anatomy of invertebrates, vol 10. Decapoda Crustacea. Wiley, New York, pp 203–294

Walker I (1994) The benthic litter-dwelling macrofauna of the Amazonian forest stream Taruma-rim: patterns of colonization and their implications for community stability. Hydrobiologia 291:75–92

Walker I, Ferreira MJN (1985) On the populations dynamics and ecology of the shrimp species (Crustacea, Decapoda, Natantia) in the Central Amazonian river Taruma-Mirim. Oecologia 66:264–270

Williams JD, Scrochi G (1994) Ofidios de agua dulce de la Republica Argentina. Fauna de agua dulce de la Republica Argentina. Vol. 42 Reptilia, fascículo 3, Oficia Lepidosauria. Centro Editor de América Latina, Buenos Aires

Williner V (2003) Ecología trófica del cangrejo *Aegla uruguayana* Schmitt 1942 (Decapoda: Anomura: Aeglidae). VIII Jornadas de Ciencias Naturales del Litoral-I Jornadas de Ciencias Naturales del NOA, pp 177

Williner V, Collins P (1999) Estudio preliminar sobre la ecología trófica del cangrejo *Trichodactylus borellianus* Nobili, 1896 (Crustacea, Decapoda, Trichodactylidae). 64° Reunión de Comunicaciones Científicas, ACNL, pp 8

Williner V, Collins P (2000) Existe jerarquización en las poblaciones de Palemónidos del valle aluvial del Río Paraná?. Natura Neotropicalis 31(1–2):53–60

Williner V, Collins P (2002a) Daily rhythm of feeding activity of a freshwater crab *Dilocarcinus pagei pagei* in National Park Río Pilcomayo, Formosa, Argentina. In: Escobar-Briones, Alvarez F (eds) Modern approaches to the study of Crustacea. Kluwer, Dordrecht, pp 171–178

Williner V, Collins P (2002b) Variación espacio-temporal de la actividad del camarón *Macrobrachium jelskii* (Miers, 1877). Ecología Austral 12:3–10

Williner V, Collins P (2003) Effects of cypermethrin upon the freshwater crab *Trichodactylus borellianus* (Crustacea: Decapoda: Braquiura). Bull Environ Contam Toxicol 71(1):106–113

Wu L, Dong S (2002) Compensatory growth responses in juvenile Chinese shrimp, *Fenneropenaeus chinensis*, at different temperatures. J Crustac Biol 22(3):511–520

Part IV
Vertebrate Animal Life

12 Fishes

Liliana Rossi[1,2,3], Elly Cordiviola[1], and María Julieta Parma[1,2]

12.1 Introduction

The middle section of the Paraná River shows a great richness of habitats, constituting an environmental mosaic, due to its spatiotemporal dynamics, inhabited by several fish species with different life strategies. The important degree of connection among environments and the displacements of fish populations in the channel–floodplain complex favor the processes of colonization and the participation of these species in complex trophic webs that link different divisions of the system. As in other large floodplain rivers, the extension of periodically flooded areas and the duration of flood pulses influence productivity and diversity of species (Junk et al. 1989; Neiff 1990). High fish diversity and productivity are observed in this section of the Paraná River (Bonetto 1986). Although the study of ichthyofauna in this area began a long time ago, there are still many aspects of their biology and ecology that should be deeply investigated.

12.2 Richness, Distribution and Abundance of Species

From the ichthyogeographical point of view, this region is part of the Subtropical domain of the Guyanese-Brazilian subregion, "Parano-platense" (e.g., Paraná River and Río de la Plata) province (Ringuelet 1975). Most fish species inhabiting the Middle Paraná River have Brazilian origin. The Paraná River could have had an essential role in their penetration, acting as a faunistic corridor and facilitating the entrance of tropical and subtropical elements of the aquatic fauna (Lopez et al. 2005).

In the "Parano-platense" province, there are currently 422 native fish species, representing 92% of Argentine continental fish and approximately 10% of

[1]Instituto Nacional de Limnología (INALI). Consejo Nacional de Investigaciones Científicas y Técnicas de la República Argentina (CONICET) y Universidad Nacional del Litoral (UNL). Santo Tomé, Argentina
[2]Facultad de Humanidades y Ciencias. Universidad Nacional del Litoral (UNL). Santa Fe, Argentina
[3]e-mail: loyrossi@arnet.com.ar

M.H. Iriondo, J.C. Paggi, and M.J. Parma (Eds.)
The Middle Paraná River: Limnology of a Subtropical Wetland

neotropical fish (Lopez 2001). Most represented orders are Characiformes (147 species) and Siluriformes (150 species). These groups also constitute the main fish component of other large Neotropical rivers (Lowe McConnel 1987). In the Middle Paraná River, there are 216 fish species (Drago et al. 2003), belonging to nine orders and 35 families (Table 12.1); this diversity is related to the important habitat richness of its great flooding area (Neiff 2001).

In these lotic systems, characterized by their great variability at different scales, the fish habitat becomes a continuous "trade-off" between that variability and life requirements (Leveque 1995). Considering life histories of the main species and their use of the habitat, Middle Paraná River populations can be grouped into three great categories, described by Welcomme (2000): species that mainly use the channel, those that mainly inhabit the lakes, and those whose life cycles involve the use of different types of habitats. Synthesis of the spatial distribution and relative abundance of species in the Middle Paraná River have been recently carried out by Drago et al. (2003) and Menni (2004). The first authors characterized different habitats and, considering different information sources, assigned values of relative abundance for each fish species. Therefore, they obtained the following information: a reduced number of species is frequent in only one habitat unit, whereas most fish use at least two out of three large units, e.g., the channel, the plain lotic environments and the lenitic area (Fig. 12.1). The most abundant species, frequently captured in the three macrohabitats, are the characid "sábalo", *Prochilodus lineatus*, and small characid species, commonly called "mojarras" (*Astyanax* (*A.*)

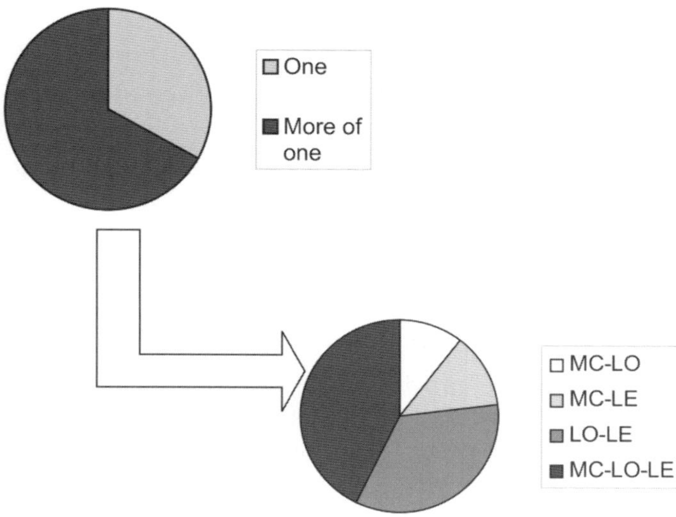

Fig. 12.1 Percentage representation of the registry of fish species per habitat of the Middle Paraná River (adapted from Drago et al. 2003). *MC* main channel; *LO* lotic environments of the flooding valley and *LE* lenitic environments

Table 12.1 Taxonomic composition of the ichthyofauna of the Middle Paraná River (taken from Drago et al. 2003)

Order	Family	No. of species
Myliobatiformes	Potamotrygonidae	6
Clupeiformes	Clupeidae	3
	Engraulidae	2
Cypriniformes	Characidae	50
	Serrasalmidae	8
	Gasteropelecidae	1
	Erythrinidae	2
	Lebiasinidae	4
	Anostomidae	8
	Hemiodidae	3
	Curimatidae	10
	Characiidae	2
	Gymnotidae	1
	Apteronotidae	3
	Rhamphichthydae	6
Silurifornes	Doradidae	8
	Auchenipteridae	4
	Aspredinidae	4
	Pimelodidae	21
	Ageneiosidae	2
	Hypophthalmidae	2
	Cetopsidae	1
	Thychomycteridae	2
	Callichthyidae	5
	Loricariidae	30
Atheriniformes	Belonidae	2
	Cyprinodontidae	1
	Jenynsidae	1
	Poecilidae	2
	Atherinidae	2
Synbranchiformes	Synbranchidae	1
Perciformes	Sciaenidae	4
	Cichlidae	12
Pleuronectiformes	Achiridae	2
Lepidosireniformes	Lepidosirenidae	1

fasciatus fasciatus and *Odontostilbe pequira*). Among the frequent but not very abundant species that also use the three mentioned units are: the jaw characin or “dorado”, *Salminus brasiliensis*, and medium-size Siluriformes, as the “moncholo” catfish, *Pimelodus albicans*, the “amarillo” catfish, *Pimelodus maculatus*, and another catfish, the “manduvé fino”, *Ageneiosus valenciennesi*. In lotic environments, large and medium-size Siluriformes are usually found and moderately abundant, such as the “surubí”, *Pseudoplatystoma coruscans*, the tiger shovelnose catfish or “ surubí atigrado”, *P. fasciatum fasciatum*, the duckbill catfish or “manduvé cucharón”, *Sorubim lima*, the granulated catfish or “armado común”, *Pterodoras granulosus*, and the ocellated river stingray o “raya” *Potamotrygon motoro*. In lenitic environments, an ichthyophagous fish, the wolf fish or “tararira”, *Hoplias malabaricus malabaricus*, is frequently and abundantly found, whereas piranhas (genera *Serrasalmus* and *Pygocentrus*), cichlids, commonly known as “chanchitas” (*Gymnogeophagus australis* and *Cichlasoma facetum*), and the pike cichlid or “San Pedro”, *Crenicichla lepidota*, are also found, although they are less abundant.

The variability of local diversity patterns is related to the structural and functional complexity of the system, influencing the availability of microhabitats and resources. The composition of fish ensembles in lotic environments has been scarcely explored in this section of the basin, mainly due to the methodological difficulties for their study. Oldani and Oliveros (1984) analyzed the composition of commercial captures in the Paraná River (at sites near the city of Paraná, 31°42′34″S and 60°29′7″W) and certain periods of presence and frequency of captures for large species. Species represented in the channel during the hydrological cycle were the “sábalo”, *P. lineatus*, the “patí” catfish, *Luciopimelodus pati*, the “moncholo” catfish, *P. albicans*, and the “surubí” catfish, *P. coruscans*, being *P. lineatus* and *Pterodoras granulosus* the most abundant ones. The use of the channel has also been documented for large fish through the study of longitudinal migrations (Bonetto et al. 1971). In relation to lower flow lotic environments, Demonte and Arias (2005) surveyed all studies carried out in the Paraná River tributaries in the province of Entre Ríos, finding a total of 66 species. The information on the structure of lenitic communities is more abundant and diverse, since there are studies in temporary and permanent environments with different degree of connection to the channels, in different hydrological cycles and with different sampling efforts.

In temporary and semi-permanent lakes, periodically connected to the Salado River, a tributary environment of the Paraná River (31°40′37′S and 60°44′22′W), Bonetto et al. (1969, 1970) analyzed the specific composition and fish biomass during an isolation phase. The species richness was 60 for a total capture in a lake, whereas a total of 75 species were found in 18 lakes of the Los Sapos Island. In these temporary environments, investigated after their connection to the river, the community was mainly integrated by juveniles of *Prochilodus lineatus*, dominating in number and biomass. Mean biomass values were near 500 kg ha^{-1} for the Los Sapos Island and 2,000 kg ha^{-1}

for environments of the El Vado Island, being the latter probably influenced by the pronounced isolation period. The important variations in structure and biomass of fish communities in these environments have been mainly attributed to fluctuations at the hydrometric level (Cordiviola 1992).

In other shallow environments of the Corrientes area (27°28′S and 58°59′W), Cordiviola de Yuan and Pignalberi (1985) also proved the dominance of *Prochilodus lineatus* in six out of 12 surveyed lakes, whereas in the others, *Plagioscion macdonaghi*, *Hypostomus robini*, *Potamotrygon* and *Apareiodon affinis* were dominant in the captures.

Several investigations have been carried out in permanent lakes of the alluvial valley, connected to the channels during a great part of the hydrological cycle. In the “La Cuarentena” Lake (31°42′S and 60°37′W), connected to the Paraná River, Tablado et al. (1988) analyzed the variations in species density and the specific composition of the community during an annual cycle. Twenty-eight species (belonging to 12 families, mainly Characidae and Pimelodidae) were registered in total in non-vegetated waters. The most abundant genus, and responsible for variations in total density, was *Prochilodus*, that generally constituted more than 50% of captures. Fish composition, despite the permanent connection of the lake, was similar to that of other valley lakes, with presence of sedentary species, as *Hoplias malabaricus*, and diverse Loricariidae. The mean density obtained was 24.8 fish/1000 m^3, the annual mean abundance was 932 fish/ha, and the capture per unit effort reached an annual mean of 21.5 kg/day. The variables that influenced the changes in density were hydrometric level and temperature.

In another lake, directly connected to the waters of the San Javier River, a secondary channel of the Paraná River, del Barco and Panattieri (1986) also proved the influence of these environmental variables in density of large fish. In studies carried out during two hydrological cycles in the Tonelli Lake (Helvecia, province of Santa Fe), they found that 15 out of 16 of the studied species were associated to fluctuations in the hydrometric level and temperature, and 10 of them evidenced a significant seasonal use of the environment (e.g., *Luciopimelodus pati*; *Prochilodus lineatus*; *Sorubim lima*; *Pseudoplatystoma f. fasciatum*; *Pterodoras granulosus*; *Ageneiosus valenciennesi* and *Raphiodon vulpinus* were associated to spring-summer; whereas *Salminus brasiliensis* and *Schizodon fasciatum* were associated to winter). Density of *Pimelodus maculatus* increased with the decrease in temperature, whereas captures of other species, as *Prochilodus*, *Raphiodon vulpinus* and *Luciopimelodus pati*, were positively correlated with both variables.

In 17 alluvial valley lakes, located along 680 km (32°05′S and 60°40′W–27°20′S and 58°35′W), Cordiviola (1980) sampled free and vegetated waters during an isolation period. Captured fish, most of them of medium to small size, were mainly Characidae, being *Odontostilbe pequira* the most abundant one. The maximum richness found in a lake was 27 species, with a record of 85 taxa for the total surveyed environments. Species from these environments were grouped considering their habitat use into: open waters fish (e.g. *Apareiodon affinis*, *Astyanax* (P) *bimaculatus*, *Pimelodella gracilis*, *Parauchenipterus galeatus*); open waters species that live in vegetated habitats (*Odontostilbe pequita*; *Pyrrhulina australis*, *Cheirodon piaba*, *Hyphessobrycon callistus*; *Aequidens vittatus*, *Rhineloricaria parva*, *Astyanax* (A) *fasciatum*), and fish closely related to vegetation (*Aphyocharax rubropinnis*, *Corydoras hastatus*, *Gymnotus carapo*, *Eigenmania virescens*, *Hypoptopoma inexpectatum*, *Crenicichla lepidota*).

Different populations use the important macrophyte development in alluvial valley lakes, of high structural complexity and abundant associated fauna (Rossi and Parma 1992). Cordiviola de Yuan et al. (1984) analyzed fish ensembles associated to marginal vegetation in the "Los Matadores" Lake, near the city of Santa Fe, connected intermittently with the Correntoso River. Samplings were carried out during one and a half years and included a period of connection with the river. A total of 71 taxa were registered; the most abundant species were *Odontostilbe pequira*, *Apistogramma corumbae* and *Characidium fasciatum*. Specific diversity oscillated between 1.25 and 3.73 bits, whereas fish density in the vegetation was correlated with hydrometric level ($r = -0.83$), showing maximum values of 232 ind.m^2 during isolation. Juveniles were frequently captured, corresponding to 59% of captured species. The use of vegetation for refuge and feeding by juvenile fish is very frequent. Agostinho et al. (2003) pointed out that 90% of High Paraná River juvenile fish were captured near the macrophytes.

12.3 Migratory Displacements

The habitat use and behavioral patterns of numerous populations are associated to the hydrological dynamics, and important longitudinal and lateral displacements have been detected in all the area from a long time ago. Extensive longitudinal migrations have been well documented in the Paraná River basin (Bonetto et al. 1971; Godoy 1975). Among the migrating species, large Siluriformes, such as two "surubí" species *Pseudoplatystoma coruscans* and *P. fasciatum fasciatum*, and Characiformes, as the "sábalo" *Prochilodus lineatus*, the "dorado" *Salminus brasiliensis*, and the "boga" *Leporinus obtusidens*, were observed. These species, that do extensive displacements upstream to reproduce, spawn in lotic environments and in high water phase, so eggs and larvae are dragged by the current downstream, colonizing the lakes connected to the alluvial valley, where their initial growth is produced (Bonetto 1976). Adult fish displacements downstream have been attributed to trophic purposes, due to the enormous richness and production of the alluvial plain. Similar migratory behavioral patterns have been registered in other Neotropical fish species (Godoy 1975; Agostinho et al. 1993).

Due to the increasing fragmentation of lotic systems and the important magnitude of their displacements, these migratory species are threatened in all large fluvial systems. Quiros and Vidal (2000) evaluated the conservation of distributional areas of migratory species in the lower section of the Paraná River, after the construction of several dams upstream, and found that these potamodromous fish maintain their position in the main channel, despite the regulation in the upper sections of the basin. Therefore, these species retain the migration patterns adjusted to the systems of pristine rivers, so that, in addition to the hydrometric level and temperature, other factors would act as triggers of the longitudinal displacements upstream.

Lateral displacements, that allow the use of lenitic environments and associated channels, have been scarcely documented in alluvial plains (Bonetto 1986; Cox Fernandez 1997). Cordiviola de Yuan (unpublished) registered lateral displacements of fish within the floodplain, through the use of a trap placed at the mouth of access to the so called "Don Felipe" watercourse (a lenitic environment associated to the Santa Fe River). The fishing art used in this study allowed to register the movement of fish (entrance to or exit from the watercourse), and abundance, sex, and gonadal developmental state were also recorded. The trap, placed in high water phase, was daily controlled for two months. The results obtained demonstrated that the highest percentage of displacing specimens corresponded to *Prochilodus*, followed by *Salminus*, *Leporinus*, and the yellow catfish, *Pimelodus maculatus*, together with different Loricariidae, and smaller fish. Captured specimens of *Prochilodus* ($n = 78$), representing lengths of 34–58 cm, and ages between 1 and 4 years, mainly corresponded to females (70%). In relation to the direction of their displacements, most of them were going inside the watercourse (95%), and were mostly spawned, whereas those that were going outside the watercourse towards the river were all mature. In the case of males, from a total of 23 specimens, 96% were captured going inside the watercourse (77% of them were mature, 18% in the post-evacuation phase and only 5% were resting), and the only specimen that was going outside the watercourse was mature. These results coincide with those obtained by Tablado et al. (1988) in relation to the detection of recently spawned fish in lenitic environments connected to a channel.

Other lateral displacements that were not much documented are the movements of young fish among different environmental units. Observations carried out in the lower section of the Salado River recorded a massive migration of juvenile "sábalos" (Rossi unpublished). In March 1988, when the river level reached 3.40 m (Santo Tomé's hydrometer) and the Los Sapos Island lakes were connected to this channel, the displacement of shoals downstream was observed for two days. Fish concentrated in circles and moved near the surface. The capture of 113 specimens showed a structure of homogeneous lengths and an optimum condition state (Fig. 12.2).

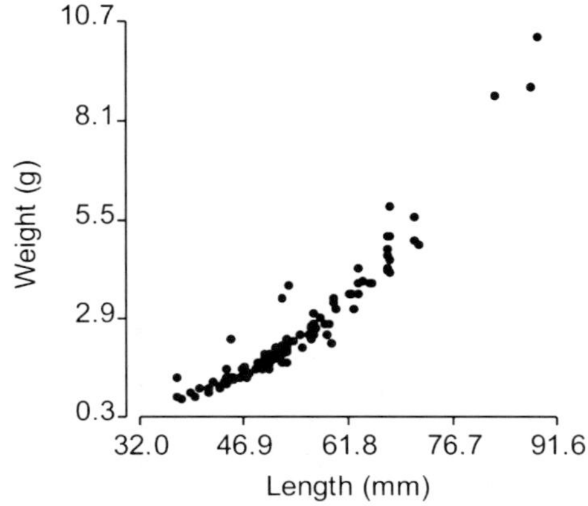

Fig. 12.2 Length–weight relationship of juveniles of *Prochilodus lineatus* captured in the Salado River during a summer migration at low waters

12.4 Reproductive Strategies and Life Histories

Floodplain rivers biota generally presents a high annual growth and high mortality rates, and develops life strategies to colonize large areas. In this way, the fauna can subsist in these systems, in which the great variability in the hydrological regime from one year to the other produces different accessible flooding areas, optimum for their growth and survival (Bayley 1995).

In large alluvial valley rivers, the synchronization between flood pulse and reproduction of migrating fish species has been recognized from a long time ago as an adaptive strategy that allows them to assure survival through the access to lenitic environments with a higher offer of refuge and food (Welcomme 1979). This interpretation, proposed by Bonetto (1976) for all migratory fish of the Paraná River, is assumed in the concept of "flood pulse", when the main function of the river channel is being the migrating route and the dispersing system for populations to have access to resources and refuges (Junk et al. 1989). The mobile littoral provides excellent breeding areas during the flood, so that many fish species anticipate those conditions when reproducing before or during the increase in flow (Bayley 1995).

As in other large rivers, in the middle and lower section of the Paraná River, the ichthyoplanktonic drift, generated by spawning of migrating fish, is produced in high water phase, during spring-summer (Oldani and Oliveros 1984; Oldani 1990; Rossi 1989; Fuentes and Espinach 1998). The increase in flow constitutes a synchronizing trigger of spawning for these species, whereas temperature and photoperiod act as predictors that set off gonadal maturation (Vazzoler et al. 1997). This synchronization, that reveals the deep adjustment of these species to the dynamics of the hydrological cycle, coincides with the pattern observed in the drift of the main neotropical

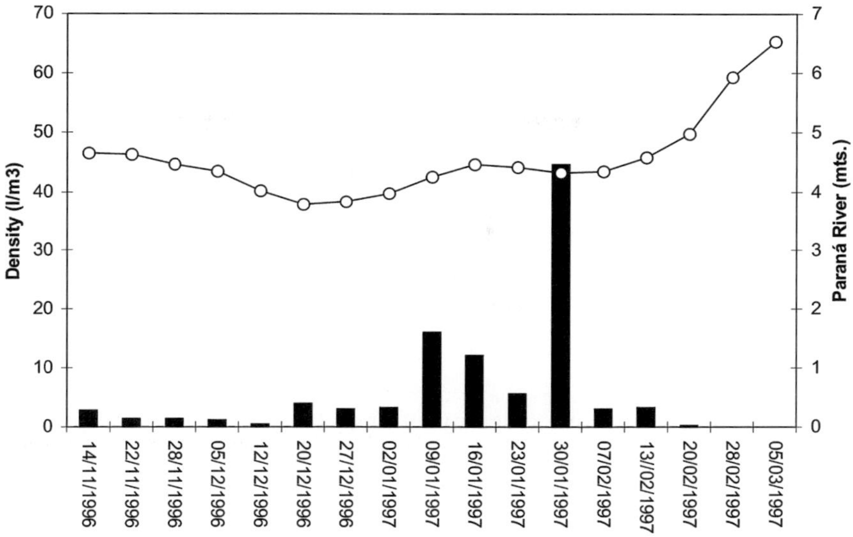

Fig. 12.3 Temporal distribution of ichthyoplankton in the Paraná River in relation to the hydrometric level (Paraná Port) during a high water phase

rheophilous species (Nakatani et al. 1997; Araujo Lima and Oliveira 1998; Merigoux and Ponton 1999; Mantero et al. 1999).

The temporal distribution of ichthyoplankton in surface waters is characterized by one or two drift pulses of higher intensity during the reproductive season, with maximum densities of 44.6 larvae/m^3 in the main channel (Fig. 12.3). The magnitude of these drifts reflects the importance of all the area for reproduction, especially if values obtained in other sections of the Paraná River are considered (Bialetzki et al. 1999; Fuentes et al. 1998). Therefore, this area has an extraordinary value for the maintenance and conservation of these populations.

In relation to ichthyoplankton composition, *Prochilodus lineatus* is the most abundant species, whereas the representation of *Leporinus obtusidens*, *Salminus brasiliensis*, *Pimelodus maculatus*, *Pseudoplatystoma coruscans* and *Sorubim lima* is generally lower and more variable from year to year. The size and age structure is heterogeneous: Characiformes *Prochilodus lineatus* and *Leporinus obtusidens* are always found at the first developmental stages, whereas *Salminus brasiliensis* and Pimelodidae are found in more advanced larval stages. This structure demonstrates the spatiotemporal variability of reproduction in the main species and the existence of spawning sites in different sections of the basin (Oldani 1990; Fuentes and Espinach 1998; Fuentes et al. 1998).

When analyzing the composition of larval drift and the gonadal developmental state, Oldani and Oliveros (1984) considered the middle section of the

Paraná River as the spawning and breeding site for *P. lineatus*, *L. obtusidens*; *P. albicans* and *Sorubim lima*, and only a breeding area for *S. brasiliensis* and *P. coruscans*. On the other hand, numerous species whose life cycles are mainly found associated to lenitic environments of the alluvial valley and that can do short displacements show complex reproductive behaviors. Some of these species have partial spawnings and external fecundity, as numerous members of the family Characidae, whereas others, as *Hoplosternum littorale* and *Hoplias malabaricus*, build nests and show parental care.

12.5 Feeding Habits

The great productivity and diversity of the alluvial plain constitutes an important source of trophic resources for fish and, since a long time ago, its importance in relation to the channels, with a lower offer of resources, has been emphasized. Among mesohabitats of lotic environments, those that show a lower current flow, as the bank areas, generally constitute more favorable feeding sites.

Fish show a great diversity of feeding habits that allow them to develop different behaviors and use a wide variety of aquatic (and even terrestrial) resources (Gerking 1994). The flexibility in the feeding behavior is another characteristic of fish species, revealed by spatial, temporal and ontogenetic changes in the diet, frequently related to changes in accessibility and availability of resources.

Spatial changes in the diet have been proven in species of lenitic environments, showing an important flexibility in relation to the availability of vegetated areas. Thus, *Astyanax abramis* behaviors as omnivorous, consuming a great amount of plants in macrophyte-abundant lakes, and as carnivorous in non-vegetated lakes; *Corydoras paleatus* presents a carnivorous diet in vegetated lakes and an omnivorous diet in non-vegetated lakes due to its high algae consumption, whereas *Odontostilbe piaba* consumes different proportions of algae and invertebrates in those environments (Oliveros 1980). Some species also show changes in their diets when they feed in the channel or in the lakes, as *Pimelodus maculatus* (Montalto et al. 1999).

Seasonal changes in the diet are observed in numerous species that, in some cases, as the *Pimelodus* catfish, can adjust rapidly to changes in availability of resources produced by flooding. Such plasticity in the feeding habits, as well as the abundance of generalist species in these lotic systems, constitutes another evidence of the adjustment of species to variations in the hydrological regime (Junk 2000). The use of terrestrial organisms is also an interesting component of fish feeding strategies in these systems. Among these organisms, frequently dragged by washout of marginal areas, we find the Formicidae, consumed by small and large species (as the characid "mojarras", *Astyanax*, *Pyrrulina australis* and "monedita", *Toracocharax stellatus*, and the catfish, *Ageneiosus valenciennesi* (Oliveros 1980, Oliveros and del Barco 1991, Arias and

Rossi 2005). Their capture is frequently carried out in surface waters and has been proven at connection sites during the flood phase (Gogniat and Rossi 2007).

The accidental introduction of the exotic mollusk *Limnoperna fortunei* (registered for the first time in the Paraná River in 1996), enriched the offer of resources and began to be used by numerous fish populations, modifying the trophic webs of the system. Montalto et al. (1999) registered their consumption in 9 out of 39 studied species, being very frequent in the diet of *Pterodoras granulosus* (with records of up to 150 mollusks in only one specimen), the characid *Leporinus obtusiden* , the catfish *Pimelodus maculatus, Pimelodus albicans, Pimelodus* sp. and *Hypostomus* cfr. *Laplatae.* Bechara et al. (1999) also registered their consumption in several omnivorous fish species downstream of Yacyretá Reservoir. The habits of this invasive species are epifaunal, in contrast with those of native mollusks, so it can colonize benthos and periphyton. This fact favors its availability for fish with different feeding habits. The great floodplain extension offers *L. fortunei* a great availability of substrates, and prolonged floods would favor its expansion and colonization (Montalto et al. 1999). The consumption of planktonic larval stages of *L. fortunei* by fish larvae of different species (mainly Pimelodidae) has been proven in different lotic environments of the Middle Paraná River (Rossi and Ezcurra de Drago unpublished).

The ontogenetic changes in the diet of several species reveal interesting behaviors related to changes in habitat and associated to the dynamics of the hydrological cycle. Thus, the early life history of *P. lineatus* appears as a good example of Bonetto's model (1976), which emphasizes the importance of the access to lakes for their first feeding. This species, during its lotic drift in the main channel and in the secondary channels, is found in early developmental stages and with a high intestinal emptiness, so it is highly probable that it would not begin its external feeding under these conditions (Oldani 1990; Fuentes et al. 1998; Rossi 2001). On the contrary, in protected and vegetated lotic habitats, as the banks of tributary rivers, characid "sábalos" exploit the littoral communities and continue their growth (Rossi 1992). In these habitats, an important ontogenetic change is produced in their diet, beginning as planktivorous fish and becoming early detritivorous fish, obtaining detritus from the roots of floating plants through profound changes in their trophic morphology. Therefore, the diversity of habitats, related to the lateral dimension of the river-floodplain system, used by this species during all its early ontogeny, constitutes a valuable gradient of conditions and resources for its survival and recruitment. The tolerance to starvation during these first life stages, that has been experimentally demonstrated (Libertelli and Espinach Ros 1994), should contribute to its survival in the channel until colonizing more favorable areas, abundant in refuge and food.

Salminus brasiliensis, one of the main piscivorous fish in the system, initially consumes zooplankton and then begins to consume fish, predating on fish larvae (mainly *Prochilodus*) from 14 mm of total length (Rossi 1989) during its lotic drift by the Middle Paraná River main and associated channels. Therefore, the channel conditions seem to be unfavorable for *Prochilodus,* not only because of its high starvation, but also because of the risk of predation by sympatric ichthyophagous fish as *Salminus*. The early ingestion of fish observed in larvae of this species allows them to acquire a rapid specialization in its feeding habits, using an abundant resource available during the summer. The early piscivory and the use of detritus and substrate during ontogeny have also been proven in fish of other Neotropical systems (Merigoux and Ponton 1998).

Other fish species of the Paraná River also use the channel as the first breeding area. Thus, larvae of Pimelodidae exploit the zooplanktonic component and accidental-drifting organisms (as insects and oligochaetes) that enrich the channel conditions during the flood (Rossi 2001). The availability of these organisms during the drift has been scarcely studied in this section of the Paraná River, although being frequently retained in ichthyoplankton nets. In other fluvial systems, Amoros and Roux (1988) have also recognized the importance of the channel enrichment with zooplankton and benthos from the alluvial plain.

The obtained information reveals that external feeding of *Sorubim lima* and other Pimelodidae, as *Pimelodus maculatus* and *Pseudoplatystoma coruscans*, can begin in the channel, and that larval development can be completed in the lotic compartment of the river-floodplain system. The time of access to the valley lakes seems to be very variable for the different species; Siluridae grow using the channel resources while they displace along them. Thus, the river constitutes the first breeding area for numerous species, in addition to a route of migration and dispersion for fish, so its role is essential in the early history of these populations (Fig. 12.4). Other authors have also

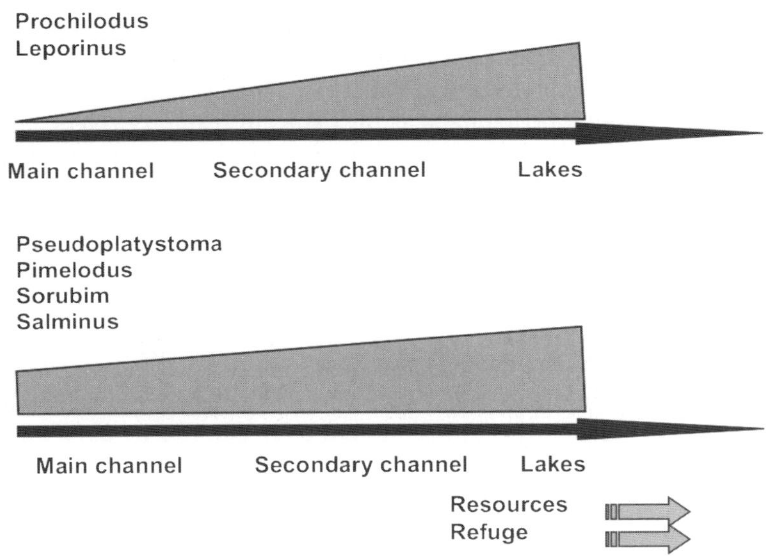

Fig. 12.4 Changes in the trophic use of lateral dimension habitats by fish larvae of different species of Paraná River

recognized the importance of the channels for fish growth and recruitment in large rivers (Schiemer and Zalewski 1992; Roux and Copp 1996).

Hoplias malabaricus, a sedentary species frequent in alluvial valley lakes, presents a sequence of ontogenetic changes in its diet similar to that of other piscivorous species of the area, with an initial ingestion of microcrustaceans, replaced by the consumption of insects and fish. Ichthyophagy is also present early in its life cycle and its importance increases rapidly with growth, so that juveniles already have a piscivorous diet (Oliveros and Rossi 1991). Adults, frequently captured with their empty intestines, have a notable capacity to bear periods of starvation that increases gradually with age. This characteristic would be related to an increasingly sedentary behavior, the development of "waiting and watching"- prey capture strategies, and a low metabolic rate (Parma de Croux 1983b). Among the available resources in the Middle Paraná River, detritus and littoral-community organisms represent important feeding sources.

In a study of the diet in 71 small and medium-size species captured in 17 lakes isolated from the alluvial valley, Oliveros (1980) found that most species showed a carnivorous diet, predominating copepods in some species (*Aequidens portalegrensis*, *Apistogramma corumbae*, *Pyrrulina australis*, *Aphyocharax alburnus* and *Roeboides paranensis*) and insect larvae in others (*Potamotrygon motoro* and *Geophagus brasiliensis*). Although the diet composition of some species revealed the exploitation of different communities (as in *Moenkhausia dichroura*, *Aphyocharax dentatus*, *Trachycorystes galeatus*, *Pachyurus bonariensis*, *Pimelodella gracilis*, and *Odontostilbe pequira*), littoral macrophyte-associated organisms constituted the main source of feeding (Fig. 12.5). This important degree of trophic participation in littoral fish has also been registered in communities of other sections of the Paraná River (Segatti Hahn et al. 1997).

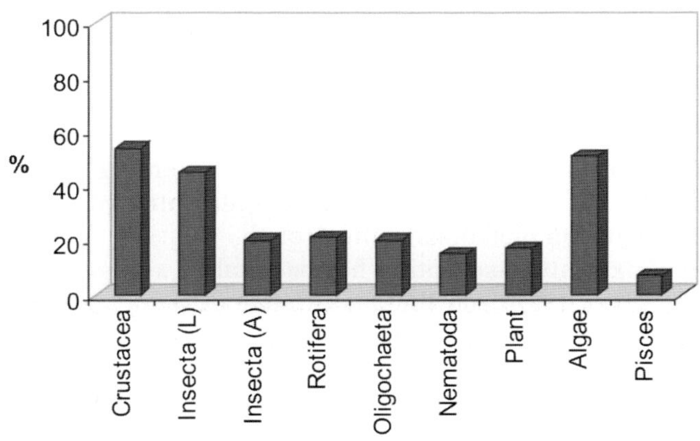

Fig. 12.5 Frequency of fish species that use different feeding resources in the lenitic environments of the floodplain valley of the Middle Paraná River L: larvae, A: adults (adapted from Oliveros 1980)

The use of resources allows grouping the species into trophic categories. In alluvial valley lakes, detritivorous, omnivorous and carnivorous (mainly insectivorous and piscivorous) species are the most represented. The importance of detritivorous species, mainly represented by *Prochilodus*, has been emphasized by Winemiller and Jepsen (1998), who pointed out that these fish increase the ecological efficiency of trophic webs by transferring biomass from the lower levels directly to species of higher levels, increasing the production of piscivorous fish through these short chains.

Among the piscivorous fish, that constitute another specialized group, we can mostly find *Salminus brasiliensis*, which mainly preys on *Raphiodon vulpinus*, *Serrasalmus* sp., *Prochilodus*, and several Siluriformes in open waters of the Paraná River (del Barco 1990). The representation of these categories in the trophic structure of the different habitats can change with the different hydrological phases. In the High Paraná River, Agostinho et al. (1997) proved that changes in the hydrological cycle strongly influence the variability of biomass in the different categories, increasing the representation of omnivorous species in flooding periods. The gradual isolation of temporary environments can also generate changes in the availability of oxygen and resources that have repercussions on the species richness and the trophic structure of these communities.

12.6 Respiratory Metabolism

12.6.1 Oxygen Consumption

There are different (physical, chemical, and biological) processes interacting in a complex way to determine the amount of dissolved oxygen in freshwater bodies. Such processes include photosynthesis and respiration, light diffusion, organic decomposition, oxygen diffusion, among others. Different oxygen requirements lead to a distinctive distribution of fish in accordance with oxygen concentration in the respective habitats (Junk et al. 1997).

Knowledge of respiratory requirements in fish is very important to evaluate natural fish production systems. Food requirements of fishes are related to their energy expenditure because the energy contained in the food consumed is expended on metabolism and growth.

The rate of oxygen consumption has been taken as a general measure of metabolic rate (Fry 1957) and the quantitative dependence on body weight was well expressed by an exponential equation. Generally, the high correlation coefficients reflect a close correlation between both variables. It is commonly expressed in milligrams of oxygen per unit of weight per hour, usually decreases as body size increases, and varies with time of day, season, temperature, activity and feeding (Doudoroff and Shumway 1979; Brett and Zala 1975).

Laboratory measurements of metabolic parameters provide information on energy requirements and uses in fish, but little information has been reported in relation to middle Paraná River ichthyofauna. It is well known that the metabolic level of fishes is related to their different degrees of activity (Lipskaya 1974), their way of life, and behavior (Newmann et al. 1981).

By comparing the metabolic rates of some Paraná River fish with different types of respiratory patterns, we can acquire a good indication of the animal needs and the way it uses O_2. Although the different metabolic rates registered in these experiences characterize the behavior of the species in their natural environment, there are only approximations, since it is not possible to describe exactly the natural conditions.

The data obtained by Parma de Croux (1981, 1983a, 1983b, 1994) and Montagna and Parma de Croux (2001) showed that the metabolic rate of *Hoplias malabaricus*, registered at different temperatures, was lower than that registered for *Pimelodus maculatus, Pimelodus albicans* and *Prochilodus lineatus* (Table 12.2). *Hoplias* consumption was 25–39% lower than *P. maculatus*, and 15–31% lower than *Prochilodus* at 15–20°C and 40% lower at 30°C. Some ecological characteristics of *Hoplias*, such as that it is considered to be a non-active and non-migratory species, typical of lentic environments (Bonetto et al. 1971; Cordiviola de Yuan 1977), solitary and sit-and-wait (Winemiller 1989a; Oliveros and Rossi 1991), may explain its low metabolic rate and situate this species in energetic advantage over other neotropical fish.

Pimelodus maculatus registered the greatest metabolic rate at the lowest temperature, but showed a significant decrease at 30°C. When compared with *Prochilodus*, it is considered to be a species with little migratory behavior,

Table 12.2 Mean metabolic rate (mgO_2/h) for different fish of the Paraná River

Species	Weight (g)	15°C	20°C	24–26°C	30°C
Hoplias	10	1.30	2.05		3.01
malabaricus	50	2.98	4.92		9.01
Pimelodus	10	1.75	2.92		4.94
maculatus	50	4.41	7.35		12.08
Prochilodus	10	1.52	2.43		5.05
lineatus	50	3.93	7.15		13.70
Leporinus	10	1.30	2.20		
obtusidens	50	3.76	6.34		
Pimelodus	10	1.64	3.27		5.22
albicans	50	3.91	7.80		14.65
Salminus	10			4.14	
brasiliensis	50			12.79	

bottom frequenter and more active in fall and winter (Ringuelet 1975; Tablado et al. 1988). *Prochilodus* makes important trophic, reproductive and thermal migrations (Bonetto et al. 1981), and appears in great shoals in spring and summer; then, the important increase in the metabolic rate between 20 and 30°C shows that this range is the most favorable for the development of its activities, mainly those concerning feeding. *Salminus brasiliensis* is considered to be an active fish with predatory and migratory habits; then, the high and continuous activity requires a higher respiratory rate than that measured for other Neotropical fish species (Parma de Croux 1997).

12.6.2 Responses to Hypoxia

Reduction in the level of available oxygen has a marked effect on many physiological, biochemical, and behavioral processes in fish. Restrictions in this supply for metabolic processes, including swimming, migrating and feeding, are likely caused by hypoxia. Adequate oxygen levels for such activities are necessary for survival of fish populations.

Fishes can develop diverse mechanisms to subsist with small amounts of oxygen, although there is controversy and doubt about how they are done. Kramer (1987) describes four response categories to low oxygen availability: (1) changes in activity; (2) increase in the aerial respiration, only possible to some fishes; (3) utilization of the oxygen-rich surface film, known as "aquatic surface respiration", and (4) horizontal and vertical changes of habitat. In behavioral ecology, the optimality theory assumes that an organism will maximize the net rate of resource uptake or minimize the cost of obtaining a required amount of that resource.

Very little has been done in terms of measurements of minimum oxygen requirements in tropical and temperate fish. The main interest has always been air-breathing fish species (Val and Almeida-Val 1995); consequently, requirements of water-breathing fish of the Paraná River are little known.

In the studied species, the data obtained showed that the lethal oxygen level oscillated between 0.3 and 0.8 ppm for *Hoplias malabaricus*, *Pimelodus maculatus*, *Pimelodus albicans* and *Prochilodus lineatus*, with no significant differences detected among species (Parma de Croux 1987, 1994).

The registry of frequency and amplitude of ventilatory movements is a good documented change of activity in response to reduced levels of dissolved oxygen (Doudoroff and Sumway 1970; Saint-Paul 1984). The normal operculum movements of these species at 20–27°C range from 16–30/min for *Hoplias*, 100/min for *Pimelodus*, and 90–100/min for *Prochilodus*, and, at critical concentrations, they were increased to 104–108 for *Hoplias*, 200–230 for *Pimelodus*, and 210–220 for *Prochilodus*, for similar temperature ranges.

There is some evidence that the decision of staying in habitats with lower oxygen or moving to one with higher oxygen availability depends on the relative costs of these alternatives. The studied species have neither air breathing nor adaptations for the aquatic surface respiration, but they showed high tolerance to low oxygen concentrations. The characteristics of *Hoplias*, such as its low metabolic rate, large respiration surface (Fernández 1985) and other physiological measurements (Hochachka et al. 1978b; Rantin and Johansen 1984) increase its ability to employ successfully aquatic respiration when the environment is under periods of hypoxia.

Prochilodus and *Pimelodus* can decrease their metabolic rates in extreme conditions. At the same time, they show changes in some activities, e.g., the increment in the frequency and amplitude of the opercular movements. These adaptations should entail the reduction of other activities, such as feeding, which is probably the most affected one. These two species can move horizontally and vertically in the aquatic environments searching for better conditions, but both may be found in places isolated from the main channel and covered with aquatic vegetation (Cordiviola de Yuan et al. 1984), where low oxygen levels can persist for several days with summer water temperatures.

The reduction in the respiration rate and the increase in the ventilatory frequency of these important species seem to be appropriate adaptations for survival to hypoxia. A valid hypothesis to explain this situation would be the existence of hematological modifications to increase oxygen carrying capacity of blood and adapt to the scarce amount of oxygen available in the environment. Much more studies will be needed in the future to confirm this hypothesis.

12.7 Conclusions

The middle area of the Paraná River, with its extensive floodplain, has an enormous value for diversity and productivity of basin fish.

The life strategies of these populations are deeply adjusted to the functioning of the system. Diverse habitats of the river-floodplain complex are used by fish during their initial growth; therefore, channels are very valuable for some populations, in addition to being "nurseries" of lenitic habitats. As in other large floodplain rivers, the fluctuations of the hydrological cycle influence the structure and dynamics of fish communities, modeling temporally and spatially their trophic webs. The conservation of these populations depends, to a great extent, on actions intended to maintain the integrity of the system.

References

Agostinho AA, Vazzoler AEA, Gomez LC, Okada EK (1993) Estratificación espacial y comportamiento de *Prochilodus scrofa* en distintas fases del ciclo de vida, en la planicie de inundación del alto río Paraná y embalse de Itaipu, Paraná, Brasil. Rev Hydrobiol Trop 26(1):79–90

Agostinho AA, Segatti Hahn N, Gomes LC, Bini LM (1997) Estructura trófica. In: Vazzoler AE, Agostinho AA, Segatti Hahn N (eds) A planicie de Inundacao do Alto Rio Paraná: Aspectos físicos, biológicos e socioeconomicos. Universidade Estadual de Maringá. Maringá

Agostinho AA, Gomez LC, Ferreira Julio Jr H (2003) Relaçoes entre macrófitas acuáticas e fauna de peixes. In: Thomaz SM, Bini LM (eds) Ecología e Manejo de Macrófitas Aquaticas. Editora da Universidade Estadual de Maringá. Maringá, pp 261–279

Amoros C, Roux AL (1988) Interaction between water bodies within the floodplains of large rivers: function and development of connectivity. Proceeding of the 2nd International Seminar of the International Association for landscape Ecology, pp 125–130

Araujo Lima CA, Oliveira EC (1998) Transport of larval fish in the Amazon. J Fish Biol 53:297–306

Arias JD, Rossi LM (2005) Dieta y estrategia alimentaria de *Pyrrhulina australe* (Pisces, Lebiasinidae). Rev FABICIB 9:197–205

Bayley PB (1995) Understanding large river-floodplain ecosystems. BioScience 45(3):153–158

Bialetzki A, Sanchez PV, Cavicchioli M, Baumgartner G, Pereira R, Nakatani K (1999) Drift of ichthyoplankton in two channels of the Paraná River, between Paraná and Mato Grosso do Sul States, Brazil. Braz Arch Biol Technol 42(1):53–60

Bechara J, Sanchez S, Roux J, Terraes J, Flores Quintana C (1999) Variaciones del factor de condición relativo de la ictiofauna del río Paraná aguas debajo de la represa de Yacyretá, Argentina. Revista de Ictiología 7:75–89

Bonetto AA (1976) Calidad de las aguas del río Paraná. Introducción a su estudio ecológico. Corrientes, Argentina, Instituto Nacional de Construcciones Portuarias y Vías Navegables Instituto de Ciencias y Técnica Hídrica

Bonetto AA, Canon Veron M, Roldan D (1981) Nuevos aportes al conocimiento de las migraciones de peces en el río Paraná. Ecosur 8(16):29–40

Bonetto AA (1986) Fish of the Paraná system. In: Davies BR, Walker KF (eds) The ecology of river systems. DW Junk Publishers, Dordrecht, pp 573–588

Bonetto AA, Cordiviola de Yuan E, Pignalberi C, Oliveros O (1969) Ciclos hidrológicos del río Paraná y las poblaciones de peces contenidas en las cuencas temporarias de su valle de inundación. Physis 29:213–223

Bonetto AA, Cordiviola de Yuan E, Pignalberi C (1970) Nuevos datos sobre poblaciones de peces en ambientes leníticos permanentes del Paraná medio. Physis 30(80):141–154

Bonetto AA, Pignalberi C, Cordiviola de Yuan E, Oliveros O (1971) Informaciones complementarias sobre migraciones de peces de la Cuenca del Plata. Physis 30(81):505–520

Brett JR, Zala CA (1975) Daily patterns of nitrogen excretion and oxygen consumption of sockeye salmon under controlled conditions. J Fish Res Bd Can 32:2479–2486

Cordiviola de Yuan E (1977) Poblaciones de peces del río Paraná. IV Fluctuaciones en la composición íctica de la laguna “Los Matadores” (Isla Clucellas), Santa Fe. Neotrópica 23(69):17–25

Cordiviola de Yuan E (1980) Campaña “Keratella I” a lo largo del río Paraná medio: Taxocenos de peces de ambientes leníticos. Ecología 4:103–113

Cordiviola de Yuan E (1992) Fish populations of lentic environments of the Paraná River. Hydrobiologia 237:159–173

Cordiviola E, Oldani N, Oliveros O, Pignalberi de Hassan C (1984) Aspectos limnológicos de ambientes próximos a la ciudad de Santa Fe (Paraná Medio): Poblaciones de peces ligadas a la vegetación. Neotrópica 30(84):127–139

Cordiviola de Yuan E, Pignalberi de Hassan C (1985) Fish populations in the Paraná River: lentic environments of Diamante and San Pedro areas (Argentine Republic). Hidrobiología 127:213–218

Cox Fernandes C (1997) Lateral migration of fishes in Amazon floodplains. Ecol Freshw Fish 6:36–44

del Barco DM, Panattieri AE (1986) Variaciones de la densidad relativa de peces en el Paraná medio en relación con factores ambientales. Rev Asoc Cien Nat Litoral 17(1):127–136

del Barco DM (1990) Alimentación de *Salminus maxillosus* Valenciennes 1840, en distintos ambientes de la Provincia de Santa fe (Pisces, Characidae). Rev Asoc Cien Nat Litoral 21(2):156–166
Demonte LD, Arias JD (2005) Ictiofauna de los ríos Paraná y Uruguay en la Provincia de Entre Ríos, Argentina. In: Temas de Biodiversidad del Litoral fluvial argentino II. INSUGEO Misceláneal 4:355–366
Drago EC, Ezcurra de Drago I, Oliveros O, Paira AR (2003) Aquatic habitats, fish and invertebrate assemblages of the Middle Paraná River. Amazoniana 27(3/4):291–341
Doudoroff P, Shumway DL (1970) Dissolved oxygen requirements of freshwater fishes. FAO Fish Tech Pap 86, pp 291
Fernándes MN (1985) Morfometría branquial do Teleosteo *Hoplias malabaricus* (Bloch). Bol Fisiol Animal Univ Sao Paulo 9:57–65
Fry FE (1957) The aquatic respiration of fish. In: Brown ME (ed) The Physiology of fishes, vol I. Metabolism. Academic Press, New York, pp 1–63
Fuentes CM, Demonte LD, Esposti MF (1998) Temporal variation of main channel ichthyoplankton at the end of Middle Paraná River. Revista de Ictiología 6 (1/2):57–64
Fuentes CM, Espinach Ros A (1998) Variación de la actividad reproductiva del sábalo *Prochilodus lineatus* (Valenciennes, 1847), estimadas por el flujo de larvas en el río Paraná inferior. Natura Neotropicalis 29(1):25–32
Hochachka PW, Guppy M, Guderley KB, Storey KB, Hulbert WC (1978b) Metabolic biochemistry of water-vs. air-breathing fishes: muscle enzymes and ultrastructure. Can J Zool 56:736–750
Gerking SD (1994) Feeding ecology of fish. Academic Press, San Diego
Godoy MP (1975) Peixes do Brasil. Suborden Characoidei. Bacia do rio Mogi Guassu. Piracicaba, Brasil. Editora Franciscana, 4 v
Gogniat D, Rossi L (2007) Dieta de *Thoracocharax stellatus* Kner (1860) (Pisces, Gasteropelecidae) en el Río Salado (Santa Fe, Argentina) Natura Neotropicalis 37
Junk WJ, Soares MG, Saint-Paul U (1997) The fish. In: Junk W (ed) The central Amazon floodplain. Springer, Berlin Heidelberg New York, pp 385–405
Junk WJ, Bayley PB, Sparks RE (1989) The flood pulse concept in river–floodplain systems. In: Doge DP (ed) Proceedings of the International Large River Symposium. Can Spec Publ Fish Aquat Sci 106:110–127
Junk WJ (2000) Mechanisms for development and maintenance of biodiversity in Neotropical floodplains. In: Gopal B, Junk WJ, Davis JA (eds) Biodiversity in wetlands: assessment, function and conservation. Backhuys Publishers, Leiden, The Netherlands, pp 119–140
Kramer DL (1987) Dissolved oxygen and fish behaviour. Environ Biol Fish 18:81–92
Leveque C (1995) L'habitat: etre au bon endroit au bon moment? Bull Fr Peche Piscic. 337/338/339:9–20
Libertelli M, Espinach Ros A (1994) Efectos del retardo en la iniciación de la alimentación externa en la supervivencia y el crecimiento de larvas de sábalo (*Prochilodus lineatus*). Tankay 1:245–247
Lipskaya NY (1974) Metabolic rates in the young of some tropical fish species. J Ichthyol 14(6):934–943
Lopez HL, Miquelarena AM, Ponte Gomez J (2005) Biodiversidad y Distribución de la Ictiofauna Mesopotámica. Temas de Biodiversidad del Litoral fluvial argentino II. INSUGEO 14:311–354
Lopez HL (2001) Estudio y uso sustentable de la biota austral: ictiofauna continental argentina. Rev Cubana Invest Pesq (Supl. Especial, e version) ISSN CUBO 138–8452
Lowe McConnel R (1987) Ecological studies in tropical fish communities. Cambridge University Press, New York
Mantero G, Martinez G, Retta S (1999) Distribución espacio-temporal del ictioplancton en el embalse de Salto Grande (Período 1988–1992). Bol Soc Zool Uruguay 10:72–82
Menni RC (2004) Peces y ambientes en la Argentina continental. Monografías del Museo Argentino de Ciencias Naturales 5 Buenos Aires, Argentina

Merigoux S, Ponton D (1998) Body shape, diet and ontogenetic diet shifts in young fish of the Sinnamary River, French Guiana, South America. J Fish Biol 52:556–569
Merigoux S, Hugueny B, Ponton D, Statzner B, Vauchel P (1999) Predicting diversity of juvenile neotropical fish communities: patch dynamics versus habitat state in floodplain creeks. Oecologia 118:503–516
Montagna M, Parma de Croux MJ (2001) Consumo de oxígeno en *Pimelodus albicans* (Pisces, Pimelodidae) en relación con la masa corporal y la temperatura. Natura Neotropicalis 32(1):49–57
Montalto L, Oliveros OB, Escurra de Drago I, Demonte LD (1999) Peces del río Paraná medio predadores de una especie invasora: *Limnoperna fortunei* (Bivalvia, Mytilidae) Revista FABICIB (3):85–101
Nakatani K, Baumgartner G, Cavicchioli M (1997) Ecologia de ovos e larvas de peixes. In: Vazzoler A, Agostinho A, Segatti Hahn N (eds) A Planicie de Inundacao do Alto Rio Paraná: Aspectos físicos, biológicos e socioeconomicos. Universidad Estadual de Maringá: Nupelia, pp 281–306
Neiff JJ (1990) Ideas para la interpretación ecológica del Paraná. Interciencia 15(6):424–440
Neiff JJ (2001) Diversity in some tropical wetland systems of South America In: Gopal B, Junk WJ, Davis JA (eds) Biodiversity in wetlands: assessment, function and conservation, vol 2. pp 157–186
Newmann DA, O'Conner JM, Sherk Jr JA (1981) Oxygen consumption of white perch (*Morone american*), striped bass (*M. Saxatilis*) and spot (*Leiostomus xanthurus*). Comp Biochem Physiol 69A:467–478
Oldani NO (1990) Variaciones de la abundancia de peces del valle del río Paraná (Argentina). Rev Hydrobiol Trop 23(1):67–76
Oliveros OB (1980) Campaña limnológica "Keratella1" en el río Paraná medio. Aspectos tróficos de los peces de ambientes leníticos. Ecología 4:115–126
Oldani NO, Oliveros OB (1984) Estudios limnológicos en una sección transversal del río Paraná. Dinámica temporal de peces de importancia económica. Rev Asoc Cienc Nat Litoral 15(2):175–183
Oliveros OB, del Barco DM (1991) Aspectos de la biología de *Ageneiosus valenciennesi* (Pisces, Ageneiosidae). Rev Asoc Cienc Nat Litoral 22(2):43–53
Oliveros OB, Rossi LM (1991) Ecología Trófica de *Hoplias malabaricus malabaricus* (Pisces, Erythrynidae). Rev Asoc Cienc Nat Litoral. 22(2):55–68
Parma de Croux MJ (1981) Metabolismo de rutina de *Pimelodus maculatus* (Lac) (Pisces, Pimelodidae). Rev Asoc Cienc Nat Litoral 12:20–26
Parma de Croux MJ (1983a) Metabolismo de rutina de *Prochilodus platensis* (Holmberg, 1889) (Pisces, Prochilodontidae) Rev Asoc Cienc Nat Litoral 14:1–6
Parma de Croux MJ (1983b) Metabolismo de rutina de *Hoplias malabaricus malabaricus* (Bloch, 1794) (Pisces, Erythrinidae) Rev Brasil Zool Sao Paulo 1(3):217–222
Parma de Croux MJ (1987) Nivel de oxígeno letal y mínimo de tolerancia en *Pimelodus albicans* (Pisces, Pimelodidae). Rev Asoc Cienc Nat Litoral 18(1):85–91
Parma de Croux MJ (1994) Metabolic rate and oxygen consumption requirements of some fish species from the middle Paraná River. Acta Biol Venez 15(2):1–10
Parma de Croux MJ (1997) Relationship between metabolic rate and body weight in *Salminus maxillosus* (Val, 1840). J Aqua Trop 12(1):17–22
Quiros R, Vidal JC (2000) Cyclic behaviour of potamodromous fish in large rivers. In: Coux I (ed) Management and ecology of river fisheries. Fishing News Books, Blackwell Science, London, pp 71–86
Rantin FT, Johansen K (1984) Responses of the teleost *Hoplias malabaricus* to hypoxia. Environ Biol Fish 11(3):221–228
Ringuelet R (1975) Zoogeografía y ecología de los peces de aguas continentales de la Argentina y consideraciones sobre las áreas ictiológicas de América del Sur. Ecosur 2(3):1–122
Rossi LM (1989) Alimentación de larvas de *Salminus maxillosus* (Val.1840) (Pisces, Characidae). Iheringia 69:49–59

Rossi LM (1992) Evolución morfológica del aparato digestivo de postlarvas y prejuveniles de *Prochilodus lineatus* (Val., 1847) (Pisces, Curimatidae) y su relación con la dieta. Rev Hydrobiol Trop 25(2):159–167

Rossi LM, Parma de Croux MJ (1992) Influencia de la vegetación acuática en la distribución de peces del río Paraná, Argentina. Ambiente subtropical 2:65–75

Rossi LM (2001) Ontogenetic diet shifts in a Neotropical catfish (*Sorubim lima*) of Paraná System. Fish Manage Ecol 8:141–152

Roux AL, Copp GH (1996) Fish populations in rivers. In: Petts G, Amoros C (eds) Fluvial hydrosystems. Chapman and Hall, London, pp 167–183

Saint-Paul U (1984) Physiological adaptation to hypoxia of a Neotropical characoid fish *Colossoma macropomum*, Serrasalmidae. Environ Biol Fish 11(1):53–62

Segatti Hahn NA, De Fátima Andrian I, Fugi R, Lescano de Almeida VL (1997) Ecología trófica. In: Vazzoler AE, Agostinho AA, Hahn NS (eds) A planicie de inundaçao do alto rio Paraná: aspectos físicos, biológicos e socioeconómicos. Editora da Universidade Estadual de Maringá, Maringá, pp 209–228

Schiemer F, Zalewski M (1992) The importance of riparian ecotones for diversity and productivity of riverine fish communities. Neth J Zool 42(2–3):323–335

Tablado A, Oldani N, Ulibarre L, Pignalberi de Hassan C (1988) Cambios Estacionales de la densidad de peces de una laguna del valle aluvial del río Paraná (Argentina). Rev Hydrobiol Trop 21(4):335–348

Val AL, AlmeidaVal VMF (1995) Fishes of the Amazon and their environment. Springer, Berlin Heidelberg New York, 224 pp

Vazzoler AE, Agostinho AA, Segatti Hahn N (1997) A planicie de inundacao do Alto Río Paraná. Aspectos físicos, biológicos y socioeconómicos Editora da Universidade Estadual de Maringá, Maringá 460 pp

Welcomme RL (1979) Fisheries ecology of floodplain rivers. Longman, London

Welcomme RL (2000) Fish biodiversity in floodplains an their associated rivers. In: Gopal B, Junk WJ, Davis JA (eds) Biodiversity in wetlands: assessment, function and conservation. Backhuys Publishers, Leiden, The Netherlands, pp 61–87

Winemiller KO (1989) Patterns of variation in life history among South American fishes in seasonal environments. Oecologia 81:225–241

Winemiller KO, Jepsen DB (1998) Effects of seasonality and fish movement on tropical river food webs. J Fish Biol 53(Suppl.A):267–296

13 Amphibians

Paola M. Peltzer[1,2,3] and Rafael C. Lajmanovich[1,2]

13.1 Introduction

The unpredictable nature of flood and flow pulses and river floodplain networks create challenges for organisms, such as a variety of anuran amphibian species, in their aquatic larval and terrestrial or semi-terrestrial life stages. These vertebrates undoubtedly form an important component of the fauna of floodplains and may contribute to the rapid recycling of detritus and mud (Welcomme 1990). Moreover, anuran species is adapted reproductively and developmentally to expansion/contraction of surface waters of riparian-derived food organisms and water volume (McDiarmid and Altig 1999).

In freshwater amphibian ecology have been emphasized the need to consider ecological processes at multiple scales. It is important to note that the complexity of river–floodplain systems results from interactions between the process that operates at spatial and temporal scales. The insight is especially relevant if we consider the diversity of waterbodies within the floodplain, which provides an interesting background for the study of the organization of amphibian communities (Joly and Morand 1994). For example, in these aquatic habitats, amphibian survivorship and breeding activities depend heavily upon large-scale processes (landscape characteristics), such as frequency and duration of hydrologic connections to waterbodies in a floodplain (Peltzer and Lajmanovich 2004). Moreover, anurans are not solely dependent upon within-pond features to determine the suitability of aquatic sites to their life requirements. Most anurans spend the bulk of their life in the upland habitats close to waterbodies, while others live in close proximity to ponds and have small home ranges during their terrestrial life stages. In this context, anuran are a sensitive indicator of environmental change and can be considered good measures of the state, quality, or condition of both terrestrial and aquatic ecosystems (US EPA 2002) due to a number of inherent biological and ecological factors (e.g., complex life cycles, thin and highly permeable

[1]Instituto Nacional de Limnología (INALI-CONICET-UNL), José Maciás 1933, 3016 Santo Tomé, Santa Fe, Argentina
[2]Facultad de Bioquímica y Ciencias Biológicas (ESS-FBCB-UNL), Ciudad Universitaria, Pje. El Pozo s/n, 3000, Santa Fe, Argentina
[3]e-mail: paolapeltzer@hotmail.com

M.H. Iriondo, J.C. Paggi, and M.J. Parma (Eds.)
The Middle Paraná River: Limnology of a Subtropical Wetland

skin at all life stages, and relatively low dispersal capabilities). While not all of the causes are clearly identifiable, it is thought that this environmental sensitivity has lead to recent worldwide declines of amphibians. In addition, concerns for global amphibian declines (Gardner 2001) have increased the urgency of understanding amphibian–habitat relationships. Thus, species-specific differences in response to (and tolerance of) different factors further complicate investigations of amphibian declines (Hayes et al. 2002). This is important since factors most likely work synergistically to negatively affect amphibian populations. For example, the nutrient-rich alluvial soils often associated with lowland floodplains have frequently been targets of intensive agriculture and livestock grazing. Due to direct impacts from agriculture and other land uses, the natural terrestrial, wetland, and aquatic habitats that are crucial for anuran reproduction and survival (e.g., forage and hibernation) are polluted, including nutrient loading, from locations anywhere within their catchments (Peltzer 2006). Consequently, these vertebrates may be exposed through their ontogeny to a variety of agricultural contaminants (pesticides and fertilizers) that are used liberally in these environments. The detrimental consequences of xenobiotics on survival and health of anuran amphibians have recently been documented for Fluvial Littoral of Argentina (Lajmanovich et al. 2005a, 2005b).

Taking into account the factors mentioned above, the objective of this chapter is to present information of 8 years (Jan. 97–March 05) regarding the diversity and natural history (tadpole and adult characteristics, habitat use, and reproductive traits) of anurans commonly found in riparian areas of the Middle Paraná River in Argentina. Moreover, we analyzed data of anuran assemblage in ponds that differed in hydroperiod and environmental features (habitat and spatial variables).

Recently, the need for landscape-management schemes and clear agricultural policies to preserve anuran and its habitats has been emphasized because several deleterious effects on health population and communities have been determined (Peltzer 2006). It is important to note that the typical landscape in the Middle Paraná River floodplain presents a complex system of islands, ponds, and permanent freshwater marshes interspersed with riparian woodland and gallery forest, making this an important region for the herpetofauna of Argentina. However, in the last 30 years, large parts of this region were gradually cultivated, urbanized, and deforested, and natural lands became severely fragmented and modified (Peltzer et al. 2003).

13.2 Amphibian Diversity and Natural History

A total of 34 anuran (Amphibia, Anura) species inhabit riparian areas of the Middle Paraná River, distributed among six families Bufonidae, Leptodactylidae, Cycloramphidae, Ceratophryidae, Hylidae, and Microhylidae (Table 13.1).

13.2.1 Reproductive Modes

Table 13.1 shows the ten reproductive modes (sensu Lavilla and Rouges 1992) per anuran species that inhabit riparian areas of the Middle Paraná River, mode 6 being the most frequent (32%), and modes 5 and 10 (2%) the less frequent ones.

13.2.2 Reproductive Pattern and Breeding Sites

Four species (12%) *Physalaemus biligonigerus Odontophrynus americanus, Scinax nasicus*, and *Hypsiboas pulchellus* are continuous breeders (Table 13.1). The first three species reproduce approximately 10 months throughout the year while *H. pulchellus* exhibit three reproductive pauses (February until the first days of March, June until the first days of July; middle October to November). Fifteen species (44%) exhibit a prolonged reproductive pattern, breeding for at least 3–4 months, and the rest of the anuran (44%) are explosive breeders, reproducing after great rains. Since most species in the Middle Paraná River floodplain prefer temporary or semi-permanent ponds for breeding ($n = 15$ species both ponds, indistinctively), the climax of the reproductive period at these unstable sites is always limited to the rainy season (Table 13.1, October to March). A total of $n = 2$ species prefer temporary, semi-permanent, or permanent breeding sites and six (n=6) anurans reproduce in either semi-permanent or permanent waterbodies. One species (*Melanophryniscus klappenbachi*) only breeds in temporary ponds.

13.2.3 Spatial Guilds

The anuran species in the Middle Paraná River floodplain exhibits a mosaic of different ecomorphological traits (Lajmanovich 2000; Peltzer and Lajmanovich 2004; Peltzer et al. 2006). The anuran tadpoles represent three ecomorphological guilds: benthic (B, 14 species), nektonic (N, 10 species), and transitional between benthic and nektonic morphotypes (B-N, 10 species) (Table 13.1). Adult anurans show differential spatial vertical position, being both terrestrial and arboreal species the most frequent guilds ($n = 12$, respectively), followed by semi-terrestrial ($n = 7$) and aquatic ($n = 3$) species.

13.3 Characterization of the Main Habitats that Provide Reproduction Sites or Refuges/Food

Temporary or semi-permanent ponds in the Paraná River floodplain provide breeding, foraging, and over-wintering sites and favorable conditions for amphibian's persistence (Peltzer and Lajmanovich 2004). Particularly, many species often breed in the same pond and, therefore, to avoid competition and

Table 13.1 Overview of biological and ecological traits of anuran species from the Middle Paraná River floodplain. Adult size (Adult: *AD*), spatial guilds (Tadpoles: *TA*, Adult: *AD* sensu Lajmanovich 2000; Peltzer et al. 2006), reproductive pattern and mode (sensu Prado et al. 2005; Lavilla and Rouges 1992), breeding sites and period. References: (*N*) Nektonic, (*B*) Benthic, (*B-N*) Benthic-Nektonic, (*TE*) terrestrial, (*ST*) semi-terrestrial, (*AQ*) aquatic, (*A*) arboreal, (*C*) continuous, (*E*) explosive, (*PR*) prolonged, (*T*) temporal, (*SP*) semi-permanent, (*P*) permanent

	Adult size	Spatial guild*		Reproductive ** Mode/***Pattern	Breeding site	Reproductive Period
ANURA	AD	TA	AD			
BUFONIDAE						
Chaunus						
C. arenarum (Hensel, 1867)	80–120	B	TE	7 [E]	T,SP,P	
C. schneideri (Werner, 1894)	210	B	TE	7 [E]	T,SP,P	
C. fernandezae (Gallardo, 1957)	56–80	B	TE	7 [E]	T,SP,P	
C. granulosus (Spix, 1824)	53	B	TE	7 [E]	T,SP,P	
Melanophryniscus						
M. klappenbachi Prigioni & Langone, 2000	51–70	B	TE	5 [E]	T	
LEPTODACTYLIDAE						
Leptodactylus (*Leptodactylus*)						
L. ocellatus (Linnaeus, 1758)	120–140	B-N	ST	8 [E]	T,SP	
L. chaquensis Cei, 1950	80–85	B-N	ST	8 [E]	T,SP	
L. gracilis (Duméril & Bibron, 1841)	45–50	B-N	TE	13 [PR]	T,SP	
L. mystacinus (Burmeister, 1861)	50–65	B-N	TE	13 [PR]	SP,P	
L. latinasus Jimenez de la Espada, 1875	32–37	B-N	TE	13 [PR]	SP,P	
L. elenae Heyer, 1978	45–50	B-N	TE	13 [PR]	T,SP	
Leptodactylus (*Lithodytes*)						
L. diptyx Boettger, 1885	27–33	B-N	TE	10 [PR]	T,SP	

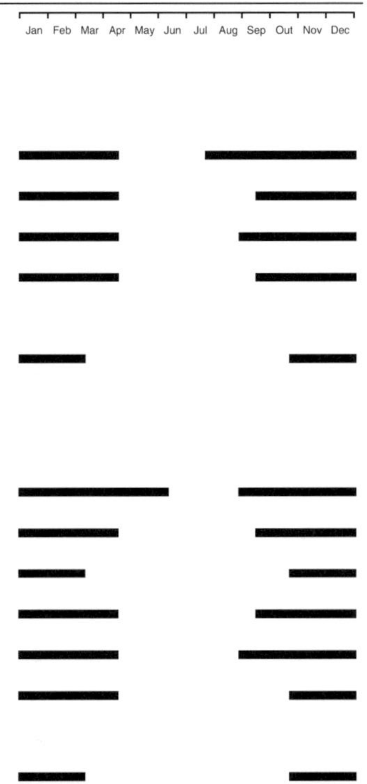

Physalaemus					
P. biligonigerus (Cope, 1861)	35–40	B	ST	8 [C]	T,SP
P. albonotatus (Steindachner, 1864)	30	B	ST	8 [E]	T,SP
P. santafecinus Barrio, 1965	28–34	B	ST	8 [E]	T,SP
P. riograndensis Milstead, 1960	25	B	ST	8 [E]	T,SP
Pseudopaludicola					
P. falcipes (Hensel, 1867)	15–20	B	ST	6 [E]	T,SP
CYCLORAMPHIDAE					
Odontophrynus					
O. americanus (Duméril & Bibron, 1841)	49–70	B-N	TE	1 [C]	T,SP
CERATOPHRYIDAE					
Ceratophrys					
C. cranwelli Barrio, 1980	110–125	B	TE	1 [E]	T,SP,P
HYLIDAE					
Phyllomedusa					
P. hypochondrialis Daudin, 1800	35–40	N	A	15 [PR]	T,SP
Trachycephalus					
T. venulosus (Laurenti, 1768)	92–114	N	A	3 [E]	T,SP
Hypsiboas					
H. pulchellus (Duméril & Bibron, 1841)	37–50	B-N	A	6 [C]	T,SP,P
H. raniceps (Cope, 1862)	70–75	B-N	A	6 [PR]	T,SP
H. punctatus (Schneider, 1799)	30–35	N	A	6 [PR]	SP,P
Dendropsophus					
D. nana (Boulenger, 1889)	15–22	B	A	6 [PR]	SP,P
D. sanborni (Schmidt, 1944)	15–21	B	A	6 [PR]	SP,P

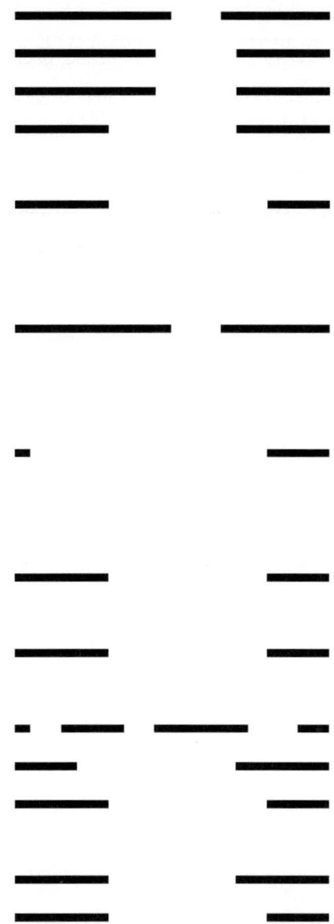

(*Continued*)

Table 13.1 Overview of biological and ecological traits of anuran species from the Middle Paraná River floodplain. Adult size (Adult: *AD*), spatial guilds (Tadpoles: *TA*, Adult: *AD* sensu Lajmanovich 2000; Peltzer et al. 2006), reproductive pattern and mode (sensu Prado et al. 2005; Lavilla and Rouges 1992), breeding sites and period. References: (*N*) Nektonic, (*B*) Benthic, (*B-N*) Benthic-Nektonic, (*TE*) terrestrial, (*ST*) semi-terrestrial, (*AQ*) aquatic, (*A*) arboreal, (*C*) continuous, (*E*) explosive, (*PR*) prolonged, (*T*) temporal, (*SP*) semi-permanent, (*P*) permanent —Continued

	Adult size	Spatial guild*		Reproductive ** Mode/***Pattern	Breeding site	Reproductive Period
Scinax						
S. nasicus (Cope, 1862)	27–35	N	A	6 C	T,SP,P	
S. squalirostris (Lutz, 1925)	23–28	N	A	6 PR	T,SP	
S. acuminatus (Cope, 1862)	19–25	N	A	6 E	T,SP,P	
S. fuscomarginatus (Lutz, 1925)	20–23	N	A	6 PR	T,SP,P	
S. berthae (Barrio, 1962)	19–28	B-N	A	6 PR	T,SP	
Pseudis						
P. paradoxus occidentalis Gallardo, 1961	35–56	N	AQ	2 PR	SP,P	
P. paradoxus platensis Gallardo, 1961	38–58	N	AQ	2 PR	T,SP,P	
Lysapsus						
L. limellum Cope, 1862	27–35	N	AQ	2 PR	T,SP,P	
MICROHYLIDAE						
Elachistocleis						
E. bicolor (Guérin-Méneville, 1838)	15–20	B	TE	3 E	T,SP,P	

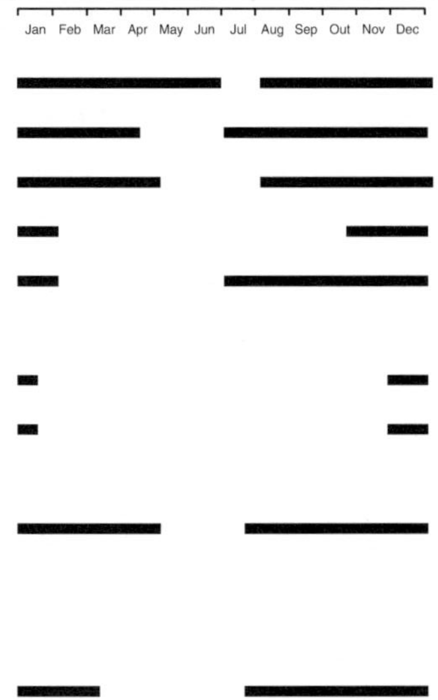

*Terrestrial: species found on the ground. Aquatic: species floating in water or inhabiting marsh. Semi-terrestrial: species inhabit terrestrial and aquatic ecotones. Arboreal: species inhabit herbaceous, shrubby, or arboreal habitats.

**Mode 1: eggs are deposited individually at the bottom in lentic water. Mode 2: eggs are individually glue in submerged vegetation. Mode 3: eggs are scattered deposited in a continuous layer in water surface. Mode 5: eggs are deposited in a mass at the bottom of the waterbody. Mode 6: eggs are deposited in spherical gelatinous clusters attached to submerged plants or objects. Mode 7: eggs are deposited in a jelly-like strand. Mode 8: eggs are deposited in a floating foam nest. Mode 10: few eggs are deposited in wet and protected burrows. Mode 13: eggs are deposited in a foam nest into land depressions, caves, subterranean cavities or mud constructions. Mode 15: eggs are deposited on leaves above water.

***Prolonged (PR): anuran species reproduces more than three consecutive months, Explosive (E): species reproduces at least for one or a few days, Continuous (C): anuran species with individuals breeding success throughout one year.

predation pressure, amphibians often show temporal segregation in the use of breeding ponds; a common phenomenon recorded in some riverine amphibian communities (e.g., Lajmanovich 2000; Snodgrass et al. 2000). We suggest that the risk of reproductive failure and/or occurrence of anuran species depend on the length of the hydroperiod and a complex network among habitat and spatial factors. In the following section, we discuss relationships between habitat ($n = 8$) and spatial ($n = 7$) variables and anuran assemblages based on multivariate analyses (principal component and canonical correspondence analyses) performed to reduce variables to a smaller number of independent components and determine which variables contributed mostly to anuran composition (diversity, evenness, and richness) variation (Lajmanovich 1997, 2000; Peltzer et al. 2003, 2006) (see Table 13.2 and Fig. 13.1).

13.3.1 Relationship among Habitat and Landscape Variables and Anuran Assemblages

13.3.1.1 Habitat Variables

Hydroperiod. The temporary ponds in riparian areas of the Middle Paraná River floodplain provide a suitable habitat for reproducing only for anuran species with shorter larval life spans, such as *Leptodactylus ocellatus* and *Physalaemus biligonigerus.* These frogs lay their eggs in a foam nest on the water's surface, giving protection from dehydration during frequent filling/drying cycles of temporary ponds, and presenting fast tadpole development that takes place in 1–2 weeks (pers. obs.) during the summer season. Relya and Werner (2000) pointed out that in response to this unpredictability, some anurans show plastic strategies on the onset of breeding or select waterbodies with a hydroperiod that is adjusted to specific requirements for larval development. Moreover, the unstable hydroperiod of temporal breeding sites, especially during the first rains, limits the occurrence of some anuran species. Alternatively, some species may breed in these habitats during heavy and continuous rains or when these sites fill with rainwater and resemble permanent habitats. For example, the hylids frogs that deposit their gelatinous eggs underwater or on the water surface have slower tadpole development (Peltzer and Lajmanovich 2004). Joly and Morand (1994) remarked the possibility that several alternative traits may serve as response to the same environmental constraint, or that one trait may be under several possibilities environmental constraint. A temporal flexibility of the breeding period appears as an adaptation to the unpredictability of water level, for example in species of *Chaunus arenarum*, *Hypsiboas pulchellus*, and *Elachistocleis bicolor*. A possible explanation for this is an adaptation to unpredictable and unstable temporal habitats (Levins 1968) that show intense variation in hydroperiod (Brasileiro et al. 2005). In other words, clutch partitioning among different sites may also be regarded as another adaptative trait used to overcome the unstable of water retention (Joly and Morand 1994).

Table 13.2 Principal component loadings for anuran diversity, habitat, and spatial variables measured at ponds from the Middle Paraná River floodplain and results of a canonical correlation analysis

Principal component analysis		Axis components	
Variables	Acronyms	PC-1	PC-2
HABITAT VARIABLES			
Air temp. (°C)	T°A	**0.45**	0.04
Water temp. (°C)	T°W	**0.45**	0.05
Turbidity (m)	Tu	0.25	–0.34
pH	pH	**0.48**	–0.02
Dissolved oxygen (mg L^{-1})	O	0.30	**–0.51**
Conductivity (μmS cm^{-1})	Cond	0.10	0.31
Rain (mm)	Rain	**0.47**	0.05
Hydroperiod	Hydr	0.26	**–0.72**
Cumulative of percentage variance explained		95	98
SPATIAL VARIABLES			
Pond area (m^2)	A	0.11	**0.56**
Distance to main channel (m)	DMC	0.15	**0.52**
Side floodplain (W–E)	SC	**0.52**	–0.21
Pond depth (m)	PD	0.38	–0.11
Surface connectivity (%)	PSC	0.12	**–0.47**
Aquatic Macrophytes (scale: 1 monospecific-5 poli-specific vegetation)	AM	**0.55**	0.36
Level of disturbances 0 natural sites-5 total degraded	LD	**0.49**	–0.23
Cumulative of Percentage variance explained		79	88
Canonical Correspondence Analysis		Axis I	Axis II
Species-environmental canonical correlation		0.87	0.86
Variables with high positive or high negative values are in bold			

Environmental factors. Heyer (1973) and Torres-Orozco et al. (2002) suggested that the presence of each anuran species and hierarchical process of habitat selection depend on both the species involved as well as environmental factors. The relationship between air and water temperature, rain, and anuran composition is observed in previous studies in tropical and subtropical

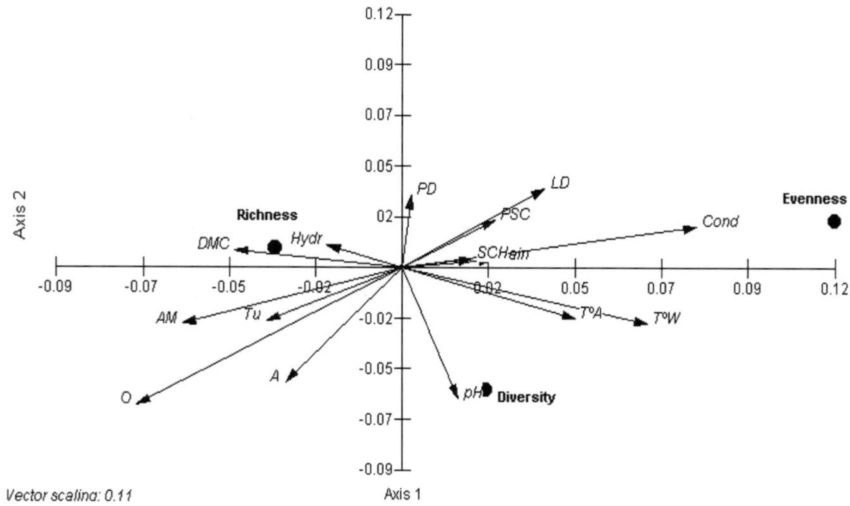

Fig. 13.1 Plot of canonical correspondence analysis by anuran composition and habitat and spatial variables. References: (*Hydr*) hydroperiod, (*T°W*) water temperature, (*T°A*) air temperature, (*Tu*) turbidity, (*pH*) pH, (*O*) dissolved oxygen, (*COND*) conductivity, (*Rain*) rainfall, (*PD*) pond depth, (*A*) area, (*DMC*) distance to main channel, (*SC*) side floodplain, (*PSC*) percentage of surface connectivity, (*AM*) aquatic macrophytes, and (*LD*) level disturbances

areas (e.g., Pombal 1997; Bertoluci 1998). It is postulated that these environmental factors may influence not only the onset of the reproductive period but also its duration. Most anuran species reproduced in a restricted period, determining a reproductive cycle. For example, rainfall may signal water availability, stimulating anuran reproductive activity and influencing the breeding success (Gascon 1991). Even though most anurans breed in the rainy season (between October and April) in riparian areas of the Middle Paraná River, some were active even during the dry season and comprises approximately 10 months, such as *P. biligonigerus*, *Odontophrynus americanus*, *H. pulchellus*, and *Scinax nasicus*. Like several other Neotropical anurans, these species can be regarded as continuous breeders. In general, in riparian areas of the Middle Paraná River, both explosive and prolonged breeders are the most important reproductive pattern, being continuous reproduction the fewer one. This pattern is commonly observed in seasonal environments, particularly in the Atlantic rainforest (e.g., Bertoluci and Rodrigues 2002; Prado et al. 2005).

pH. With respect to relationships between pH and anuran assemblages, several studies demonstrated that increasing the acidity of waterbodies has major deleterious effects on an amphibian tadpole's survival. Particularly, we observed that in some study ponds surrounded by agricultural, agroecosystems decreased pH two fold and had lower anuran composition (Peltzer et al.

2006; Peltzer 2006) not only affecting tadpoles but also adults dynamics. This result is of particular interest because some organochlorine pesticides used extensively in agricultural practices, such as Endosulfan, can increase the half-time life in this acidic aquatic environment and can persist for more than 3 months. Recently, the deleterious and genotoxic effect of this insecticide on anuran tadpoles (*H. pulchellus*) was demonstrated by Lajmanovich et al. (2005a).

Dissolved oxygen. We observed that dissolved oxygen was inversely related to anuran composition in ponds of riparian areas of the Middle Paraná River. Especially the low level of dissolved oxygen was recorded in some ponds located near or surrounded by agroecosystems and/or cattle presence (Peltzer et al. 2006). High stocking rates of cattle have contributed to declines in water quality and to the generally poor condition of many floodplain wetlands (Jansen and Healey 2003).

Aquatic vegetation. Several studies demonstrated that aquatic vegetation provides spatial heterogeneity, places for attaching eggs, and protection against pesticide contamination better than the open water (see Peltzer and Lajmanovich 2004; Lajmanovich et al. 2003). Moreover, the availability of aquatic vegetation as a structure in ponds can reduce the ability of fish to observe, pursue, and capture tadpoles (Jansen and Healey 2003). In addition, the phytophagous diets of some anuran tadpoles (Lajmanovich 2000) also suggest that in the aquatic vegetation zones they may largely find their optimal food. The high density of macrophytes makes the availability of support for epiphytic algae highest in littoral zone of lentic waterbodies (Joly and Morand 1997). Moreover, taking into account that livestock grazing was one of the major disturbances detected on ponds of Middle Paraná River floodplain, Jansen and Healey (2003) suggested that the removal of vegetation by cattle not only decreases water quality but also reduces an amphibian's reproductive activity.

Disturbance to the aquatic and surrounding terrestrial habitats. Riparian forest have been described as important anuran habitats that increase the connectivity in a landscape, providing niches for hibernation and foraging and routes for migration (Hazell et al. 2001). We found that the anuran composition (diversity, evenness, and richness) was highest in temporary and semi-permanent ponds without evidence of natural (flood) or anthropic (urbanization, bridge, roads, livestock grazing, deforestation, and agriculture) perturbations (Peltzer and Lajmanovich 2004). First, we suggest that these patterns can be explained as a function of disturbance by flooding and drying, with aquatic sites experiencing floods or with higher connectivity to the main basin having the lowest anuran species diversity, evenness and richness. In such highly unstable sites, it is likely that the entrance of major predators (particularly fishes) of anuran tadpoles is higher, making colonization of ponds by potential competitors. Secondly, an important decrease in anuran composition was observed in ponds surrounding or nearby agriculture or urban areas, altered by domestic livestock grazing or deforestation of riparian vegetation (Peltzer 2006). When the breeding sites are surrounded by crops, agrochemicals may accumulate in the vegetation where some adult anuran live or in nearest waterbodies by pluvial runoff, producing several detrimental effects on its health and survival (e.g., Lajmanovich et al. 2004; Cabagna et al. 2005). Furthermore, cleared areas experienced greater temperature extremes than areas with native vegetation and consequently anuran species' response may be magnified, affecting exposure to daily and seasonal climatic extremes (Hazell et al. 2001).

13.3.1.2 Landscape Variables

Area. The area of the suitable aquatic ponds constitutes one of the main regulating factors in population dynamics, acting both on larval and adult stages (Peltzer 2006). In other words, the carrying capacity of one aquatic site is closely correlated to the pond size as well as the physiological/life history traits. For example, at higher densities of anuran tadpoles, both exploitation and interference competition interact with environmental stochasticity (e.g., small ponds) in increasing extinction risk (Joly and Morand 1997). The competition results in subtle microhabitat partitioning among the tadpoles of different anuran species (McDiamird and Altig 1999), which is consistent with differential location of microhabitat of tadpoles find in different ponds. In accordance with Joly and Morand (1994), the highest values of anuran composition is reached at intermediate spatial and high temporal variability. Townsend (1989) mentioned the patch dynamic concept, assuming that at the highest spatial and intermediate temporal variability more species can co-exist. However, co-existence includes species other than anurans, and generally these non-anuran organisms are predacious competitors, a further reason for the low anuran composition, particularly larval stages, which we find in permanent riparian pond and ponds that are most frequently disturbed by floods. We confirmed this in field and lab investigations (e.g., Peltzer and Lajmanovich 2003, 2004; Peltzer 2006), finding that anuran abundance and richness were negatively correlated with the presence of potential predators. In this sense, the presence of potential tadpole predators are more frequently recorded in permanent ponds, particularly the presence of fishes (e.g., *Hoplias malabaricus*, *Astyanax* spp.); however, the presence of small predacious invertebrates (aeshnid odonate naiads: *Rhionaeschna* sp., hemipterans: *Belostoma* spp., hemipterans-heteropterans: *Notonecta glauca*, crabs *Dilocarcinus pagei*, and snails: *Pomacea canaliculata*) are more variable in temporal or semi-permanent ponds (Peltzer 2006).

Pond distance to the main channel and floodplain side (west or east). Within aquatic networks of alluvial floodplains of the Middle Paraná River the anuran composition from ponds increases with decreasing distance to the main channel and surface connectivity. One explanation for this may be that when ponds are connected (total or partially), larval, juveniles, and adult fishes may be exported to them, increasing the tadpole predation risk. In general, as mentioned previously, high-quality amphibian breeding sites are free of fish. However, certain fish species such as *Pterodoras granulosus, Astyanax* spp. and *Hoplias malabaricus* do colonize ponds in low-lying areas that have a connection to other water bodies during partially flooding, while large predatory fish, such as *Salminus maxillosus, Serrasalmus* spp., *Raphiodon vulpinus* and *Pseudoplatystoma coruscans* may invade these isolated ponds at times of abundant rainfall and/or during drastic floods, becoming an important threat for anuran. Moreover, in the west side of floodplain more anuran diversity, evenness and richness are frequently recorded (Peltzer 2006). A possible explanation for this is the influence of the Wet Chaqueña eco-region, and, consequently, its associated anuran lineages, for example the presence of *C. granulosus, M. klappenbachi*, and *L. diptyx*, species that were not found on the east side.

13.4 Final Remarks

The factors that determine the presence of anuran species in ponds of the Middle Paraná River floodplain are part of a complex network of relationships that work together. These dynamics are strongly influenced not only by habitat and spatial constraint but also on the species involved (analysis of the natural history of each species, reproductive traits, functional types, and spatial distribution).

Acknowledgements. We thank Adolfo Beltzer, Néstor Basso, Alfredo Berduc, Pedro Cacivio, Lucía Federico, Diego Baldo, Julián Faivovich, Maximiliano Attademo, Laura Sánchez, Raquel Tardivo, Germán Bock, and Alejandro Giraudo for their advice and collaboration during this 8-year period.

References

Cabagna MC, Lajmanovich RC, Stringhini G, Sanchez-Hernandez JC, Peltzer PM (2005) Hematological parameters of health status in the common toad *Bufo arenarum* in agroecosystems of Santa Fe Province, Argentina. App Herpetol 2:373–380

Bertoluci J (1998) Annual patterns of breeding activity in Atlantic Rainforest anurans. J Herpetol 32:607–611

Bertoluci J, Rodrigues MT (2002) Utilização de habitats reprodutivos e micro-habitats de vocalização em uma taxocenose de anuros (Amphibia) da Mata Atlântica do sudeste do Brasil. Papéis Avulsos de Zoología 42:287–297

Brasileiro CA, Azuaya RJ, Kiefer MC, Martins M (2005) Amphibians of an open Cerrado fragment in southeastern Brazil. Biota Neotrop 5:93–109

Gardner T (2001) Declining amphibian populations: a global phenomenon in conservation biology. Anim Biodiv Conserv 24:25–44

Gascon C (1991) Populations and community level analyses of species occurrences of Central Amazonian rainforest tadpoles. Ecology 72:1731–1746

Hayes TB, Collins A, Lee M, Mendoza M, Noriega N, Stuart AA, Vonk A (2002) Hermaphroditic, demasculinized frogs after exposure to the herbicide atrazine at low ecologically relevant doses. Proceedings of the National Academy of Sciences of the United States of America 99:5476–5480

Hazell D, Cunnungham R, Lindenmayer D, Mackey B, Osborne W (2001) Use of farm dams as frog habitat in an Australian agricultural landscape: factors affecting species richness and distribution. Biol Conserv 102:155–169

Heyer WR (1973) Ecological interactions of frog larvae at a seasonal tropical location in Thailand. J Herpetol 7:337–361

Jansen A, Healey M (2003) Frog communities and wetland condition: relationships with grazing by domestic livestock along an Australian floodplain river. Biol Conserv 109:207–219

Joly P, Morand A (1994) Theoretical habitat templates, species traits, and species richness: amphibians in the Upper Rhöne River and its floodplain. Freshw Biol 31:455–468

Joly P, Morand A (1997) Amphibian diversity and land–water ecotones. In: Lachavanne JB, Juge R (ed) Biodiversity in land–inland water ecotones, vol 18. Man Biosphere Ser Geneva, pp 161–182

Lajmanovich RC (1997) Alimentación de larvas de anuros en ambientes temporales del sistema del río Paraná, Argentina. Doñana Acta Vertebrata 24:191–202

Lajmanovich RC (2000) Interpretación ecológica de una comunidad larvaria de anfibios anuros. Interciencia 25:71–79

Lajmanovich RC, Lorenzatti E, de la Sierra P, Marino F, Stringhini G, Peltzer P (2003) Reduction in the mortality of tadpoles (*Physalaemus biligonigerus*, Amphibia: Leptodactylidae) exposed to cypermethrin: uptake by aquatic ferns. Freshw Environ Bull 12:1558–1561

Lajmanovich RC, Cagabna M, Peltzer PM, Stringhini G, Attademo A (2005)a Micronucleus induction in erythrocytes of the tadpoles of *Hyla pulchella* (Amphibia: Hylidae) exposed to insecticide endosulfan. Mut Res 587:67–72

Lajmanovich RC, de la Sierra P, Marino F, Peltzer PM, Lenardon A, Lorenzatti E (2005b) Determinación de residuos de organoclorados en vertebrados silvestres del litoral fluvial de Argentina. In: Aceñolaza FG (ed) Temas de la Biodiversidad del Litoral Fluvial Argentino II, vol 14. INSUGEO, Tucumán, pp 389–398

Lajmanovich RC, Sanchez-Hernandez J, Stringhini G, Peltzer PM (2004) Levels of serum cholinesterase activity in the Rococo toad (*Bufo paracnemis*) in an agrosystem of Argentina. Bull Environ Cont Toxicol 72:548–591

Lavilla EO, Rougès M (1992) Modos de Reproducción de Anuros Argentinos. Serie Divulgación, Asociación Herpetológica Argentina 5:1–66

Levins R (1968) Evolution in changing environments. Princeton University Press, Princeton, 120 pp

McDiarmid RW, Altig R (1999) Tadpoles: the biology of anuran larvae. The University of Chicago Press, Chicago

Peltzer PM (2006) La fragmentación de hábitat y su influencia en la diversidad y distribución de anfibios anuros de áreas ecotonales de los dominios fitogeográficos amazónico y chaqueño. Doctoral Thesis, Universidad Nacional de La Plata, Buenos Aires, Argentina, 276 pp

Peltzer PM, Lajmanovich RC (2003) *Hyla pulchella* (NCN). Predation. Herpetol Rev 34:231

Peltzer PM, Lajmanovich RC (2004) Anuran tadpole assemblages in riparian areas of Paraná River (Argentina). Biodiv Conserv 13:1833–1842

Peltzer PM, Lajmanovich RC, Beltzer AH (2003) The effects of habitat fragmentation on amphibian species richness in the floodplain of the Middle Paraná River. Herpetol J 13:95–98

Peltzer PM, Lajmanovich RC, Attademo MA, Beltzer AH (2006) Anuran diversity across agricultural pond in Argentina. Biodiv Conserv 15:3499–3513

Pombal JP (1997) Distribuição espacial e temporal de anuros (Amphibia) em uma poça permanente na Serra de Paranapiacaba, sudeste do Brasil. Rev Brasil Biol 57:583–594

Prado ACP, Uetanabaro M, Haddad C (2005) Breeding activity patterns, reproductive modes, and habitat use by anurans (Amphibia) in a seasonal environment in the Pantanal, Brazil. Amph Rep 26:211–221

Relyea RA, Werner EE (2000) Morphological plasticity in four larval anurans distributed along an environmental gradient. Copeia 2000:178–190

Snodgrass JW, Komoroski MJ, Bryan AL, Burger J (2000) Relationship among isolated wetland size, hydroperiod, and amphibian species richness: implications for wetland regulations. Conserv Biol 14:414–419

Torres-Orozco R, Jimenez–Sierra CL, Vogt RC, Villareal Benitez JL (2002) Neotropical tadpoles: spatial and temporal distribution and habitat use in a seasonal lake in Veracruz, Mexico. Phyllomedusa 1:81–91

Townsend CR (1989) The patch dynamic concepts of stream community ecology. J North Am Benthol Soc 8:36–50

U.S. Environmental Protection Agency (2002) Methods for evaluating wetland condition: using amphibian in bioassessments of wetlands. Washington, DC, EPA-822-R-02-022

Welcomme RL (1990) Status of fisheries in South American rivers. Interciencia 15:337–345

14 Reptiles

Alejandro R. Giraudo[1,2,3], Vanesa Arzamendia[1,2], and Soledad M. López[1]

14.1 Introduction

Aquatic ecosystems and their species have traditionally been studied separately from the terrestrial systems that surround them. This tendency is being modified in the last decades, and the study of the great South American rivers and their floodplains had a central role in recognizing the importance of the interaction between aquatic and terrestrial phases to understand their functioning (Junk 1997a). However, studies on terrestrial fauna associated with wetlands are even more scarce (Gibbons 2003). There are animal groups that have the function of relating aquatic and terrestrial ecosystems: reptiles are one of them. Various groups and species of reptiles have adapted to live in the water, although most of them still depend on the mainland to complete their life cycle, developing several of their vital functions, as reproduction and hibernation, in terrestrial ecosystems that surround the wetlands (Gibbons 2003; Roe et al. 2003). Such species are called semi-aquatic, and the peripheral aquatic habitats and corridors that connect wetlands (e.g., flooding valley lakes) are vital for these reptile communities (Gibbons 2003; Roe et al. 2003). The great South American rivers, including the Middle Paraná, have extensive floodplains where aquatic and terrestrial phases alternate, and flood pulses constitute an important regulating factor in the structure and functioning of their biota (Junk 1997b; Neiff 1990). However, the Amazon and Orinoco basins are located in intertropical regions and their main rivers run in a west–east direction, generally surrounded by tropical forests. One particularity of the Paraná basin is that its main collectors, the Paraná and Paraguay Rivers, begin in tropical latitudes and run in a north–south direction into temperate latitudes (37° latitude S) until the Río de la Plata (Giraudo and Arzamendia 2004). The effect of these rivers as biogeographical corridors of tropical biotas towards temperate latitudes is a pattern indicated by diverse authors (Menalled and Adámoli 1995; Giraudo and Arzamendia

[1]Instituto Nacional de Limnología, CONICET-UNL, José Maciá 1933 (3016) Santo Tomé, Santa Fe, Argentina
[2]Facultad de Humanidades y Ciencias, Universidad Nacional del Litoral (UNL). Maestría en Ecología, Facultad de Ciencia y Tecnología, UADER, Entre Ríos, Argentina
[3]e-mail: alejandrogiraudo@hotmail.com

M.H. Iriondo, J.C. Paggi, and M.J. Parma (Eds.)
The Middle Paraná River: Limnology of a Subtropical Wetland

2004). Consequently, the Middle Paraná River biota is dominated by tropical elements that contrast with xerophilous forests and temperate steppes that go across it. This characteristic, added to the high heterogeneity of habitats modeled by flood pulses, make the Middle Paraná River an area with high species richness in relation to adjacent non-flooded lands (Arzamendia and Giraudo 2004). Some reptiles are conspicuous fauna in the great South American rivers (Junk and Silva 1997). The famous naturalist Alcides d'Orbignyi mentioned in 1828: "I also saw numerous caimans and I was told that they live further south", when he navigated through sectors of the Middle Paraná River. With the incorporation of reptile leather into contemporary fashion, all crocodilians have been hunted by the thousands during the 20th century and all species declined in numbers (Busack and Pandya 2001, see Medem 1983 for a revision on their decline in Argentina). Although the commercial capture pressure has decreased in the last decades and caimans are again observed in the area, they are not so frequent now in human densely populated sectors, as was indicated by d'Orbignyi in the 19th century. As with amphibians, the global decline of reptiles has been postulated, and it has been proven that many populations of semi-aquatic and aquatic reptile populations declined due to wetland disappearance, canalization, pollution and dam constructions (Gibbons et al. 2000). Reptiles are among the most important predators of the Middle Paraná River, since large mammals, as the jaguar (*Phanthera onca*) or the giant otter (*Pteronura brasiliensis*), are locally extinct. Crocodilians and large snakes (e.g., *Eunectes notaeus*, that reaches 5 m in total length) feed on all kinds of upper vertebrates and, in the Amazon floodplain lakes, crocodilians and other reptiles can have a key function in nutrient recycling, e.g., nitrogen (Fittkau 1973). This chapter aims to analyze the reptile community of the Middle Paraná River floodplain, including aspects of species richness, its main ecological traits (abundance, habitat and substrate use, reproductive and feeding strategies), and the influence of the flood-pulse and other environmental variables in the reptile community.

14.2 Species Richness and Ecological Traits

A total of 71 reptile species were recorded in the Middle Paraná River floodplains (Table 14.1), including five turtles, two crocodilians, nine lizards, six amphisbaenians and 49 snakes. The list of species recorded in the Middle Paraná River floodplains is probably not yet complete. In the 1992–2006 period, 3,934 reptile specimens were recorded, corresponding to 56 species, and the additional taxa are reliable from literature records (Table 14.1). Continued samples will probably add rare or secretive reptile species inhabiting surrounding localities (e.g., *Anisolepis longicauda, Liophis flavifrenatus, Pseudoeryx plicatilis*, see Cei 1993 and Giraudo 2001 for a revision of expected species). Reptile composition in the Middle

Paraná River is still not completely known, as evidenced by the recent description of two new species: the aquatic snake *Hydrops caesurus* (Scrocchi et al. 2005) and the amphisbaenid *Amphisbaena hiata* (Montero and Céspedez 2002).

14.2.1 Turtles

The five turtle species that inhabit the Middle Paraná River floodplain have aquatic habits (Table 14.1). *Phrynops hilarii, Mesoclemmys vanderhaegei, Acanthochelys pallidipectoris* and *A. spixii* belong to the family Chelidae and *Trachemys dorbigni* belongs to the family Emydidae. *Phrynops hilarii* is one of the most abundant reptiles in the Middle Paraná River (see Sect.14.4.2), and reaches the largest size within the turtles (400 mm of carapace length). It is frequent in ponds and lakes of the flood valley and in larger lotic channels. The other four species are scarce or rare in the Middle Paraná River. *Mesoclemmys vanderhaegei* is known by a unique reference in the Malabrigo Port (Iverson 1986). *Trachemys dorbigni* is a frequent species in the Uruguay River, although with scarce records in the Middle Paraná River (Freiberg 1969; Cabrera 1998). *Acanthochelys spixii* inhabits the Middle Paraná River marginally, in riverbank lakes of the province of Corrientes, connected with the Paraná River during great floods. Similarly, *Achantochelys pallidipectoris* was recorded near Resistencia (province of Chaco) (Waller 1988), in lakes occasionally connected with the Middle Paraná River during great floods. All mentioned turtle species are carnivorous and include fish in their diets, in addition to vertebrates and invertebrates and, in the case of *T. dorbigni,* it also consumes plants (Table 14.1, Cei 1993; Cabrera 1998).

14.2.2 Crocodilians

Two crocodilian species inhabit the Middle Paraná River (Table 14.1). *Caiman latirostris* inhabits throughout the extension of the Middle Paraná River. It is the species with the southernmost distribution in the New World, reaching 33° of latitude S in temperate areas of the Middle Paraná River (Waller 1987; Scott et al. 1990). *Caiman yacare* only inhabits the northern sector of the Middle Paraná River, between the mouth of the Paraguay River and 29° of latitude S (Micucci and Waller 1995). Both species have been recorded occasionally in higher latitudes due to transport by the Paraná and de la Plata rivers produced by extraordinary floods (Micucci and Waller 1995). In coincidence with what was indicated by Scott et al. (1990) for these species in the Paraguay River, they are in sympatry in the northern sector of the Middle Paraná River, showing differences in their habitat use. *Caiman latirostris* is mainly found in smaller and more ephemeral wetlands of the Aquatic/Terrestrial Transition Zone (ATTZ; Junk et al. 1989), such as

Table 14.1 Reptiles from Middle Paraná River floodplains with some morphological and ecological traits

Species	Length[c] mean	Maximum or range	Substrate use	Food[d]	Reproductive modality	Number of eggs or offspring
TURTLES						
Trachemys dorbigni	–	250	Aquatic	fi, am, bir, m, inv, pl	Oviparous	11–16
Phrynops hilarii[a,b]	–	400	Aquatic	mo, fi, am, bir	Oviparous	22–23
Mesoclemmys vanderhaegei[b]	–	270	Aquatic	Car	Oviparous	3–6
Acanthochelys spixii	–	170	Aquatic	fi, am, inv	Oviparous	4–9
Acanthochelys pallidipectoris	–	175	Aquatic	FI, am	Oviparous	5
CROCODYLIANS						
Caiman latirostris chacoensis[a,b]	–	3,000	Aquatic	inv (mo, ins, cr, ar) fi, am, sn, bir, m	Oviparous	23–41
Caiman yacare[a,b]	–	2,500	Aquatic	inv (mo, ins, cr, ar) FI, bir, m	Oviparous	30–45
LIZARDS						
Tropidurus torquatus[a,b]	–	105	Rupicola and Arboreal	INV, re, bir, pl	Oviparous	7
Cercosaura schreibersii[a]	36	29–43	Terrestrial	INV	Oviparous	2
Kentropix viridistriga[a]	–	107	Terrestrial	INV	Oviparous	4–5
Teius oculatus[a]	–	120	Terrestrial	INV (co, him, isop)	Oviparous	9
Tupinambis merianae[a]	–	800	Terrestrial	Omn	Oviparous	20–54
Mabuya dorsivittata[a]	66	62–70	Terrestrial	INV (co)	Viviparous	–
Mabuya frenata[a,b]	–	42	Arboreal	INV (ar, or, sp, ins)	Viviparous	–
Homonota borelli[a]	–	42	Terrestrial–Arboreal	INV	Oviparous	–

Ophiodes intermedius[a]	134	110–153	Terrestrial	INV (ins, sp)	Viviparous	–
AMPHISBAENIANS						
Amphisbaena angustifrons	–	90–420	Fossorial	INV (ins)	Oviparous	–
Amphisbaena heterozonata[a]	189	118–255	Fossorial	INV (ins)	Oviparous	2–8
Amphisbaena hiata	173	146–204	Fossorial	INV (ins)	Oviparous	–
Amphisbaena mertensi[b]	–	131–410	Fossorial	INV (ins)	Oviparous	–
Anops kingii[a]	–	76–223	Fossorial	INV (ins)	Oviparous	2–4
Leposternon microcephalum[a,b]	–	121–524	Fossorial	INV (ins)	Oviparous	2–8
SNAKES						
Liotyphlops ternetzii[b]	240	108–334	Fossorial	INV	Oviparous	2–7
Typhlops brongersmianus[a]	148	85–287	Fossorial	INV	Oviparous	6
Leptotyphlops albipunctus[a]	195	83–283	Fossorial	INV	Oviparous	2
Leptotyphlops unguirostris	–		Fossorial	INV	Oviparous	3
Eunectes notaeus[a,b]	1,846	546–4,000	Aquatic	bir, am, re, fi, ca, egbir, m	Viviparous	4–37
Boiruna maculata[a]	364	296–1,590	Terrestrial	SN, liz, bir	Oviparous	8–17
Clelia bicolor[a]	430	292–825	Terrestrial	am, liz, sn	Oviparous	–
Clelia clelia[a,b]	977	335–1,830	Terrestrial	SN	Oviparous	10–14
Clelia rustica[a]	805	296–1,378	Terrestrial	liz, sn, m	Oviparous	7–8
Chironius maculoventris[a]	829	400–890	Arboreal	AM	Oviparous	–
Echinanthera occipitalis[a]	293	139–470	Terrestrial	liz, sn, am	Oviparous	–
Helicops infrataeniatus[a,b]	486	146–795	Aquatic	FI, am	Viviparous	1–28

(*Continued*)

Table 14.1 Reptiles from Middle Paraná River floodplains with some morphological and ecological traits—Continued

Species	Length[c] mean	Maximum or range	Substrate use	Food[d]	Reproductive modality	Number of eggs or offspring
Helicops leopardinus[a,b]	317	104–755	Aquatic	FI, am	Viviparous	15–26
Hydrodynastes gigas[a,b]	1,092	106–2,360	Aquatic	am, fi, sn, m, egfi	Oviparous	18–29
Hydrops caesurus[b]	253	91–484	Aquatic	fi?	Oviparous	9
Leptodeira annulata pulchriceps[a]	490	183–585	Arboreal	am, liz	Oviparous	8–12
Leptophis ahaetulla marginatus[a,b]	652	263–1,832	Arboreal	AM, bir, egbir	Oviparous	6–9
Liophis almadensis[a,b]	364	228–540	Semi-aquatic	AM, liz, m	Oviparous	6–10
Liophis anomalus[a]	377	136–580	Semi-aquatic	AM	Oviparous	6–11
Liophis dilepis[a]	303	144–440	Semi-aquatic	AM	Oviparous	3–5
Liophis jaegeri coralliventris[a,b]	323	115–500	Semi-aquatic	AM	Oviparous	5–14
Liophis meridionalis[a,b]	458	179–605	Semi-aquatic	AM	Oviparous	–
Liophis semiaureus[a,b]	558	170–1,147	Aquatic	FI, am	Oviparous	8–20
Liophis poecilogyrus[a]	375	162–623,5	Semi-aquatic	AM, fi	Oviparous	4–11
Lystrophis dorbignyi[a]	335	12–795	Fossorial	am, fi, liz, inv	Oviparous	3–7
Lystrophis pulcher	–		Fossorial	am, liz	Oviparous	–
Mastigodryas bifossatus triseriatus[a,b]	1,042	263–1,960	Terrestrial	am, liz, bi, m	Oviparous	8–12
Oxyrhopus guibei[a,b]	728	206–1,038	Terrestrial	LIZ	Oviparous	–
Oxyrhopus rhombifer[a]	473	230–810	Terrestrial	liz, bir	Oviparous	10–15
Phalotris bilineatus[a]	396	137–695	Fossorial	RE	Oviparous	1
Philodryas aestiva subcarinata[a]	624	208–950	Terrestrial	AM, m	Oviparous	4–19

Philodryas mattogrossensis[a,b]	535	291–778	Terrestrial	am, liz, bir, m	Oviparous	–
Philodryas olfersii latirostris[a,b]	646	282–1,125	Arboreal	am, liz, bir, m	Oviparous	3–8
Philodryas patagoniensis[a]	665	175–1,315	Terrestrial	RE, am, bir, m	Oviparous	1–26
Phimophis guerini	796	735–852	Terrestrial	LIZ	Oviparous	–
Pseudablabes agassizii	308	214–411	Terrestrial	INV	Oviparous	6–10
Psomophis obtusus[a]	238	95–370	Terrestrial	AM, inv	Oviparous	1–18
Sibynomorphus turgidus[a]	331	129–509	Terrestrial	MO	Oviparous	7
Thamnodynastes chaquensis[a]	425	170–619	Terrestrial	AM	Viviparous	–
Thamnodynastes hypoconia[a,b]	342	140–513	Semi–aquatic	AM, liz, sn, egrep	Viviparous	8–12
Thamnodynastes strigatus[a,b]	619	311–795	Aquatic	AM, fi, m, re	Viviparous	1–17
Tomodon ocellatus	405	375–446	Terrestrial	MO	Viviparous	4–26
Waglerophis meremii[a]	574	315–950	Terrestrial	AM	Oviparous	7–35
Micrurus altirostris[b]	545	225–959	Fossorial	SN	Oviparous	1–7
Micrurus baliocoryphus[a]	664	253–818	Fossorial	SN	Oviparous	–
Micrurus pyrrhocryptus[a]	683	214–1,175	Fossorial	SN	Oviparous	14
Bothrops alternatus[a]	750	118–1,360	Terrestrial	M	Viviparous	7–32
Bothrops diporus[a]	601	234–1,020	Terrestrial	M, am	Viviparous	7–8
Crotalus durissus terrificus[a]	–	1,800	Terrestrial	M	Viviparous	20–30

[a]Species recorded by us in the Middle Paraná River.
[b]Tropical species restricted to the Middle Paraná River floodplain.
[c]In turtles, carapace length is indicated; in crocodilians, total length is indicated; in lizards and snakes, the snout-vent length is indicated.
[d]When an item-prey occupies quantitatively more than 50% of the species diet, it is indicated in *capital letters.*
ca carrion, *car* carnivorous, *inv* invertebrates (*co* cockroaches, *ins* insects, *cr* crustaceans, *mo* mollusks, *him* Himenoptera, *isp* Isoptera, *or* Orthoptera, *sp* spiders), *omn* omnivorous, *pl* plants, *ver* vertebrates (*am* amphibians, *bir* birds, *egbir* bird eggs, *egfi* fish eggs, *egrep* reptile eggs, *fi* fish, *liz* lizards, *m* mammals, *re* reptiles, *sn* snakes).

marshes, generally highly vegetated temporary ponds, and small lotic channels. *Caiman yacare* inhabits permanent waters, occupying main streams, secondary and principal channels of the Paraná River and surrounding permanent marshes, generally with wide open waters. According to Scott et al. (1990), *C. latirostris* seemed to inhabit a wider range of habitats when it occurred by itself, than when it occurred in sympatry with *C. yacare*. Moreover, when both species are in sympatry, generally *C. yacare* is more abundant. Scott et al. (1990) suggested, based on local people comments, that *C. yacare* was aggressive toward *C. latirostris* and would displace it when they came into contact. These authors suggested that "*C. latirostris* has been subjected to more intense hunting pressure", and that "the observations of replacement of *C. latirostris* by *C. yacare* could be the result of a disproportionate reduction of the former by hidden hunters". However, we have observed that *C. yacare* is still more abundant than *C. latirostris* in protected areas located in the Paraná River basin, where populations were protected for decades, which does not support the hypothesis mentioned before. We consider that probably *C. latirostris* can be underestimated in number since it mainly inhabits densely vegetated sectors where it basks in the sun, which does not make it very conspicuous and observable. On the contrary, *C. yacare* occupies open waters and environments, and basks in the sun in beaches and visible sites (Medem 1983). Both species can subsist in sectors with a high human presence, although they can reach very high densities in sectors with low anthropic pressure or in protected areas. In less accessible areas of the Middle Paraná River floodplain, we have observed up to 16 specimens of *C. yacare* in 3 h of distance covered. In the Pilcomayo National Park, a close protected area with similar habitats to those of the Middle Paraná River, a population of 2,122 caiman specimens was estimated in a 700-ha lake during a dry period (Giraudo et al. in prep.).

14.2.3 Amphisbaenians

Six species of these insectivore and fossorial habits reptiles were registered. They occupy the highest lands in the flood valley, occasionally flooded by extraordinary events. During these events, it is possible to find numerous specimens of rare and secretive fossorial reptiles (amphisbaenians and wormsnakes) that very probably suffer high mortalities in such periods. It is probable that some specimens could subsist or that such areas could be recolonized during the terrestrial phases.

14.2.4 Lizards

Lizards have only nine species in the Middle Paraná River (Table 14.1). Most are terrestrial species (six), two of them inhabit the trees, and one of them inhabits rocks and trees. None of them has special adaptations to live in the water, although terrestrial and arboreal species inhabit the islands of the Middle Paraná River. *Tupinambis merianae* can swim and move among the islands. Most lizard species are small and feed on arthropods. *Tupinambis merianae* is the largest lizard in the Middle Paraná River, with a maximum total length of 1,500 mm. It has an omnivorous diet, including arthropods, small vertebrates, eggs, carrion, and fruits (Cei 1993).

14.2.5 Snakes

Snakes are the richest reptile group, with 49 taxa. They occupy all type of habitats and substrates in the Middle Paraná River (Table 14.1), including various species adapted to live in the water, with a relatively depressed body and more dorsal eyes and nostrils (*Eunectes, Helicops, Hydrops* and *Liophis semiaureus*), modifications in the cephalic scales (*Hydrops* and *Helicops*), and a reduction in ventral plates (*Eunectes*). Viviparity in *Eunectes, Helicops* and *Thamnodynastes* is advantageous to avoid the egg loss in floods and to favor the retention of embryos until favorable conditions appear (see Sect. 14.4.3). The yellow anaconda (*Eunectes notaeus*) is a very large Boidae that reaches up to 4 m in total length (occasionally 5 m, Scrocchi and Giraudo 2005). Because of its size, it is one of the largest predators of the Middle Paraná River, feeding on a wide range of vertebrates, frequently including aquatic birds (*Phalacrocorax, Euxenura, Egretta, Chauna, Aramus* and *Vanellus*), reptiles (*Tupinambis* and crocodilians), medium to large mammals (*Lontra, Hydrochoerus, Euphractus, Procyon* and *Cerdocyon*), and fish (*Cichlasoma, Pygocentrus* and Siluriformes) (Scrocchi and Giraudo 2005).

14.2.6 Species Richness Patterns of the Middle Paraná River in a Regional and Landscape Context

The river system, with a wide floodplain, is exceptionally diverse environmentally, and, hence, it normally supports high species richness, sometimes the highest in the landscape. Riparian corridors in temperate areas have been called "linear oases" and contain many rare or uncommon species (Forman 1995). Many animals in the surrounding matrix also depend on these corridors for water, food or refuge (Forman 1995; Roe et al. 2003). We compared reptile richness in 1° geographical cells, crossed by the Middle Paraná River, and contiguous cells located in interior high lands of the Paraná River western sector. A significantly higher mean richness was observed in the cells crossed by the Middle Paraná River (means difference $t = 2.90$, d.f. = 14, $P < 0.012$). The gradual loss of reptile species was observed when latitude increased (Pearson correlation $r^2 = -0.81$, $n = 8$, $P = 0.01$) in the Middle Paraná River (Fig. 14.1). The northern extreme cell had 51 species and the southern extreme cell only had 22 species.

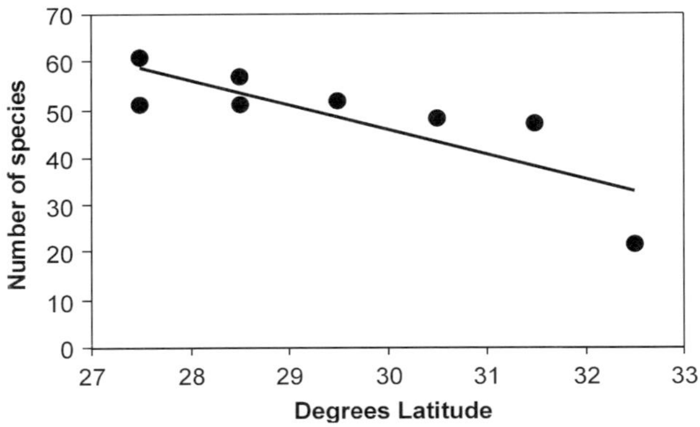

Fig. 14.1 Relationship between reptiles species richness and latitude in the Middle Paraná floodplain (Pearson correlation, $r^2 = -0.81$, $n = 8$, $P = 0.01$)

14.3 The Paraná River Floodplain as Biogeographical Corridor of Tropical Species

The rivers of the Paraná basin run from tropical regions to temperate latitudes (Giraudo and Arzamendia 2004; Arzamendia 2006). The Middle Paraná River extends between 27° and 32° of latitude S, crossing regions of subtropical and temperate xerophilous forests, savannahs and steppes. Thirty-eight percent of reptiles (27 species, Table 14.1) registered in the Middle Paraná River are tropical species that inhabit exclusively in its floodplain and surrounding sectors, and are not found in high lands of the region. These species reach higher latitudes through the Paraná River and 64% of them show the southernmost limit of their geographical distribution in the Middle Paraná River.

The Middle Paraná River begins in the confluence of two great collectors: the Paraguay and Paraná rivers. The Paraguay River, that contributes with the lowest water flow and the highest amount of sediments to the Middle Paraná River (Paoli et al. 2000), begins in the Pantanal of Mato Grosso (Brazil), where tropical forests and savannahs, in the transition of the Amazonian and "Cerrado" biogeographical provinces, develop (Cabrera and Willink 1980). The Paraná River upper basin, that provides the highest flow, although with a lower amount of sediments, receives an extensive net of tributaries from a mountainous area of southeastern Brazil (Paoli et al. 2000), where tropical forests of the "Atlántic" and "Paranaense" (i.e., Paraná River) biogeographical provinces are found (Cabrera and Willink 1980). The influence of these two great collectors as a faunistic corridor in the Middle Paraná River is not equal. The Middle Paraná River shows a higher faunistic influence than the Paraguay River, with a higher proportion of tropical species distributed in the southern Amazonia, Central South America, the Pantanal and the Humid "Chaco" ($\chi^2 = 13.56$, d.f. = 2, $P = 0.0011$,

18 species, 67% of tropical species, e.g., *Phrynops hilarii, Caiman yacare, Eunectes notaeus, Helicops leopardinus, Hydrodynastes gigas, Hydrops caesurus, Liophis meridionalis, Philodryas mattogrossensis, P. olfersii latirostris*), and a lower proportion of species with Atlantic "Paranaense" distribution (five species, 19% of tropical species, *Amphisbaena mertensi, Helicops infrataeniatus, Oxyrhophus guibei, Thamnodynastes strigatus* and *Micrurus altirostris*) and a wide tropical distribution (four species, 15%, *Tropidurus torquatus, Leposternon microcephalum, Clelia clelia* and *Liophis almadensis*). 54% of tropical species are aquatic species and 36% of species inhabit in humid riparian forests of the ATTZ (Giraudo and Arzamendia 2004). Only 10% are generalistic species inhabiting rocks or trees. This shows the importance of aquatic habitats and riparian forests for tropical fauna in temperate latitudes of the Middle Paraná River.

14.4 Influence of Flood Pulses and other Environmental Factors on the Reptile Community

14.4.1 Seasonal Variation in Abundance

As ectotherms, reptiles are highly dependent on the physical exchange with the environment. Most obviously, activity in cold-climate reptile populations is often constrained by ambient temperatures or basking opportunities (i.e., incident solar radiation) (Brown and Shine 2002). Conspicuous seasonal peaks in encounter rates are often associated with significant biological events, such as hibernation, mate-searching by adult males, egg-laying migrations by gravid females, and hatching of eggs (Bonnet et al. 1999, Brown and Shine 2002). The Middle Paraná River is subjected to a higher climatic seasonality (e.g., temperature, precipitations) than other great neotropical rivers, as the Amazon or the Orinoco; however, similarly to these rivers, there is an additional factor that can be preponderant in the activity patterns of animals, as the variations in the hydrological cycle (Junk 1997b). We observed significant differences in the mean number of reptiles registered in the different seasons of the year (Kruskal Wallis test, $KW = 14.55$, $P = 0.0022$) in 5,038 km covered between 1997 and 2004 in the Middle Paraná River floodplains ($n = 887$ individuals), with a significantly lower mean in winter (0.07 ind./km) (*a posteriori* Dunn's comparison, $P < 0.05$), and with a higher mean quantity of reptiles in spring (0.30 ind./km), autumn (0.22 ind./km) and summer (0.17 ind./km), although without significant differences among these last three seasons (Fig. 14.2). The number of reptiles registered showed a positive and significant correlation (Spearman) with the hydrometric level ($r^2 = 0.24$; $P = 0.05$), temperature ($r^2 = 0.36$, $P = 0.0042$) and precipitations ($r^2 = 0.44$, $P = 0.0026$). In the Paraná River floodplain, the activity of reptiles is related not only with temperatures and precipitations, as in high land communities, but also with variations at the hydrological level (Fig. 14.3).

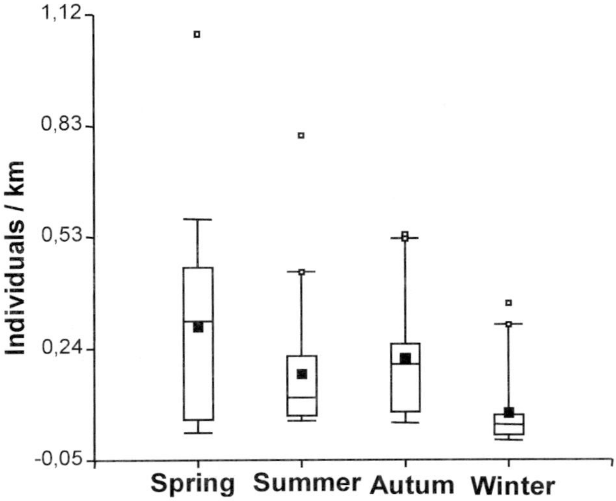

Fig. 14.2 Box-plot and comparison of mean number of reptiles (ind./km) per each year seasons (1997–2004 period)

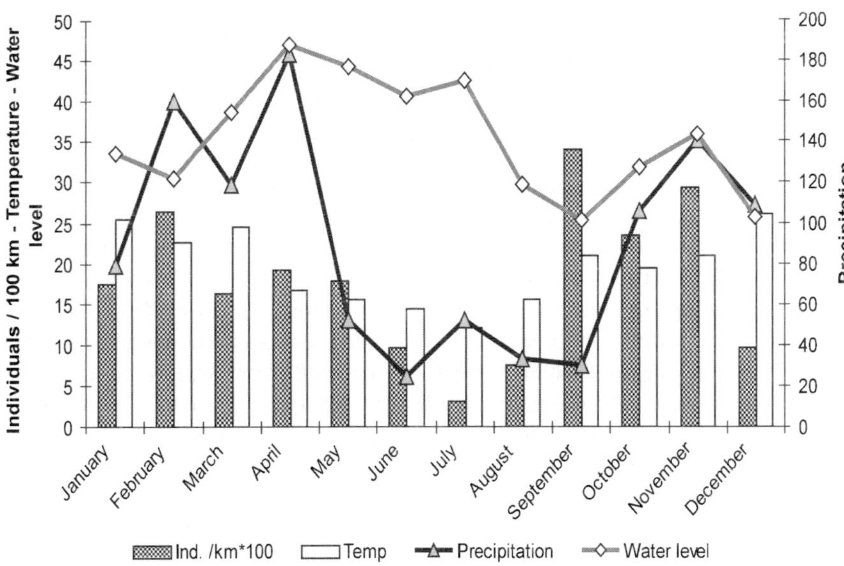

Fig. 14.3 Monthly variations in the mean number of reptiles registered (n = 887 ind., 5,038 km covered between 1997 and 2004) in the Middle Paraná River floodplain in relation to temperature, precipitations, and river level (in decimeters) in the Santa Fe Port between 1997–2004

14.4.2 Substrate and Habitat Use

Due to recurrence of floods, we could expect the reptile community of the Middle Paraná River to show a higher abundance in aquatic and semi-aquatic species, with adaptive advantages to survive to these situations, and a lower abundance of fossorial, arboreal and terrestrial species. Data obtained in this study indicate that 586 individuals from a total of 887 belong to aquatic and semi-aquatic species (66%), with a significantly higher frequency ($\chi^2 = 1{,}028.52$, d.f. = 3, $P < 0.0001$) than terrestrial species (285 specimens, 32%), fossorial species (11.1%) and arboreal species (5.1%). Among the most abundant species we found four aquatic species, including *Helicops leopardinus* (196 specimens), *Liophis semiaureus* (88 specimens), *Phrynops hilarii* (87 specimens) and *Hydrodynastes gigas* (74 specimens), and one terrestrial species *Philodryas patagoniensis* (193 specimens), with a total of 71% from 887 sampled reptiles.

14.4.3 Reproductive Strategies

Reproduction in reptiles is influenced by ecological, environmental, phylogenetic and geographical factors (Cadle and Greene 1993; Gregory and Larsen 1993; Di-Bernardo 1998), most of which began to be studied recently in species of the Neotropical subtropics. Middle Paraná River reptiles show two types of reproduction: oviparity and viviparity (Table 14.1). The former is the most extended one among reptiles in general; however, viviparity could be favorable in certain variable environments, as the Paraná River floodplains. This type of reproduction can provide an effective solution to the difficulty constituted by egg-laying in dry places and with variable temperatures, and to avoid the loss of eggs in floods. In proportion, the Middle Paraná River reptile community has twice viviparous species (18%) (13 viviparous taxa from a total of 71 species) than reptile viviparous species in Argentina (9%) (28 viviparous taxa from a total of 313 species). In subtropical and temperate zones, the greater extent of reproductive cycles in viviparous species may be related to the lower dependence of embryos to the environmental temperature (Aguiar and Di-Bernardo 2005). For oviparous species, soil humidity can be an important factor. Shine and Brown (2002) showed that seasonal variations in soil water content in a natural nest can generate strong phenotypic variations in hatching snakes (as snout-vent length, body mass, strength and swimming speed). Reproductive tactics are clearly linked to features of the environment or of the species' morphology and ecology (Shine 2003). Reproductive phenology of most reptiles of the Middle Paraná River starts at the beginning of spring and extends throughout the summer. However, there are some particularities in the reproductive strategies of species in relation to the environments they frequent and their type of reproduction. When comparing reproduction in two of the

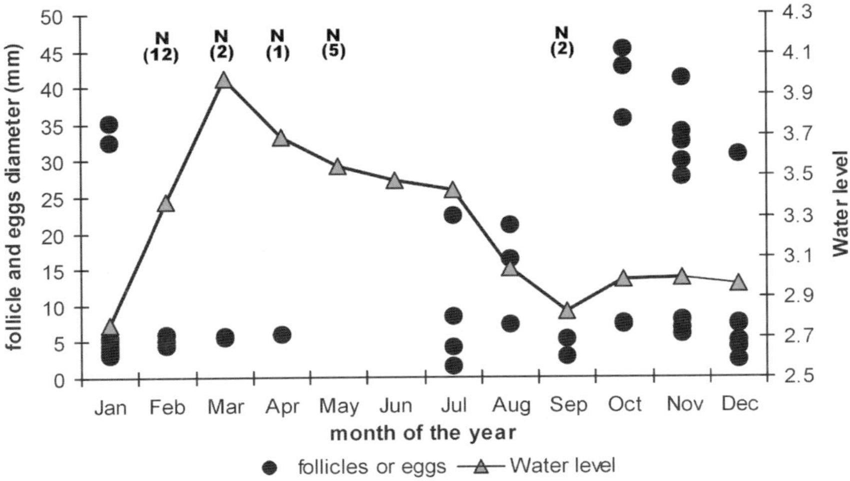

Fig. 14.4 Reproductive cycle of *Philodryas patagoniensis*, seasonal variation in the diameter of ovarian follicles (smaller than 25 mm) and eggs (larger than 25 mm) in relation to the monthly mean hydrometric levels (in meters) in the Paraná Port between 1905 and 2003. The number of neonates (N) registered are indicated in *parentheses*

most frequent snakes in the Middle Paraná River, a terrestrial and oviparous snake, *Philodryas patagoniensis*, and an aquatic and viviparous snake, *Helicops leopardinus*, we observed some differences in their reproductive strategies (Figs. 14.4, 14.5). In *P. patagoniensis*, vitellogenesis begins earlier in July (winter), and oviposition is mainly carried out between spring and the beginning of summer (October to January), when the river level is generally low. When the river begins to grow, most neonates already hatched (Fig. 14.4). In *H. leopardinus*, the development of vitellogenic follicles begins lately in spring (October), and the development of embryos can even be extended from November to June. Most births and neonates are found in March, coinciding with the maximum historical values in the hydrometric and precipitation level (Fig. 14.5). It is probable that newborns and gravid females of this aquatic snake could find a higher availability of aquatic environments in floodplain lakes and marshes, optimizing the possibilities to find refuge and feeding areas. Reproduction of *H. leopardinus* is similar to that of *H. infrataeniatus*, whose reproductive cycle extends even until June in southern Brazil (Aguiar and Di-Bernardo 2005). We observed that oviposition of the aquatic turtle *Phrynops hilarii* in the Middle Paraná River is concentrated in March (six nests registered, a turtle digging the nest, a run-over turtle with 13 well-developed eggs in its body) and April (three nests recorded), also coinciding with the maximum historical river levels. Turtles lay their eggs in the riverbank at a height that oscillates between approximately 2 and 20 m above the water level. It is logical

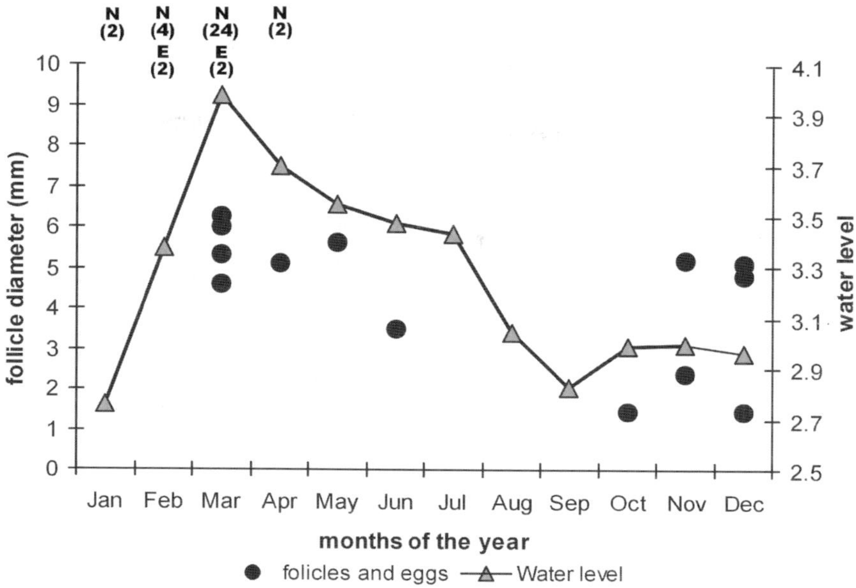

Fig. 14.5 Reproductive cycle of *Helicops leopardinus*, seasonal variations in the diameter of ovarian follicles (*black circles*) in relation to the monthly mean hydrometric levels (in meters) in the Paraná Port between 1905 and 2003. The number of neonates (N) and embryos (E) registered are indicated in *parentheses*

to assume that if turtles go out of the water to oviposite with maximum water levels, it is less probable that their nests be lost by the effect of floods. Caimans construct their nests in floating "embalsados" (floating accumulation of aquatic vegetation, plant rests and muddy material) or in high sectors of the riverbanks. Nests of *C. latirostris* were recorded up to 2,000 m away from water bodies (Larriera 1995). These strategies can minimize nest losses during floods.

14.4.4 Feeding Strategies

Middle Paraná River reptiles feed on a wide variety of terrestrial and aquatic invertebrates and vertebrates, playing an important role as predators (Table 14.1, Fig. 14.6). Insect larvae that develop partly in the water (e.g., Ephemeroptera, Odonata, Plecoptera, Trichoptera) are among the main feeding resources of aquatic turtles, also including a wide range of vertebrates, other invertebrates, and carrion (Souza 2004). Lizards and amphisbaenids mainly feed on terrestrial invertebrates. Ontogenetic variations in feeding are frequent in reptiles; for example, crocodilians feed on insects

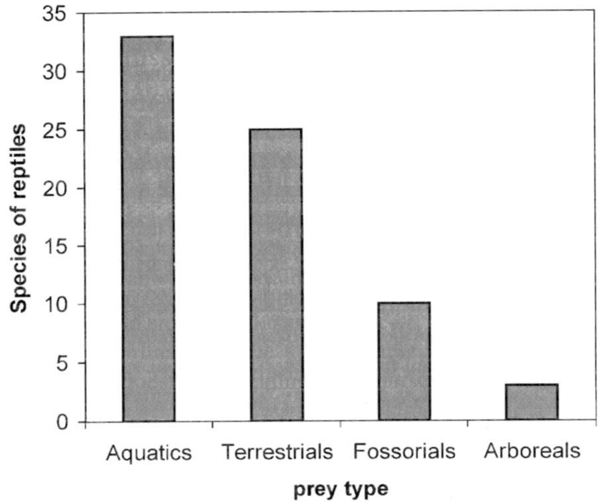

Fig. 14.6 Number of reptile species that eat aquatic, terrestrial, fossorial and arboreal preys

and small frogs when they are juveniles and, as they grow, they incorporate fish and snails, and large specimens can feed on all types of vertebrates, including medium and large mammals (Medem 1983). In aquatic reptiles, there are generalist species, as turtles, alligators, and some snakes as *Hydrodynastes gigas*, although aquatic or semi-aquatic preys are an important component of their diet. 72% of preys of *Hydrodynastes gigas* are aquatic or semi-aquatic vertebrates (López and Giraudo 2004). This species shows specialized behaviors to search for amphibians, using the tail, in shallow water refuges (Strussmann and Sazima 1990). Snakes that show morphological adaptations to live in the water, as *Helicops leopardinus*, *H. infrataeniatus* and *Liophis semiaureus*, mainly feed on fish (more than 60% of their diet, Aguiar and Di-Bernardo 2004, pers. obs.), very abundant preys, at least seasonally, in the Middle Paraná River floodplain. Semi-aquatic snakes, of genera *Liophis* and *Thamnodynastes*, mainly feed on amphibians, and occasionally on fish (Table 14.1). Around 50% of Middle Paraná River snakes (25 out of 49 species) include amphibians in their diet. Ecological factors, as the high abundance of amphibians in South American wetlands, or phylogenetic factors, as the preponderance of frog-eating species of the subfamily Xenodontine in the South American snake fauna, have been mentioned to explain this pattern (Cadle and Greene 1993; Strussmann and Sazima 1993). The few arboreal snake species of the Middle Paraná River, as *Leptophis ahaetulla marginatus* and *Philodryas olfersii latirotris*, include preys from such habitats, as arboreal amphibians (Hylidae), birds, and bird eggs in their diets (López et al. 2003, Table 14.1).

14.5 Discussion and Conclusions

14.5.1 Species Richness

The Middle Paraná River floodplain, although it is located in a subtropical-temperate region, shows high reptile richness (71 species), comparable with values registered in tropical sectors of South America, considered as megadiverse areas. For example, Cunha et al. (1985) mentioned 77 species for the Carajas region, Pará, in the Brazilian Amazonia, and Dsuelmann (1978) registered 91 reptile species in the Ecuadorian Amazonia. This would be explained by the following factors: (1) Riparian ecosystems are between the most diverse, dynamic, and complex biophysical habitats of the world's surface (Tiegs et al. 2005). Flood pulses generate a wide spatiotemporal heterogeneity of habitats and resources, which can be used by terrestrial and aquatic species, increasing considerably plant and animal richness in the system (Junk et al. 1989). (2) The Middle Paraná River is a biogeographical transition area where tropical faunistic elements, that reach higher latitudes through the river, and faunistic elements from subtropical and temperate biogeographical regions (e.g., Chaquenian and Pampean biogeographical provinces) come together (Arzamendia 2006). (3) Some of the Middle Paraná River habitats, as riparian humid forests, are structurally complex and provide numerous microhabitats for reptiles. (4) The extraordinary floods of the Paraná River could interrupt locally the processes of competitive exclusion, allowing the maintenance of a high diversity (Connell 1978; Junk et al. 1989). (5) The wide food offer, at least seasonal, for reptiles in the Middle Paraná River (e.g., aquatic invertebrates, fish, amphibians). (6) Adaptations of reptiles that can favor their survival in pulsatile and variable environments: high metabolic efficiency to transform food in biomass, prolonged capacity of starvation in moments of food scarcity (crocodilians and snakes can starve for more than a year), capacity of adjusting their reproduction to environmental variations and food abundance (Junk and da Silva 1997).

The decrease in species richness with the increase in latitude in the Middle Paraná River is probably related with the decrease in temperature, as has been widely discussed for different organisms (e.g., Willing et al. 2003; Hawkins 2004). The Middle Paraná River extends for 620 km in a north–south direction, with a marked gradient in temperature, varying from 22°C at the north to 18°C at the south (Iglesias de Cuello 1982).

14.5.2 The Paraná River Floodplain as Biogeographical Corridor of Tropical Species

The existence of a microclimate on the riverbanks and the development of riparian humid forests and numerous wetlands facilitate the survival of tropical floristic and faunistic species in temperate latitudes of Paraná River basin large rivers (Cabrera 1994; Menalled and Adamoli 1995). This pattern has been mentioned in different aquatic and terrestrial organisms. The higher faunistic influence of the Paraguay River on the Middle Paraná River can be due to the interaction of ecological factors (higher similarity between the Middle Paraná and Paraguay rivers floodplains in relation to the High Paraná River) and historical factors (hydrological isolation between High and Middle Paraná River) (Giraudo and Arzamendia 2004; Arzamendia 2006).

14.5.3 Influence of Flood Pulses and other Environmental Factors on the Reptile Community

Several aquatic turtle, crocodilian, and snake species (e.g., *Eunectes* and *Helicops*) are abundant in the Middle Paraná River, as in the Amazon River, and many of their life cycles (activity, reproduction and feeding) are strongly influenced by the flood pulses (Junk and da Silva 1997, see Sect. 14.4). However, there is a scarce influence of other climatic variations (e.g., temperature) in the activity and reproduction of reptiles in the Amazonia and other tropical areas of South America (Duellman 1978; Pizzato and Marques 2006) due to the low climatic seasonality. On the contrary, Middle Paraná River reptiles were strongly influenced by the marked seasonality in temperature and precipitations, and their activity and reproductive patterns were coincident with the warmest and rainiest months of the year. This seasonal pattern coincides with what was indicated for other reptiles in subtropical and temperate regions of South America (e.g., southern Brazil, Schmid de Aguiar and Di-Bernardo 2005). Particular adjustments were observed in the reproductive types and phenology when species of aquatic and terrestrial snakes were compared. Viviparity, frequent in aquatic species (e.g., *Helicops leopardinus*), can represent advantages in environments that show high changes through time, as the Paraná River floodplain; for example, the adjustment between the peak of births and the historical peaks (modal) of floods and precipitations to favor the juveniles a wide offer of aquatic habitats. Viviparous snakes may retain the embryos and give birth only when the environmental conditions are appropriate (Girons 1985; Shine 2003; Aguiar and Di-Bernardo 2005). In viviparous species, the long reproductive cycle can be a consequence of a low metabolism and low food ingestion (Bizerra et al. 2005). According to feeding data of *H. leopardinus* in the region, this species shows low food ingestion (only 11 out of 106 individuals (10.3%) showed stomach content). In consequence, another reason for a late reproductive cycle in *H. leopardinus* could be that feeding and reproduction are restricted to periods with favorable conditions (e.g., higher river level). In the case of *P. patagoniensis*, 84 individuals out of 184 (46%) showed stomach content (López and Giraudo, in prep.), which evidences a marked difference in the energy incorporated to food and that potentially can be invested in reproduction. This last terrestrial and oviparous species is reproduced earlier, laying the eggs when the Paraná

River is generally in low water. The percentage of reproducing adult females reached 40% in *H. leopardinus* and 50% in *P. patagoniensis*, indicating that only part of the mature females in both species is reproduced every year. This would allow a higher plasticity, so an important proportion of females are reproduced in favorable years (e.g., with adequate hydrological and pluviometric levels). Extraordinary intense floods, mostly in duration (Neiff 1990), and generally not predictable, can have catastrophic consequences for the physical and biological environment (Junk et al. 1989; Bó and Malvarez 1999). In two periods of extraordinary floods (1992–1993, 1997–1998), a high mortality of reptiles in routes and marginal paths to the Middle Paraná River was observed, in addition to a great amount of live reptiles accumulated in high lands and killed by local settlers. These extreme flood events act on individuals directly (elimination of animals by predation or drowning, destruction of nests) or indirectly (changes in habitats, availability of preys and refuges) (Bó and Malvarez 1999). Terrestrial reptile species would possibly be more affected than those adapted to the water, and it is possible that some aquatic reptiles would benefit from a higher availability of preys and carrion.

14.5.4 Function and Conservation of Reptiles in the Middle Paraná Floodplain

It has been hypothesized that in Amazon River floodplain lakes, reptiles, e.g., crocodilians, would have a key function in the storage and recycling of allochthonous nutrients, increasing the production of phytoplankton and other organisms of aquatic trophic chains through the contribution of nutrients when these lakes are disconnected with the river (Fittkau 1973). Various species of very abundant aquatic reptiles, as *Phrynops, Caiman, Helicops, Hydrodynastes* and *Liophis*, could have the same function in the Middle Paraná River. As in the Amazon River (Junk and da Silva 1997), flood pulses have favored overexploitation and elimination of reptiles in the Middle Paraná River, since they have to abandon their refuges during large floods, or concentrate in the remaining aquatic systems during low waters, being more vulnerable to hunting. Changes in drainage levels in the Paraná River, that provoked the increase in historical mean floods and changes in their seasonal distribution, have been mentioned since 1970. Deforestation, agriculture, dam constructions in the basin, and global climatic changes are among their main causes (García and Vargas 1998; Giaccosa et al. 2000). This could have serious consequences on reptile populations that have adjusted their reproduction and activity to historical and more predictable river cycles. The accelerated modifications occurring in the high lands near the river (because of urbanization, agriculture, cattle raising and deforestation) can eliminate key habitats for aquatic, semi-aquatic and exclusive reptiles of the Middle Paraná River, used in essential periods of their life cycles, especially in extreme floods. According to Roe et al. (2003), as Middle Paraná

River reptiles are animals that require wetlands as well as adjacent high lands, the classic conservation strategies applied separately in wetlands or in high lands are not enough: a landscape-level approach is necessary to protect and manage Middle Paraná River floodplain valley wetlands and their surrounding lands.

Acknowledgements. We thank Maria Ofelia de Emiliani for inviting us to participate in this project; Tito Paggi, Susana Paggi, Julieta Parma, Ramón Regner, Esteban Creus, Lionel Mehaudy, Estela Torres del Sel, Elena Córdoba, Gabriel Tourn, Lucrecia Valldeneu, Cristian De Bonis and Adolfo Beltzer from the National Institute of Limnology are thanked for their collaboration and support in carrying out this study; the Hydrology Department of the FICH, National University of the Littoral, for providing hydrological and climatic data, and José Pekerman, for his dignity upon injustice. This investigation was supported by the following institutions and projects: PEI 6129 (CONICET), CAID 2005 PE 249 (UNL), PICTO 2005 N° 23191 (UNL, ANPCyT) and PICT 2002 N° 01-12831 (ANPCyT).

References

Aguiar LFS, Di-Bernardo M (2004) Diet and feeding behavior of *Helicops infrataeniatus* (Serpentes: Colubridae in southern Brazil. Studies on Neotropical Fauna and Environment 39:7–14

Aguiar LFS, Di-Bernardo M (2005) Reproduction of the water snake *Helicops infrataeniatus* (Colubridae) in southern Brazil. Amphibia-Reptilia 26:527–533

Arzamendia V (2006) Las serpientes (Reptilia) de los ríos Paraná y Uruguay: patrones de diversidad y rol de los ríos como corredores faunísticos. Tesis Doctoral. Universidad Nacional de Córdoba, Córdoba, Argentina

Arzamendia V, Giraudo AR (2004) Usando patrones de biodiversidad para la evaluación y diseño de áreas protegidas: las serpientes de la provincia de Santa Fe (Argentina) como ejemplo. Rev Chil Hist Nat 77:335–348

Bizerra A, Marques OAV, Sazima I (2005) Reproduction and feeding of the colubrid snake *Tomodon dorsatus* from south-eastern Brazil. Amphibia-Reptilia 26:33–38

Bó R, Malvárez AI (1999) Las inundaciones y la biodiversidad en humedales: un análisis de eventos extremos sobre la fauna silvestre. In: Malvárez AI (ed) Tópicos sobre humedales subtropicales y templados de Sudamérica. MAB, UNESCO, Buenos Aires, pp 147–168

Bonnet X, Naulleau G, Shine R (1999) The dangers of leaving home: dispersal and mortality in snakes. Biol Conserv 89:39–50

Brown GP, Shine R (2002) Influence of weather conditions on activity of tropical snakes. Austr Ecol 27:596–605

Busack SD, Pandya S (2001) Geographic variation in *Caiman crocodilus* and *Caiman yacare* (Crocodylia: Alligatoridae): systematic and legal implications. Herpetologica 57(3):294–312

Cabrera AL (1994) Regiones fitogeográficas argentinas. Encicl arg Agric Jard 2(1):1–85

Cabrera MR (1998) Las tortugas continentales de Sudamérica austral. Edición del Autor, Córdoba, Argentina

Cabrera AL, Willink A (1980) Biogeografía de América Latina. OEA, Ser Biol Monog (13):1–122

Cadle JE, Greene HW (1993) Phylogenetic patterns, biogeography, and the ecological structure of Neotropical snake assemblages. In: Rickelfs RE, Schluter D (eds) Species diversity in ecological communities: historical and geographical perspectives. University Chicago Press, Chicago, pp 281–293

Cei JM (1993) Reptiles del noroeste, nordeste y este de la Argentina. Herpetofauna de las Selvas subtropicales, Puna y Pampas. Monografía 15, Mus Reg Sc Nat Torino, Italia

Connell JH (1978) Diversity in tropical rainforest and coral reefs. Science 199:1302–1310

Cunha OR, Nascimento FP, Ávila-Pires TCS (1985) Os repteis da área de Carajás, Pará, Brasil (testudines e Squamata). I. Publicações Avulsas do Museu Paraense Emilio Goeldi 40:9–92
Di-Bernardo M (1998) História natural de uma comunidade de serpentes da borda oriental do planalto das araucárias, Rio Grande do Sul, Brasil. Unpubl. MS Thesis, Universidade Estadual Paulista, São Paulo, Brazil
Duellman WE (1978) The biology of an Equatorial herpetofauna in Amazonian Ecuador. The University of Kansas, Museum Natural History, Lawrence, Misc Publ 65:1–352
d'Orbignyi A (1998) Viaje por América meridional I. memoria argentina. Emece, Buenos Aires
Fittkau EJ (1973) Crocodiles and nutrient metabolism of Amazonian waters. Amazoniana 4:103–133
Forman RTT (1995) Land mosaics: the ecology of landscapes and regions. Cambridge University Press, Cambridge
Freiberg MA (1969) Una nueva subespecies de *Pseudemys dorbignyi* (Duméril et Bibron) (Reptilia, Chelonia, Emydidae). Physis 28:299–314
García NO, Vargas WM (1998) The temporal climatic variability in the "Río de la Plata" Basin displayed by the river discharges. Clim Change 38:359–379
Giacosa R, Paoli C, Cacik P (2000) conocimiento del régimen hidrológico. In: Paoli C, M Schreider (eds) El río Paraná en su tramo medio. Contribución al conocimiento y prácticas ingenieriles en un gran río de llanura. Universidad Nacional del Litoral, Santa Fe, Argentina, pp 70–103
Gibbons JW (2003) Terrestrial habitat: a vital component for herpetofauna of isolated wetlands. Wetlands 23:630–635
Gibbons JW, Scott DE, Ryan JT, Buhlmann KA, Tuberville TD, Metts BS, Greene JL, Mills T, Leiden Y, Poppy S, Winne CT (2000) The global decline of reptiles, Déjà Vu Amphibians. BioScience 50(8):653–666
Giraudo AR (2001) La diversidad de serpientes de la Selva Paranaense y del Chaco Húmedo: Taxonomía, biogeografía y conservación. Literatura of Latin América, Buenos Aires, Argentina
Giraudo AR, Arzamendia V (2004) ¿Son las planicies fluviales de la Cuenca del Plata, corredores de biodiversidad? Los vertebrados amniotas como ejemplo. In: Neiff JJ (ed) Humedales de Iberoamérica, Editado por el CYTED, Programa Iberoamericano de Ciencia y Tecnología para el Desarrollo, Cuba, pp 157–170
Girons HS (1985) Comparative data on Lepidosaurian reproduction and some time tables, Chap. 2. In: Gans C, Billet (eds) Biology of the reptilia. Wiley, New York, pp 33–58
Gregory PT, Larsen KW (1993) Geographic variation in reproductive characteristic among Canadian populations of the common garter snake (*Thamnophis sirtalis*). Copeia (4):946–958
Hawkins BA (2004) "Latitude" and geographic patterns in species richness. Ecography 27:268–272
Iglesias de Cuello A (1982) Atlas físico de la República Argentina, vol. 2. Atlas total de la República Argentina, Centro Editor de América Latina, Argentina
Iverson JB (1986) A checklist with distribution maps of the turtles of the world. Privately printed, Richmond, IN
Junk WJ (1997a) General aspects of floodplain ecology with special reference to Amazonian floodplains. In: Junk WJ (ed) The central Amazon floodplain. Ecol Studies 126:4–20
Junk WJ (1997b) Synthesis and discussion. In: Junk WJ (ed) The central Amazon floodplain. Ecol Studies 126:453–472
Junk WJ, Bayley PB, Sparks RE (1989) The flood pulse concept in river–floodplain systems. In: Dodge DP (ed) Proc Int Large River Symp (LARS). Can Spec Publ Fish Aquat Sci 106:110–127
Junk WJ, da Silva VM (1997) Mammals, reptiles and amphibians. In: Junk WJ (ed) The central Amazon floodplain. Ecol Studies 126:409–417
Larriera A (1995) Áreas de nidificación y momento óptimo de cosecha de huevos de *Caiman latirostris* en Santa Fe, Argentina. In: Larriera A, Verdade LM (eds) La conservación y el manejo de caimanes y cocodrilos de América Latina. Fundación Banco Bica, Santo Tomé Santa Fe, Argentina, pp 221–232
López MS, Giraudo AR (2004) Diet of the large water snake *Hydrodynastes gigas* (Colubridae) in Argentina. Amphibia-Reptilia 25:178–184

López MS, Giraudo AR, Arzamendia V (2003) *Leptophis ahaetulla marginatus*. Diet. Natural History Notes. Herpetol Rev 34:68–69

Medem F (1983) Los Crocodylia de Sur América. Volumen II. Univ. Nac. Colombia, Col Ciencias, Bogotá, Colombia

Menalled FD, Adámoli JM (1995) A quantitative phytogeographic analysis of species richness in forest communities of the Paraná River Delta, Argentina. Vegetatio 120:81–90

Micucci PA, T Waller (1995) Los yacarés en Argentina: hacia un aprovechamiento sustentable. In: Larriera A, Verdade LM (eds) La conservación y el manejo de caimanes y cocodrilos de América Latina. Fundación Banco Bica, Santo Tomé Santa Fe, Argentina, pp 81–112

Montero R, Céspedes J (2002) New two-pored Amphisbaena (Squamata: Amphisbaenidae) from Argentina. Copeia (3):792–797

Neiff JJ (1990) Ideas para la interpretación ecológica del Paraná. Interciencia 15(6):424–441

Paoli CM, Iriondo, Garcia N (2000) Características de la cuencas de aporte. In: Paoli C, Schreider M (eds) El río Paraná en su tramo medio. Contribución al conocimiento y prácticas ingenieriles en un gran río de llanura. Universidad Nacional del Litoral, Santa Fe, Argentina, pp 29–68

Pizzatto L, Marques AV (2006) Interpopulational variation in sexual dimorphism, reproductive output, and parasitism of *Liophis miliaris* (Colubridae) in the Atlantic forest of Brazil. Amphibia-Reptilia 27:37–46

Roe JH, Kingsbury BA, Herbert NR (2003) Wetland and upland use patterns in semi-aquatic snakes: implications for wetland conservation. Wetlands 23:1003–1014

Scott NJ Jr, Aquino AL, Fitzgerald L.A (1990) Distribution, habitats and conservation of the *Caimans* (Alligatoridae) of Paraguay. Vida Silvestre Neotropical 2:43–51

Scrocchi GJ, Ferreira, VL, Giraudo AR, Avila RW, Motte M (2005) A new species of *Hydrops* (Serpentes: Colubridae: Hydropsini) from Argentina, Brazil and Paraguay. Herpetologica 61:468–477

Scrocchi GJ, Giraudo AR (2005) Reptiles de la Reserva El Bagual. In: Di Giacomo A, Krapovickas SF (eds) Historia natural y paisaje de la Reserva EI Bagual, provincia de Formosa, Argentina. Inventario de la fauna de vertebrados y de la flora vascular de un área protegida del Chaco Húmedo. Temas de naturaleza y conservación 4, Asociación Ornitológical del Plata, Buenos Aires, Argentina, pp 155–198

Shine R, Brown GP (2002) Effects of seasonally varying hydric conditions on hatching phenotypes of keelback snakes (*Tropidonophis mairii*, Colubridae) from the Australian wet-dry tropics. Biol J Linnean Soc (76):339–347

Shine R (2003) Reproductive strategies in snakes. Proc R Soc Lond B 270:995–1004

Souza FL (2004) Uma revisão sobre padrões de atividade, reprodução e alimentação de cágados brasileiros (Testudines, Chelidae). Phyllomedusa 3:15–27

Strussmann C, Sazima I (1990) Esqudrinhar com a cauda: uma tática de caça da serpente *Hydrodynastes gigas* no pantanal, Mato Grosso. Mem Inst Butantan 52:57–61

Strussmann C, Sazima I (1993) The snakes assemblage of the Pantanal at Poconé, Western Brazil: faunal composition and ecological summary. Studies Neotrop Fauna Environ 28(3):157–168

Tiegs SD, O'leary JF, Pohl MM, Munill CL (2005) Flood disturbance amd riparian species diversity on the Colorado River Delta. Biodiversity and Conservation 14:1175–1194

Waller T (1987) Registro de localidades de distribución de las especies del género *Caiman* (Crocodilia, Alligatoridae) en Argentina. Amph Rept (conserv) 1:68–75

Waller T (1988) Notas sobre la distribución de *Acanthochelys spixii* (D. y B., 1835) y *Acanthochelys pallidipectoris* (Freiberg, 1945) en la Argentina (Testudines: Chelidae). Amph Rep (conserv) 3:93–98

Willing MR, Kauffman DM, Stevens RD (2003) Latitudinal gradients of biodiversity: Pattern, Process, Scale, and Synthesis. Ann Rev Ecol Evol Syst 34:273–309

15 Birds

Adolfo H. Beltzer[1] and Martin A. Quiroga

15.1 Introduction

Aquatic birds usually exploit unstable or fluctuating environments along their cycle and among different years. These environments also show high productivity, resource-partitioning mechanisms, as well as ecological isolation, partial ecological specialization, prey captures of different size, type and shape, at different times of the day, and in different microhabitats, are observed in these species.

An additional factor that also facilitates the coexistence of aquatic birds is the great diversity of prey they can find. Potentially, each one of the species differs from the others in relation to their visual capacity, hours of activity, escape behavior, abundance or non-abundance in certain seasons of the year, depth in which it lives, etc., and will respond differently to this mosaic of possibilities.

Fishing birds possibly constitute the most complex aquatic birds. In many species, the term "fishing" does not imply an exclusively ichthyophagous diet, since they also frequently capture quite an amount of insects, crustaceans, arachnids, and mollusks, among others. Among the fishing birds, waders constitute a continuous series of similar shape species, which only differ in their body size. Almost all of them capture their preys while they walk or stalk standing in the water bodies. Thus, the feeding depth in each case is limited by the feet length. In this group, a sequence in the beak size and the body weight is observed, so larger or heavier species can capture and consume larger preys. Moreover, larger birds move at a lower velocity while they localize their large preys and they can even remain static for hours until the moment of capture. In contrast, generally small and medium birds move more and faster.

Birds are important consumers of aquatic systems (Martínez 1993), so the interrelation among their body size, the energetic requirements and activity levels are significant (Margalef 1983). On the other hand, they can obtain food in different environmental units of the aquatic system (differential spatial use

Instituto Nacional de Limnología (INALI-CONICET-UNL), José Maciá 1933, 3016 Santo Tomé, Argentina, [1]e-mail: adolfohec2001@yahoo.com.ar

M.H. Iriondo, J.C. Paggi, and M.J. Parma (Eds.)
The Middle Paraná River: Limnology of a Subtropical Wetland

of the environment or eco-space, Dobhanzky et al. 1983), so they can exploit the resources along the annual cycle and at different times of the day (differential models of temporal activity, Pianka 1982; Martínez 1993).

For birds, food, water, and refuge are essential requisites. Feeding is related to the first two ones. However, an adequate place to be protected is important to assure permanent feeding without disturbances (Nagy and Haufler 1987). The scientific study of feeding and feeding habits in wild fauna is only a comparatively recent direction (Korschgen 1987).

For all these reasons, and because the simple localization of species in great environmental and vegetation units (GUVAS, Beltzer and Neiff 1992) is not enough to determine their degree of dependence, and considering the important role that birds perform in energy transference, feeding biology studies are important. In general, there is not enough information for the area, since it is based on occasional observations and under the level of precision in relation to the taxonomic groups used as food.

Therefore, to get adequate information on feeding ecology of birds associated with the middle section of the Paraná River, a qualitative-quantitative analysis of stomach contents began in 1979 to establish the composition of the trophic spectrum, constituting the first records for the area. Seasonal variations in the spectrum composition, trophic niche amplitude, circadian rhythm of feeding activity, preference rate and dietary selectivity, niche overlapping, feeding efficiency, degrees of fidelity and trophic participation, area of influence, nuclear area, and habitat selection could be establish through quantitative analysis of the obtained information for each feeding category.

15.2 Trophic Groups

Trophic guilds constitute one of the most significant ecological aspects to take into account in the study of an ensemble within a very complex ecosystem as the Paraná River. A functional group or guild can be considered as a subgroup of the community in which individuals use similar resources in an analogous way (Root 1967). The analysis of guilds allows one to know the most ecologically related syntopic species, with respect to the trophic and behavioral aspect, the interspecific interactions, type of food they consume, and how, when and where they consume it (Kirkconnel et al. 1992). Using this approach, and due to the importance of this issue, it has been studied by numerous specialists throughout the world (Orians 1969; Cody 1974; Wilson 1974; Diamond 1975; Feisinger 1976; Eckhardt 1979; Hirth and Marion 1979; Holmes et al. 1979; Thiollay 1979; Williams and Batzli 1979; Desgranges 1980; Faaborg 1982; Karr 1982; Blake 1983; Landres and MacMahon 1983; MacNally 1983; Nudd 1983; Osborne et al. 1983; Poysa 1983; Siegfried 1983; Adams 1985, among others). According to Wilson (1974), the three main components of the functional group are the type of food, the way of consuming it, and the site.

Moreover, according to Johnson (1978), there are variations in the diets of organismic units when they have extensive geographical distributions due to the differences in the availability of preys. Other aspects to consider are the size of the foraging group, i.e., if they tend to search for food through solitary behavioral patterns, in pairs, in unispecific or mixed flocks and, finally, if they do it at daily, nightly, or crepuscular hours. Another aspect to be studied is the different foraging heights (Kirkconnel et al. 1992) that establish different structural subniches for the species and allow ecological segregation. Osborne et al. (1983) mentioned three subdivisions according to the foraging heights: (1) low, between 0 and 1 m; (2) medium, between 1 and 6 m, and (3) high, higher than 6 m. Landres and MacMahon (1983) also analyzed the portion used (for example, external or internal portion of branches).

It is not easy to gather the rich group of birds associated to the aquatic system of the Paraná River into guilds, since some populations are strictly aquatic, while others are terrestrial but are also related to the river valley, so they are considered as well. The proposed classification considers the following guilds or functional groups:

- The birds that use rapier pecking as a feeding strategy basically feed on aquatic invertebrates and vertebrates captured by rapid pecking: *Ardea cocoi* (cocoi heron), *Ardea alba* (great egret), *Egretta thula* (snowy egret), *Butorides striatus* (striated heron) and *Nycticorax nycticorax* (black-crowned night heron). Other species feed on aquatic invertebrates and vertebrates using non-superficial pecking (by pecking under the water surface): *Plegadis chihi* (white-faced ibis), *Phimosus infuscatus* (bare-faced ibis), etc.
- The pecking phytophagous-carnivorous birds feed on floating and rooted macrophytes and on invertebrates associated to vegetation by pecking on the surface film, the emerging vegetation, or under it: *Fulica leucoptera* (white-winged coot), *Fulica rufifrons* (red-fronted coot), *Gallinula chloropus* (common moorhen), *Porphyrio martinicus* (purple gallinule) and *Gallinula melanops* (spot-flanked gallinule). Another group is composed of carnivorous birds, basically diving piscivorous birds that feed on aquatic vertebrates and invertebrates by diving under the water: *Podicephorus major* (great grebe), *Rollandia rolland* (white-tufted grebe), *Podylimbus podiceps* (pied-billed grebe), *Phalacorrax brasilianus* (neotropical cormorant), *Anhinga anhinga* (anhinga), etc. Filtering walkers feed while walking and filtering water by moving their heads from one side to the other to catch algae, invertebrates, and even small fish: *Phoenicopterus chilensis* (Chilean flamingo).
- The surface filterer birds feed on macrophytes (leaves, stems, and seeds of floating and rooted plants) and invertebrates (pleuston fauna): *Anas versicolor* (silver teal), *Anas platalea* (red shoveler), *Dendrocygna bicolor* (fulvous whistling-duck), *Dendrocygna viduata* (white-faced whistling-duck), *Netta peposaca* (rosy-billed pochard), etc.

- In some species, the diet is basically composed of aquatic and some terrestrial invertebrates they capture by walking and pecking on shallow waters, in the sand, or in muddy beaches. They are riverside walking carnivorous birds (coasts and banks), such as: *Calidris fuscicollis* (white-rumped sandpiper), *Calidris melanotos* (pectoral sandpiper), *Tringa flavipes* (lesser yellowlegs), *Tringa melanoleuca* (greater yellowlegs), *Tringa solitaria* (solitary sandpiper), *Calidris alba* (sanderling), *Charadrius collaris* (collared plover), *Pluvialis dominica* (golden plover), etc. The vegetated riverside walking carnivorous-omnivorous guild (coasts and banks) has a diet composed by floating and rooted macrophyte seeds and fauna associated to aquatic vegetation (pleuston fauna) that they incorporate by walking on the plant carpet, facilitated by their anatomical adjustments, in the case of the *Jacana jacana* (wattled jacana), whereas some Rallidae swim and walk on the riverside, for example, *Pardirallus sanguinolentus* (plumbeous rail), *Pardirallus maculatus* (spotted rail), *Aramides ypecaha* (giant wood-rail), etc.
- Other carnivorous birds basically feed on fish and some invertebrates (insects and palaemonid crustaceans) they obtain through visual perception from perches or from the air and by pecking plunges: *Megaceryle torquata* (ringed kingfisher), *Chloroceryle amazona* (amazon kingfisher), *Chloroceryle americana* (green kingfisher), *Sterna phaetusa* (large-billed tern), *Sterna supercilliaris* (yellow-billed tern), etc. There are also those that feed on aquatic invertebrates and vertebrates (basically fish) at the surface water while they swim or float, although they usually use the strategy of pecking plunges, constituting the surface aquatic carnivorous birds, as for example, *Croicocephalus maculipennis* (brown-hooded gull), *Croicocephalus cirrocephalus* (grey-hooded gull), etc.
- In some cases, the behavioral patterns reveal specific adjustments which, in the case of the Middle Paraná River, are represented by only one species, as in the case of the flying surface filter bird or *Rynchops níger* (black skimmer).

In contrast, there are many species that feed by insect, fruit and seed gleaning, whose guilds are grouped into:

- Gleaning insectivorous-frugivorous birds: they mainly feed by gleaning insects, fruits, and seeds they obtain in the bark of trunks, branches, and leaves, as well as of fruits, seeds and buds: *Saltator coerulescens* (greyish saltator), *Saltator aurentiirostris* (golden-billed saltator) and *Agelaioides badius* (bay-winged cowbird).
- Pecking soil insectivorous birds: they basically feed by pecking on the surface, generally on the pasture and, in occasions, on the humid surface next to the muddy, sandy water: *Machetronis risoxus* (cattle tyrant), *Cinclodes fuscus* (bar-winged cinclodes), etc.
- Pecking and gleaning soil and foliage insectivorous birds: they mainly consume insects and larvae by pecking on the grass (pasture) and gleaning on

the foliage (branches, trunks, and leaves): *Paroaria coronata* (red-crested cardinal), *Paroaria capitata* (yellow-billed cardinal), *Taraba major* (great antshrike), *Polioptila dumicola* (masked gnatcatcher), etc.

- Fluttering and hunting perch insectivorous-omnivorous birds: they basically consume insects, arachnids and fruits, even fish and small vertebrates. This strategy is used by some species as: *Fluvicola albiventer* (black-backed water-tyrant), *Pitangus sulphuratus* (great kiskadee), *Tyrannus melancolicus* (tropical kingbird), *Pyrocephalus rubinus* (vermilion flycatcher) and *Xolmis irupero* (white monjita).
- Aerial insectivorous birds: they feed on insects they trap in the air while they fly: *Tachycineta leucorrhoa* (white-rumped swallow) and other migratory species, as *Progne chalybea* (grey breasted martin) and *Hirundo rustica* (barn swallow).
- Nocturnal aerial insectivorous birds: they basically feed on insects they capture in flight: *Podager nacunda* (nacunda nighthawk) and *Hydropsalis brasiliana* (scissor-tailed nightjar).
- Trunk insectivorous birds: they feed on insects and larvae they extract from the bark by pecking trunks and branches: *Picoides mixtus* (checkered woodpecker), *Melanerpes candidus* (white woodpecker), *Colaptes campestris* (camp flicker) and *Colaptes melanocloros* (royal woodpecker).
- Rapier pecking soil insectivorous birds: their diet basically consists of terrestrial invertebrates (orthopterous insects and arachnids), although they also consume amphibians at surface waters: *Bubulcus ibis* (cattle egret).
- Soil insectivorous-frugivorous birds: they obtain their food (insects and soil seeds) by moving about the dead leaves on the ground: *Turdus rufiventris* (rufous-bellied thrush) and *Turdus amaurochalinus* (creamy-bellied thrush).
- In the case of the sifting flying nectarivorous-insectivorous birds, they feed by using the hanging or suspended flight, as is the case of *Chlorostilbon aureventris* (glittering-bellied emerald), *Hylocharis chrysura* (gilded sapphire), *Polytmus guainumby* (white-tailed goldenthroat) and *Heliomaster furcifer* (blue-tufted starthroat).
- A large group is constituted by the soil granivorous birds, among which we find the pigeons and doves, which basically feed on seeds: *Zenaida auriculata* (eared dove), *Columba picazuro* (picazuro pigeon), *Columba maculosa* (spot-winged pigeon), *Leptotila verreauxi* (white-tipped dove) and *Columbina picui* (picui ground-dove).
- Soil and foliage pecking and gleaning omnivorous birds feed on seeds and invertebrates by pecking on the ground and gleaning in the tree formations of the woodland and gallery forest: *Crotophaga ani* (smooth-billed ani), *Guira guira* (guira cuckoo), *Tapera naevia* (striped cuckoo), *Coccyzus melacoryphus* (dark-billed cuckoo) and *Coccyzus cinereus* (ash-colored cuckoo).
- Aerial predators include those that feed on vertebrates and invertebrates they trap from perches or from the air through pecking falls: *Buteo magnirostris* (roadside hawn), *Falco sparverius* (American kestrel) and *Circus buffoni* (long-winged harrier).

- The nocturnal aerial predator guild basically traps rodents, as well as birds and insects, from perches or from the ground: *Tyto alba* (common barn owl), *Othus coliba* (common alicucu), *Glaucidium brasilianus* (ferruginous pygmy owl), and *Bubo virginianus* (great horned owl).
- The malacophagous birds basically feed on mollusks, so some of them have structural adjustments that allow them to be very efficient to obtain and use the food: *Rosthramus sociabilis* (everglade kite) and *Aramus guararuna* (limpkin).
- Finally, ghouls are those species that basically include carrion in their diet: *Caracara plancus* (crested-caracara), *Coragyps atratus* (black vulture) and *Cathartes aura* (turkey vulture).

15.3 Seasonal Dynamics

Aquatic birds constitute a characteristic feature of rivers and aquatic environments throughout the world. Their specific composition indicates the state of the water body, their productivity at the different trophic levels and the characteristics of their structure and function (Reichholf 1981). However, an environment dominated by the presence of herons (Ardeidae) (Beltzer 1995; Beltzer et al. 2005) requires abundant fish, and the specific composition of a certain group not always indicates the availability and accessibility of the resource (Beltzer 1991).

Undoubtedly, the "hydrometric level-vegetation dynamics-associated fauna" complex constitutes a macrofactor with different weight according to the characteristics of the biotic complex, to which we should add such concepts as ordinary-extraordinary flood, energy-matter pulses, environmental units, and functional groups (Beltzer and Neiff 1992).

A preliminary exam of richness and abundance provides important information on the nature of aquatic systems.

Therefore, each water level at a given moment is responsible for the higher or lower offer of habitats and therefore for the participation of the different guilds.

According to the magnitude and variety, it can occur that no guild could use the environments. In contrast, the 'walking birds' guild is the first one conditioned to make displacements, since their structural adaptations limit the exploitation of resources. Changes in the water level determine variations in habitats and, consequently, in the spatial heterogeneity. Therefore, the simultaneous presence of all groups can be observed, or also their sequential replacement or absence.

Thus, due to their flight capacity, this group can respond to each situation, moving to alternative areas so they can adjust efficiently to the great variability of these types of systems.

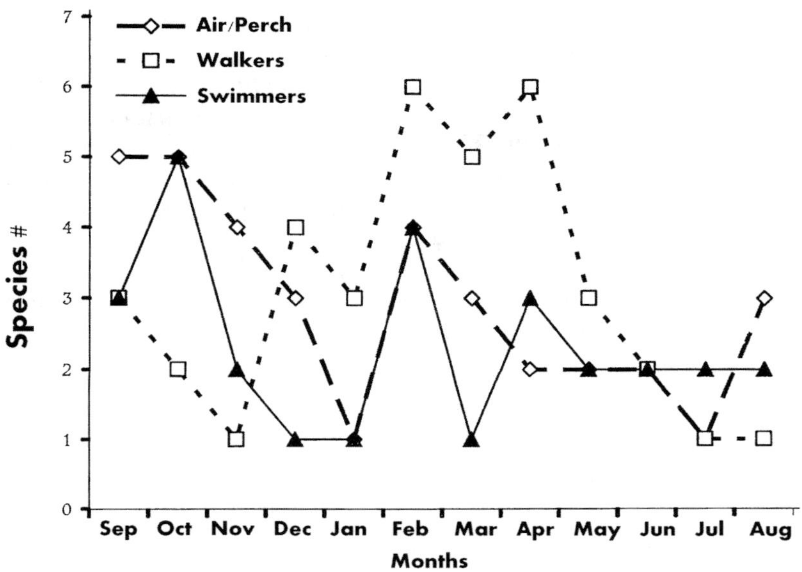

Fig. 15.1 Dynamics of three guilds during an annual cycle (Beltzer et al. 2004) in which walkers predominated from December to May (i.e., *Ardea cocoi, Casmerodius albus, Plegadis chihi*), while those that obtain their food from the air or perches (*Rosthramus sociabilis, Pitangus sulphuratus*, etc.) predominated in October and November. On the other hand, those that swim and plunge (i.e., *Amazonetta brasiliensis, Netta peposaca, Dendrocygna bicolor* and *D. viduata*) did not show a clear tendency

In a study carried out by Bossisio and Beltzer (2004), 99 species were identified, a number that contrasts with the results by Beltzer et al. (2004), in which only 29 species were identified. These values would indicate that this type of systems has a very marked seasonal behavior in relation to the species change over.

Figure 15.1 shows the dynamics of three guilds during an annual cycle (Beltzer et al. 2004), in which walkers predominated between December and May (i.e., *Ardea cocoi, Casmerodius albus, Plegadis chihi*), while those that obtain their food from the air or perches (*Rosthramus sociabilis, Pitangus sulphuratus*. etc.) predominated in October and November. On the other hand, those that swim and plunge (i.e., *Amazonetta brasiliensis, Netta peposaca, Dendrocygna bicolor* and *D. viduata*) did not show a clear tendency.

Figure 15.2 represents the fluctuation in the hydrometric level, allowing the interpretation of guild variations in Fig. 15.1.

The descriptive study of bird communities as a factor to understand other aspects relative to the functioning of ecosystems is one of the strongest fields in current ornithology, with a very recent development in Argentina. These studies are becoming more important since the integration between diversity and function can lead to very useful interpretations, as mentioned by Lawton (1994).

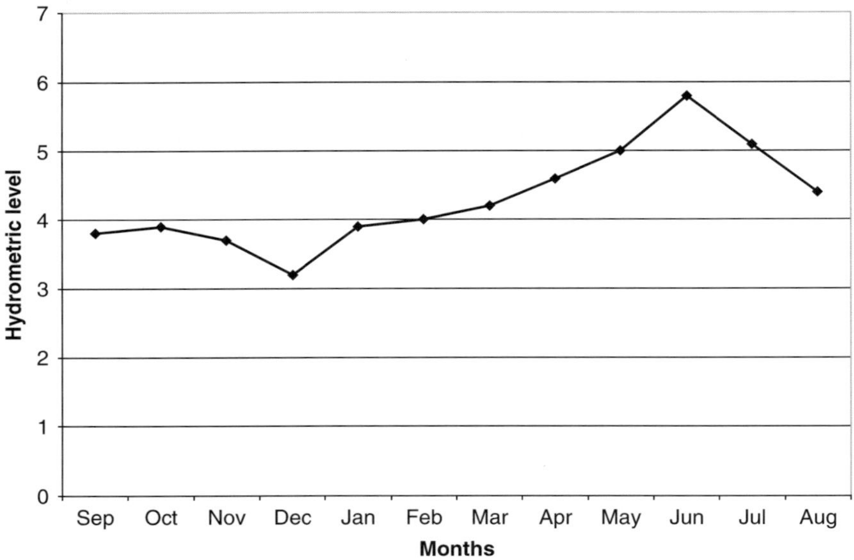

Fig. 15.2 Fluctuation of the hydrometric level, that allows to interpret variations in guilds in Fig. 15.1

15.4 Reproductive Biology

Little is known about the reproductive habits of species in this area, although this is a relevant issue. Reproductive biology of a particular species is the result of extensive evolutionary processes that were derived in an "optimum" reproductive strategy for each species in the habitat where it is found. Undoubtedly, the formation of pairs and mating systems reflect the availability of key ecological resources (space, food and protection), as well as the availability of pairs and the possibility of uniparental care (Gill 1995).

Reproduction involves a great quantity of complex behaviors and processes that go from the evolution of ornaments and selection of pairs, through the courtship behavior, until the spermatic competition and the establishment of a mating system that maximizes the quantity of obtained descendants.

Although many of the questions necessary for reproduction are genetically inherited (as the courtship rituals, ornaments, etc.), the optimum reproductive strategy of a species is definitively defined by the interaction with the environment in which it is found. In a widely variable environment such as the Paraná River, where the hydrosedimentological regime generates periods of low waters, with a higher mainland surface and numerous water bodies that contain a great variety of feeding resources, and others of high waters, where emerging lands are scarce and food is more difficult to obtain, it would not be strange to see different mating strategies and systems for different species, or for a same species in different situations.

Variations in water level in addition to modifying the conditions of the land severely affects the access to food, which is one of the main determinants of mating systems, so species can go from monogamy to polygamy with control of the resource (i.e., Alcock 1997; Gill 1995). Moreover, the availability of nesting sites also suffers variations leading to different mating systems.

Knowledge of reproductive parameters of any species, as for example the size of the layer or the reproductive success, is interesting to understand the questions related to its ecology and evolution (Rotenberry and Wiens 1989), as well as for species management, conservation and repopulation programs (Caro 1998).

Reproductive success has been widely used as a measurement of the fitness of an animal (Lack 1954; Fisher 1958; Williams 1966). The total fitness of an individual could be defined as the sum of the reproductive success throughout its life, and it is easy to define reproductive success as the number of chicks per year (in spite of the difficulty to define "success") (Burger 1982).

This is observed in that small fluctuations in some variables of the reproductive biology, as the size of egg laying, can influence the population dynamics, and the differences in the reproductive success can provide basic material for evolutionary changes (Rotenberry and Wiens 1989).

The factors related to the environment can reduce the reproductive success of most bird species. Important environmental events that cause problems in birds are mainly of two types: daily changes in temperature and precipitation cause the death of embryos or chicks, while catastrophic events (as storms, strong winds, extreme temperatures, etc.) can cause the death of embryos or chicks when they are exposed, even for a short period of time (Nisbet 1975; Hand et al. 1981).

Finally, we will refer to one of the less studied subjects in the area: the reproductive biology of birds from the Paraná River. In the study area, few advances have been done in this subject, especially if they are referred to quantitative contributions. The first contributions were carried out by de la Peña (1976, 1977, 1978, 1994, 1995, 1996), and consist of qualitative data, including descriptions of nests, eggs and quantity of chicks of most species in the area.

A few years ago a project for the study of the reproductive biology of birds from the Paraná River was begun to obtain data on some variables, as for example: birth and mortality rates (gross and caused) of chicks, reproductive success and yield, growth and development of chicks, etc.

Until now, there are data on four species of Passeriformes: *Sicalis flaveola* (saffron finch) (Quiroga et al. 2003), *Trogodytes aedon* (house wren) (Quiroga et al. 2007) and *Zonotrichia capensis* (rufous-collared sparrow), which have been studied since 1999 until now, and *Tachycineta leucorroha* (white-rumped swallow), that has been monitored since 2003 until now (Quiroga et al. unpublished).

Within the non-Passeriformes, the reproductive biology has been studied in *Butorides striatus* (striated heron) (Beltzer 1991, 1995; Mosso and Beltzer

1992) from 1986 to 1991. The interest for this species was based on the fact that it is the only heron in the flood valley of the Paraná River that shows patterns of migratory behavior, being a spring-summer resident.

This heron did not construct its nests simultaneously, producing overexposed stages in which incubating pairs, nests with chicks, and others with juveniles that were ready to abandon them, were observed. The incubation was begun with the laying of the first egg, so hatching is not synchronous. This does not coincide with the patterns observed in the Passeres, which begin the incubation with the penultimate egg.

The average number of eggs per nest was 3 ± 1. During the studied years, a notable increase in breeding pairs and, thus, in the number of nests per area was observed. The birth rate averaged 0.17 and 17% hatched and 73.08% of chicks reached the stage of juveniles in conditions to fly. The mean time of incubation was 21 days, with a permanence of chicks in nests of 14 days. Gross mortality rate was 48%, while mortality per ages was: eggs 48.09%, and juveniles 21.1% (Fig. 15.3). The net growth rate (R) was 1.61 and the reproductive yield was (n) 42.2%.

Lays of *Sicalis flaveola* mostly consisted of four eggs (± 1) to one egg/day, with some nests with five eggs and others with three eggs.

The longest egg axis was 18.91 cm. (± 1.09), the shortest axis was 13.95 cm. (± 0.48), the mean weight was 1.92 g (± 0.2) and the mean volume was 3.67 mm^3 (± 0.2).

The incubation period was 16 days (± 0.71). Apparently, the female would be the only one participating in the incubation.

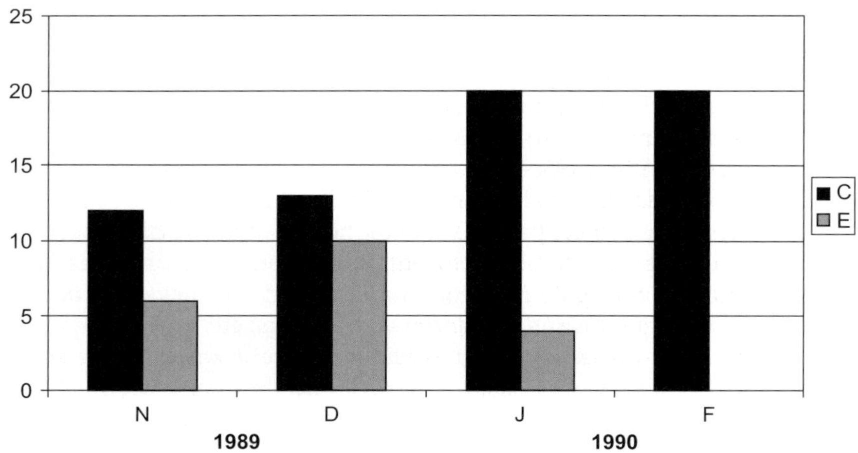

Fig. 15.3 Percentage values of the number of eggs and chicks of *Butorides striatus*. *E* eggs; *C* chicks. *Letters* indicate months: *N* Nov.; *D* Dec.; *J* Jan; *F* Feb. *Numbers* on the bar express absolute values

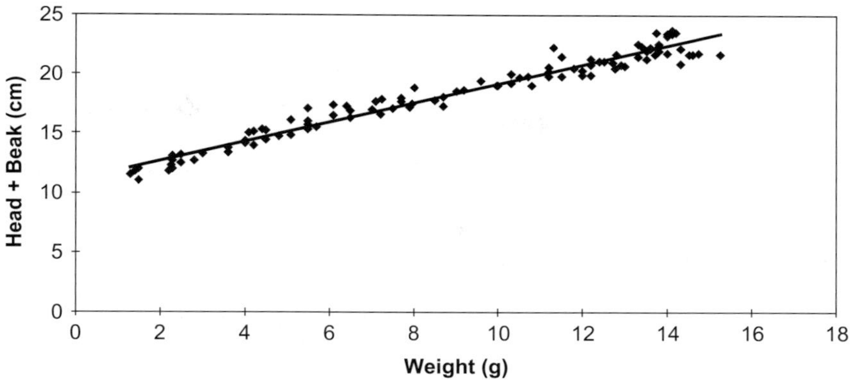

Fig. 15.4 Body weight and Head + Beak regression. Coefficient of multiple correlation: 0.97952477

The birth rate was 0.031, while mortality of chicks increased to 20%. Most of them died by starvation or by climatic conditions.

Regression analysis showed a strong dependence among the variables of growth, bird weight, wing length, and size of head + beak. All correlations were positive (Fig. 15.4).

As expected, the three variables (body weight, wing length, and size of head plus beak) increased as the age of the chicks increased. Wing length and body weight increased rapidly, while head plus beak size increased only gradually.

In relation to developmental indicators, chicks opened their eyes between the sixth and seventh day of development, and primary and secondary feathers appeared around the same time.

In the case of the house wren, the nest construction was not synchronous, beginning at mid September and continuing at least until mid February (when samplings ended). Their construction took between 3 and 14 days, being more frequent between 4 and 8 days. Some pairs of *Troglodytes aedon* took the nests of *Sicalis flaveola* just after they finished their construction. Once usurped, they added typical *T. aedon* nest material and immediately began egg laying. In this way, a type of mixed nest was generated, but, anyway, it was successful (Fig. 15.5, 15.6).

The size of the egg laying (four eggs on average) and the incubation period were similar to those described for *Sicalis flaveola*. The length of the largest axis was 17.09 mm (± 0.92), the length of the smallest axis was 12.99 mm (± 0.567), the weight was 1.54 g (± 0.201), and the volume was 2.88 mm^3 (± 0.285). For this species, the birth rate was 0.0156, while the mortality rate was 29.9%. Correlations carried out between measured variables were strong, and growth patterns and chick development followed the same tendencies of most studied species of Passeriformes.

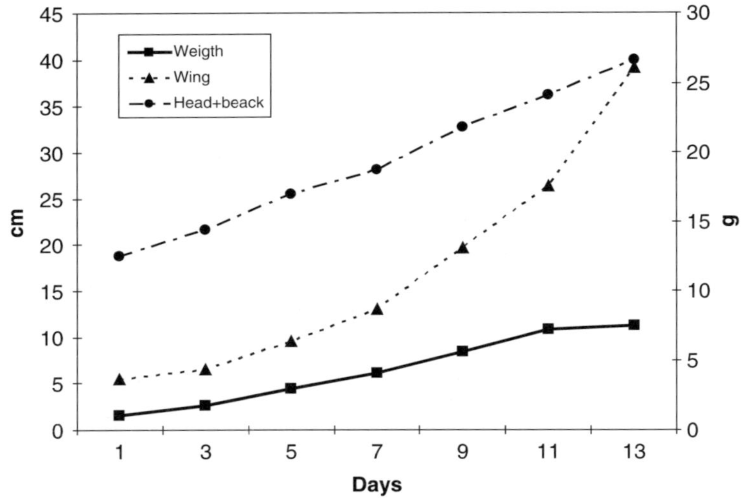

Fig. 15.5 Evolution of variables through time of *Troglodytes aedon* chicks development in Santa Fe, Argentina

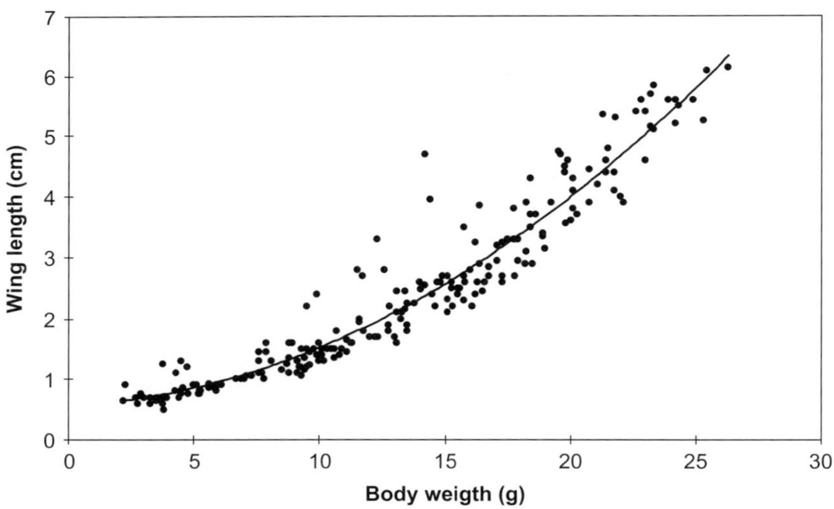

Fig. 15.6 Evolution of variables through time of *Troglodytes aedon* chicks development in Santa Fe, Argentina. Body weight and wing length regressions. Coefficient of multiple correlation: 0.974

In the case of the rufous-collared sparrow (*Zonotrichia capensis*), a sequential egg laying of an egg per day was also registered, and each sitting was not greater than three eggs. The incubation period demanded between 10–11 days, remaining a total of 11 chicks. The birth rate was 82% and the mortality rate was 18%, while net birth rate was 2.37.

For *Tachycineta leucorroha*, the reproductive activity extended from the beginning of September to mid December, reaching maximum activity in November, coinciding with a peak in the river flood, which inundated the area. This could have generated a higher abundance of insects, basic diet of this species, which would justify the low rate of mortality of chicks in the studied periods.

The average egg-laying size was four eggs, but layers with three, four, and six eggs were registered. For this species, the average egg length was 20.21 (± 0.843), the width was 14.31 (± 0.753), and the weight was 2.152 (± 0.221). The birth rate was 81.76% and mortality rate was 7.5%, and there were strong correlation indices among the measured variables.

Although future investigations are necessary, these contributions provide knowledge of a group that constitutes an indicator of the state of water bodies, which justifies their continuation to reach an adequate management of the resource in the framework of a sustainable development.

References

Adams J (1985) The definition and interpretation of guild structure in ecological communities. J Anim Ecol UK 54:43–59

Alcock J (1997) Animal behavior. Sinauer, Sunderland, MA

Beltzer AH (1991) Aspects of the breeding biology and the death rate of the *Butorides striatus* (Aves: Ardeidae). Rev Asoc Cienc Nat Litoral Argentina 22(1):35–40

Beltzer AH (1995) Natalidad y mortalidad de la Garcita azulada (*Butorides striatus*) (AVES: Ardeidae) en las proximidades de la ciudad de Santa Fe período 1989–1990 Argentina. Rev Ecol Lat Am Venezuela 4(1–3):11–14

Beltzer AH, Oliveros OB (1982) Alimentación del macá grande (*Podiceps major*) en el valle aluvial del río Paraná medio (Podicipediformes: Podicipedidae). Rev Asoc Cienc Nat Litoral Argentina 13:5–10

Beltzer AH, Neiff JJ (1992) Distribución de las aves en el valle del río Paraná. Relación con el régimen pulsátil y la vegetación. Ambiente Subtropical Argentina 2:77–102

Beltzer AH, Quiroga MA, Bortoluzzi AL (2004) Variación estacional de la comunidad de aves en un ambiente lenítico del valle de inundación del río Paraná. Natura Neotropicalis Argentina 34–35(1–2):33–37

Beltzer AH, Quiroga MA, Schnack JA (2005) Algunas ardeidas del valle de inundación del río Paraná: consideraciones sobre el nicho ecológico y mecanismos de aislamientos. In: Aceñolaza FG (ed) Temas de la Biodiversidad del Litoral Fluvial Argentino II. Miscelaña 14, ISUGEO, Tucumán, pp 499–526

Blake JG (1983) Trophic structure of birds communities in forest parches in east-central Illinois. Wilson Bull USA 95(3):416–430

Bossisio AC, Beltzer AH (2004) Dinámica primavera estival de un ensamble de aves en el área de inundación del río Paraná. Natura neotropicalis Argentina 34–35(1–2):51–61

Burger J (1982) An overview of proximate factors affecting reproductive success in colonial bids: concluding remarks and summary panel discussion. Colonial Waterbirds USA 5:58–65

Caro T (1998) The significance of behavioural ecology for conservation biology. In: Caro T (ed) Behavioural ecology and conservation biology. Oxford University Press, Oxford, pp 3–30

Cody ML (1974) Competition and the structure of birds' communities. Princeton Univ Press, Princeton

De la Peña MR (1976) Aves de la Provincia de Santa Fe. Castellví, Santa Fe, Argentina

De la Peña MR (1977) Aves de la Provincia de Santa Fe. Castellví, Santa Fe, Argentina
De la Peña MR (1978) Enciclopedia de las aves argentinas. Colmegna, Santa Fe, Argentina
De la Peña MR (1994) Nueva guía de flora y fauna del Río Paraná. Martín de la Peña (ed) Santa Fe, Argentina
De la Peña MR (1995) Ciclo Reproductivo de las Aves Argentinas. UNL, Santa Fe, Argentina
De la Peña MR (1996) Ciclo Reproductivo de las Aves Argentinas, part 2, UNL, Santa Fe, Argentina
Desgranges JL (1980) Avian community structure of six forests stands in La Maurice National Park. Quebec. Can Wildl Serv 41:3–32
Diamond JM (1975) Assembly of species communities. In: Cody ML, Diamond JM (eds) Ecology and evolution of communities. Harvard Univ Press, Boston, pp 342–444
Dobhanzky T, Ayala F, Stebins JL, Valentine JW (1983) Evolución. Ormega, Barcelona
Eckhardt RC (1979) The adaptative syndrome of two guilds of insectivorous birds in the Colorado Rocky Mountains. Ecol. Monogr USA 49:129–149
Faaborg JR (1982) Trophic size structure of west Indian Birds communities. Proc Natl Acad Sci USA 79:1563–1567
Feinsinger P (1976) Organization of a tropical guild of nectarivorous birds. Ecol Monogr USA 46:257–291
Fisher RA (1958) The genetical theory of Natural selection, 2nd edn. Dover Publ, New York
Gill FB (1995) Ornithology. Freeman, New York
Hand JL, Hunt JR, Warner M (1981) Thermal stress and predation: influences on the structure of a gull colony and possibly on breeding distributions. Condor USA 83:193–203
Hirth DH, Marion WR (1979) Birds communities of a South Florida flatwoods. Fla Sci USA 42(3):142–151
Holmes RT, Bonney RE, Pacala SW (1979) Guild structure of the Hubbard Brook bird community a multivariate approach. Ecology USA 60(3):512–520
Johnson DR (1978) The study of raptors populations. Univ Press, Idaho
Karr JR (1982) Avian extinction on Barro Colorado Island, Panama: a reassessment. Am Nat USA 119:220–239
Kirkconnel A, Garrido O, Posada RM, Cubillas SO (1992) Los grupos tróficos en la avifauna cubana. Poeyana, Cuba 415:1–21
Korschgen LJ (1987) Procedimientos para el análisis de los hábitos alimenticios. In: Rodríguez Tarrés R (ed) Manual de Técnicas de Gestión de la Vida Silvestre. The Wildlife Society, Bethesda, MD
Lack D (1954) The natural regulation of animal numbers. Clarendon Press, Oxford
Landres PB, MacMahon JA (1983) Guilds and community organization: analysis of an oak woodland in Sonora, Mexico. Auk USA 97:351–365
Lawton JH (1994) What do species do in ecosystems? Oikos Denmark 71:367–374
MacNally R (1983) On assessing the significance on interspecific competition to guild structure. Ecology USA 64(6):1643–1652
Margalef R (1983) Limnología. Omega, Barcelona
Martinez MM (1993) Las aves y la limnología. In: Boltosvskoy A, Lopez H (eds) Conferencias de limnología. Instit. Limnol R. Ringuelet, La Plata, pp 127–142
Mosso ED, Beltzer AH (1992) Nuevos aportes a la biología reproductiva de la garcita azulada *Butorides striatus* (Aves: Ardeidae). Hornero, Argentina 13:236–237
Nisbet ICT (1975) Selective effects of predation in a tern colony. Condor USA 77:221–226
Nudd TD (1983) Niche dynamics and organization of waterfowl guild in variable environments. Ecology USA 64:319–330
Orians GH (1969) The number of bird species in some tropical forests. Ecology USA 50(5):783–796
Osborne DR, Beissinger RS, Bourne GR (1983) Water as an enhancing factor in bird community structure. Carib J Sci Puerto Rico 19(1):35–38
Pianka ER (1982) Ecología evolutiva. Omega, Barcelona
Poysa H (1983) Resource utilization pattern and guild structure in waterfowl community. Oikos Denmark 40:317–326

Quiroga MA, Del Barco O, Agostinelli F (2003) First Approaches to the Reproductive Biology of *Sicalis flaveola* (Birds: Emberizidae) at the Alluvial Valley of Paraná River, Argentina. FAVE Argentina 2(1):35–40
Quiroga MA, Beltzer AH, Auce L, Lorenzón L (2007) First approaches to the reproductive biology of *Troglodytes aedon* (Birds: Trogloditidae) at the alluvial valley of Paraná river, Argentina. FABICIB Argentina (in press)
Reichholf JH (1981) Aves. In: Hurlbert S, Rodríguez G, Dias Dos Santos N (eds) Aquatic biota of tropical South America. Part 2. Anarthropoda. San Diego State University, San Diego, pp 262–266
Root RB (1967) The niche exploitation pattern on the bluegray gnatcatcher. Ecol Monogr USA 37:317–350
Rotenberry JT, Wiens JA (1989) Reproductive biology of shrubsteppe passerine birds: geographical and temporal variation in clutch size, brood size, and fledging success. Condor USA 91(1):1–14
Siegfried WR (1983) Trophic structure of some communities of Fynbos birds. J S Afr Bot South Africa 49 (1):1–43
Thiollay JM (1979) Structure et dynamique du peuplement aviend d'un mattoral aride (Bolson de Mapimi, Mexique). Terre et Vie, France 33:563–589
Williams GC (1966) Natural selection, the costs of reproduction and refinement of Lack's principle. Am Naturalist USA 100:687–692
Williams JB, Batzli GO (1979) Interference competition and niche shifts in the bark foraging guild in central Illinois. Wilson MF (1974) Avian community organization and habitat structure. Ecology USA 55:1017–1029

Subject Index

Printing: Krips bv, Meppel
Binding: Stürtz, Würzburg